VOLUME NINETY FIVE

CURRENT TOPICS IN DEVELOPMENTAL BIOLOGY

Forces and Tension in Development

VOLUME NINETY FIVE

CURRENT TOPICS IN DEVELOPMENTAL BIOLOGY

Forces and Tension in Development

Edited by

MICHEL LABOUESSE
Development and Stem Cells Program
Institut de Génétique et de Biologie Moléculaire et Cellulaire (IGBMC)
CNRS-UMR7104, INSERM-U964
Université de Strasbourg
Illkirch, France

ELSEVIER

AMSTERDAM • BOSTON • HEIDELBERG • LONDON
NEW YORK • OXFORD • PARIS • SAN DIEGO
SAN FRANCISCO • SINGAPORE • SYDNEY • TOKYO
Academic Press is an imprint of Elsevier

Academic Press is an imprint of Elsevier
525 B Street, Suite 1900, San Diego, CA 92101-4495, USA
225 Wyman Street, Waltham, MA 02451, USA
32, Jamestown Road, London NW1 7BY, UK
Linacre House, Jordan Hill, Oxford OX2 8DP, UK

First edition 2011

ISBN: 978-0-12-385065-2
ISSN: 0070-2153

For information on all Academic Press publications
visit our website at elsevierdirect.com

Printed and bound by CPI Group (UK) Ltd, Croydon, CR0 4YY

Transferred to Digital Print 2011

Contents

Contributors

Lance A. Davidson
Department of Bioengineering and Developmental Biology, University of Pittsburgh, Pittsburgh, Pennsylvania, USA

Emmanuel Farge
Mechanics and Genetics of Embryonic and Tumoral Development Group, UMR168 CNRS, Institut Curie, Paris, France

Carl-Philipp Heisenberg
Institute of Science and Technology Austria, Klosterneuburg, Austria

S. F. Gabby Krens
Institute of Science and Technology Austria, Klosterneuburg, Austria

Karsten Kruse
Theoretical Physics, Saarland University, Saarbrücken, Germany

Pierre-François Lenne
IBDML, UMR6216 CNRS-Université de la Méditerranée, Campus de Luminy, Case 907, 13288 Marseille Cedex 09, France

Charles B. Lindemann
Department of Biological Sciences, Oakland University, Rochester, Michigan, USA

Claire M. Lye
Department of Physiology, Development, and Neuroscience, University of Cambridge, Cambridge, United Kingdom

Matteo Rauzi
IBDML, UMR6216 CNRS-Université de la Méditerranée, Campus de Luminy, Case 907, 13288 Marseille Cedex 09, France

Daniel Riveline
Laboratory of Cell Physics, Institut de Science et d'Ingénierie Supramoléculaires (ISIS, UMR 7006) and Institut de Génétique et de Biologie Moléculaire et Cellulaire (IGBMC, UMR 7104), Université de Strasbourg, France

Bénédicte Sanson
Department of Physiology, Development, and Neuroscience, University of Cambridge, Cambridge, United Kingdom

Willy Supatto
Laboratory for Optics and Biosciences, Ecole Polytechnique, Centre National de Recherche Scientifique (CNRS) UMR 7645, and Institut National de Santé et de Recherche Médicale (INSERM) U696, Palaiseau, France

Julien Vermot
Institut de Génétique et de Biologie Moléculaire et Cellulaire (IGBMC), Institut National de Santé et de Recherche Médicale (INSERM) U964, Centre National de Recherche Scientifique (CNRS) UMR 1704, and Université de Strasbourg, Illkirch, France

Preface

> "Over the last 20 years, progress in developmental biology has been so dramatic that developmental biologists may be excused for having the view, possibly an illusion, that the basic principles are understood, and that the next 20 years will be devoted to filling in the details."
>
> Lewis Wolpert (1994) Do we understand development? *Science* **266**, 571–572

Biologists, and philosophers alike, have long tried to understand what makes embryos grow to reach their mature form. Answering this tough question has been contingent on the technologies and concepts available at a given point in history. This has been particularly true for the study of embryonic development. During the nineteenth century, embryology arguably represented a central discipline in biology. The then recent improvements in microscopy and the identification of cytological reagents labeling subcellular structures certainly contributed to the success of embryology. However, the study of embryonic development progressively declined through much of the twentieth century, being sidelined by the emergence of genetics, biochemistry, and subsequently molecular biology. It all changed with the pioneering work of Nusslein-Volhard and Wieschaus (1980), and the possibility to molecularly identify genes through molecular cloning. Embryogenesis then became heralded as developmental biology. The imprint and power of molecular biology strongly influenced the studies and thinking in developmental biology with a strong emphasis on the notion of information provided by DNA. The model of selector and realizator genes put forward by Garcia-Bellido (1975) exemplifies the belief that genes and proteins, particularly transcription factors and signaling cascades, control all events, with little room for other input.

Attention for physics in biology, particularly in morphogenesis, has witnessed an almost parallel curve. The notion that mechanical forces might influence the development of embryos was well accepted a century ago, in particular, through the pioneering work of Wilhelm His, who proposed that differences in pressure could account for the various forms of our organs. Wilhelm Roux was also influential by introducing mechanical alterations to the embryo and coining the term "developmental mechanics" (with more than one meaning). By far, the most influential scientist to have emphasized the central role of physical laws and mechanics in morphogenesis is d'Arcy Thompson with his book *On Growth and Form*

(Thompson, 1917). Although well received, Thompson's lessons did not bear fruit until recently, for the scientific community was not equipped with the right tools to address the influence of physics in development. Except for a few pioneers (Beloussov *et al.*, 1975; Keller, 1980; Odell *et al.*, 1981; Steinberg, 1962) and a few notable theoreticians (see Keller, 2002), the interest for physics waned until the late 1990s.

Several factors contributed to a resurgence in the interest for physical forces in development. First, developmental biologists slowly turned their attention away from the identification of signaling pathways controlling embryonic patterning to study morphogenesis, which is primarily a biomechanical process (Peifer and Wieschaus, 1990; Priess and Hirsh, 1986; Tepass *et al.*, 1990). Meanwhile, cell biologists and physiologists had discovered that cells respond to forces by opening channels or by strengthening their junctions (Choquet *et al.*, 1997; Guharay and Sachs, 1984; Hudspeth, 1985; Palecek *et al.*, 1997; Riveline *et al.*, 2001) and could measure the force produced by motors (Finer *et al.*, 1994; Gelles *et al.*, 1988; Kishino and Yanagida, 1988; Kuo and Sheetz, 1993). These studies also introduced physical methods to probe the role of forces either on single molecules or entire cells. Cell biologists had also predicted, based in part on the phenotype of *Drosophila* integrin mutants, that integrin is a mechanochemical transducer (Ingber, 1991; Leptin *et al.*, 1989). Third, the achievements of the various genome programs prompted the field to embrace less gene-centered approaches of developmental biology. Last but not least, progress in time-lapse imaging progressively enabled all of us to see cells alter their shapes before our eyes. Together these changes paved the way to a less deterministic view of embryonic development and to the study of forces and tension as a major parameter in driving morphogenesis.

We have reached a stage where it becomes possible to reconcile two philosophically opposing views of embryonic morphogenesis, a purely gene-centric view and a mechanical-only view. At the subcellular level, numerous biophysical studies now provide rules to account for the behavior of many proteins in response to forces (Moore *et al.*, 2010). What is missing at one end of the spectrum is an understanding of the physical laws accounting for soft matter behavior, which could, for instance, explain self-organization, and an understanding of hydrodynamics involving microflows. At the other end of the spectrum, biologists need to gain an integrated view of the different forces that sculpt the embryo in a series of linked events, and their relationship to "classical" cellular processes.

This volume of *Current Topics in Developmental Biology* entitled *Forces and Tension in Development* hopes to touch on both aspects. One of its goals is to cover through a series of six reviews the most recent findings in the field of embryonic morphogenesis for which the role of forces has been teased apart. Chapter 2 explores how microtubule-based cilia generate microscopic flows and how these flows influence development. Chapters 4–7 discuss actomyosin- and adhesion-based forces and their role in cell shape changes, cell sorting, and the emergence of novel structures; Chapter 8 discusses the mechanotransduction processes downstream of tension. In parallel, Chapters 1 and 3 explore the theoretical basis for the mechanism of ciliary beating and actomyosin oscillatory behaviors. A second goal of this volume is to present some of the methods used to measure the effects and magnitude of forces *in vivo*.

Microtubule-Generated Flows

Chapter 1 by Charles B. Lindemann discusses conceptual and experimental data supporting the "geometric clutch hypothesis" to account for ciliary beating, which results from cycles of dynein attachment and detachment from microtubules as microtubules bend. Willy Supatto and Julien Vermot in turn explore the physical features of flow fields generated by ciliary beating as they pattern left–right asymmetry or control otolith formation. Their contribution also provides interesting insights into the methods used to image microscopic flows in embryos.

Actomyosin-Based and Adhesion-Based Forces

In a chapter particularly accessible to nonphysicists, Karsten Kruse and Daniel Riveline provide some basic elements to understand the physics behind the spontaneous mechanical oscillations of the actomyosin cytoskeleton. Next, Matteo Rauzi and Pierre-François Lenne discuss how actomyosin-based cortical forces can drive cell shape changes in embryos, with particular emphasis on *Drosophila*. Their contribution also reviews the pros and cons of laser-based methods to measure the effects of forces in vivo. Claire M. Lye and Bénédicte Sanson focus their attention on the integration of cell intrinsic and extrinsic forces shaping *Drosophila* embryos. Gabby Krens and Carl-Philipp Heisenberg examine the different models that have been proposed to account for cell sorting *in vivo*, with a focus on vertebrate embryos. Their chapter also reviews the different methods that have been used to exert forces on isolated tissues. In a complementary and lively contribution, Lance A. Davidson examines how tissues integrate forces and stress to generate new structures during morphogenesis. He also introduces methods for quantifying embryo-mechanics, outlining the limitations of each method. In a final chapter, Emmanuel Farge looks at mechanosensing and at the mechanotransduction pathways induced by forces during development in different systems, and how tension can feedback to the nucleus.

Undoubtedly, understanding how forces mould the embryo will require the input of physicists to make sense of the hydrodynamics of microscopic flows or to define the principles behind the behavior of soft matter. Time has come for biologists to interact with scientists from an altogether different background. After all, it is perhaps appropriate to recall that one of the most recent influential figures in developmental biology, Lewis Wolpert, who put forward the concept of the French flag for positional information[1] (Wolpert, 1969), was trained as an engineer.

Additional Foreword to Physicists

While I anticipated that most readers of this volume of *Current Topics in Developmental Biology* will most likely be biologists interested by the role of physical forces, I do hope it will prove useful and accessible to physicists. The references mentioned above to trace back the evolution of ideas in biology might go beyond the background of physicists, but suffice it to say that until recently, except for a few rare pioneers, biologists did not consider the influence of mechanical forces in embryonic development. Not only is this not the case anymore, but also an increasing number of biologists are using approaches and concepts imported from physics.

The main difficulty for a fruitful interaction between scientists with different backgrounds is to overcome the barrier of language. Biologists are generally not comfortable with many concepts in physics such as potential, degree of freedom, or phase transition (to name a few) and have generally not used their college thermodynamics since a long time. Likewise for physicists, the anatomy of embryos, all the knowledge hidden behind gene names, or recent concepts of induction/signaling can represent serious deterrents.

From this respect, physicists should find the chapters in this volume quite accessible. In particular, Chapters 5 by Lye and Sanson, 6 by Krens and Heisenberg and 7 by Davidson include remarkably simple, yet detailed enough, presentations of the anatomy and developmental stages of *Drosophila* and *Xenopus* embryos, two of the most heavily studied models in developmental biology. These two chapters should help newcomers get a sense for the organization of these embryos and of their morphogenetic movements. All chapters have been contributed either by physicists by training or by biologists now quite familiar with physical methods. These chapters

[1] The French flag model suggests that a diffusible molecule made at one end of the flag would diffuse in a graded manner and impart at each position of the flag (vertical bands) different information that would in turn dictate cell behavior/destiny. This model was inspired in part by theories on the role of gradients in biology put forward by the mathematician Alan Turing in the 1950s.

introduce the key background information for reading the chapter, and none of them relies too heavily on prior knowledge of classical biology.

REFERENCES

Beloussov, L. V., Dorfman, J. G., and Cherdantzev, V. G. (1975). Mechanical stresses and morphological patterns in amphibian embryos. *J. Embryol. Exp. Morphol.* **34,** 559–574.

Choquet, D., Felsenfeld, D. P., and Sheetz, M. P. (1997). Extracellular matrix rigidity causes strengthening of integrin-cytoskeleton linkages. *Cell* **88,** 39–48.

Finer, J. T., Simmons, R. M., and Spudich, J. A. (1994). Single myosin molecule mechanics: Piconewton forces and nanometre steps. *Nature* **368,** 113–119.

Garcia-Bellido, A. (1975). Genetic control of wing disc development in Drosophila. *Ciba Found. Symp.* 161–182.

Gelles, J., Schnapp, B. J., and Sheetz, M. P. (1988). Tracking kinesin-driven movements with nanometre-scale precision. *Nature* **331,** 450–453.

Guharay, F., and Sachs, F. (1984). Stretch-activated single ion channel currents in tissue-cultured embryonic chick skeletal muscle. *J. Physiol.* **352,** 685–701.

Hudspeth, A. J. (1985). The cellular basis of hearing: The biophysics of hair cells. *Science* **230,** 745–752.

Ingber, D. (1991). Integrins as mechanochemical transducers. *Curr. Opin. Cell Biol.* **3,** 841–848.

Keller, R. E. (1980). The cellular basis of epiboly: An SEM study of deep-cell rearrangement during gastrulation in Xenopus laevis. *J. Embryol. Exp. Morphol.* **60,** 201–234.

Keller, E. F. (2002). Making Sense of Life: Explaining Biological Development with Models, Metaphors, and Machines. Harvard University Press, Cambridge.

Kishino, A., and Yanagida, T. (1988). Force measurements by micromanipulation of a single actin filament by glass needles. *Nature* **334,** 74–76.

Kuo, S. C., and Sheetz, M. P. (1993). Force of single kinesin molecules measured with optical tweezers. *Science* **260,** 232–234.

Leptin, M., Bogaert, T., Lehmann, R., and Wilcox, M. (1989). The function of PS integrins during Drosophila embryogenesis. *Cell* **56,** 401–408.

Moore, S. W., Roca-Cusachs, P., and Sheetz, M. P. (2010). Stretchy proteins on stretchy substrates: The important elements of integrin-mediated rigidity sensing. *Dev. Cell* **19,** 194–206.

Nusslein-Volhard, C., and Wieschaus, E. (1980). Mutations affecting segment number and polarity in Drosophila. *Nature* **287,** 795–801.

Odell, G. M., Oster, G., Alberch, P., and Burnside, B. (1981). The mechanical basis of morphogenesis. I. Epithelial folding and invagination. *Dev. Biol.* **85,** 446–462.

Palecek, S. P., Loftus, J. C., Ginsberg, M. H., Lauffenburger, D. A., and Horwitz, A. F. (1997). Integrin-ligand binding properties govern cell migration speed through cell-substratum adhesiveness. *Nature* **385,** 537–540.

Peifer, M., and Wieschaus, E. (1990). The segment polarity gene armadillo encodes a functionally modular protein that is the Drosophila homolog of human plakoglobin. *Cell* **63,** 1167–1176.

Priess, J. R., and Hirsh, D. I. (1986). Caenorhabditis elegans morphogenesis: The role of the cytoskeleton in elongation of the embryo. *Dev. Biol.* **117,** 156–173.

Riveline, D., Zamir, E., Balaban, N. Q., Schwarz, U. S., Ishizaki, T., Narumiya, S., Kam, Z., Geiger, B., and Bershadsky, A. D. (2001). Focal contacts as mechanosensors: Externally applied local mechanical force induces growth of focal contacts by an mDia1-dependent and ROCK-independent mechanism. *J. Cell Biol.* **153,** 1175–1186.

Steinberg, M. S. (1962). Mechanism of tissue reconstruction by dissociated cells. II. Time-course of events. *Science* **137,** 762–763.
Tepass, U., Theres, C., and Knust, E. (1990). Crumbs encodes an EGF-like protein expressed on apical membranes of Drosophila epithelial cells and required for organization of epithelia. *Cell* **61,** 787–799.
Thompson, D. W. (1917). On Growth and Form. Cambridge University Press, Cambridge, England.
Wolpert, L. (1969). Positional information and the spatial pattern of cellular differentiation. *J. Theor. Biol.* **25,** 1–47.

MICHEL LABOUESSE

CHAPTER ONE

Experimental Evidence for the Geometric Clutch Hypothesis

Charles B. Lindemann

Contents

Abstract

The cilia and flagella of eukaryotic cells are complex filamentous organelles that undulate rapidly and produce propulsive force against the fluids that surround the living cell. They provide a number of important functions in the life cycle of higher organisms including humans. A flagellum propels the spermatozoa to the site of fertilization and cilia move the egg through the oviduct to the uterus and have a role in left–right asymmetry in the developing embryo and contribute to normal brain morphology. The geometric clutch hypothesis is a mechanistic explanation of how the repetitive bending of cilia and flagella is generated. This chapter recounts the events leading to the development of the geometric clutch hypothesis, explores the conceptual framework of the hypothesis as it relates to properties of the axoneme, and considers the experimental support for the existence of such a mechanism in real cilia and flagella.

Department of Biological Sciences, Oakland University, Rochester, Michigan, USA

Current Topics in Developmental Biology, Volume 95
ISSN 0070-2153, DOI: 10.1016/B978-0-12-385065-2.00001-3

1. Introduction

Cilia and flagella are slender filamentous organelles of eukaryotic cells present in a remarkable variation of sizes and perform many useful functions. Historically, they were named "cilia" if there were many of them, and "flagella" if they were longer and occurred singly or in pairs on a cell. The variation in length is extreme. They can be long, up to 6 cm in length, like the sperm tails of certain *Drosophila* (fruit fly) species, or very short, like the respiratory cilia in humans and the cilia of many protozoa. A drop of pond water visualized microscopically will yield single-celled organisms using cilia to swim, to feed, to crawl, and to attach to each other for mating.

The pattern of movement of cilia and flagella is quite diverse. Some appear to do a paddle stroke, somewhat like the breast stroke of a human swimmer, while others execute an almost perfectly sinusoidal pattern of propagating waves. There are flagella that reverse their beating direction (e.g., *Trypanosomes*) and there are flagella that can change from beating symmetrically to a paddling motion depending on external stimuli (e.g., *Chlamydomonas)*. Dr. Michael Sleigh published a wonderful book (Sleigh, 1962) on cilia and flagella examining the types of ciliary beating and the many biological applications that cilia and flagella carry out. In the summer of 1968, it was my good fortune to be assigned as Dr. Sleigh's summer assistant to help him film ciliary beating in the lab of my Ph.D. mentor, Dr. Robert Rikmenspoel. It was my introduction to the diversity of cilia and flagella.

Cilia are typically short, and a single cell can have more than a hundred of them, while flagella are usually longer and occur singly, or in pairs. A cilium typically displays a two-phase beat comprised of a stiff-looking effective stroke and a recovery stoke that rolls along the cilium like a whip lash, as illustrated in Fig. 1.1. Long flagella ordinarily have multiple undulations propagating along the flagellum in symmetry reminiscent of traveling waves (Fig. 1.1). It is understandable why early observers thought that cilia and flagella were different from each other. The advent of electron microscopy (EM) established that cilia and flagella all possess a common internal structure of tubular filaments arranged in a characteristic 9 + 2 pattern and are, in fact, just variations of the same cell organelle. This internal structure, called the axoneme, is illustrated in Fig. 1.2.

In the late 1960s, owing to the pioneering work on excitation–contraction coupling in skeletal muscle, there was much interest in the role of electrical signals in activating contractile events in cells. A study from Rikmenspoel's lab determined that there was no detectable variable electrical voltage that correlated with the flagellar beat cycle of bull sperm (Lindemann and Rikmenspoel, 1971). However, a key observation during

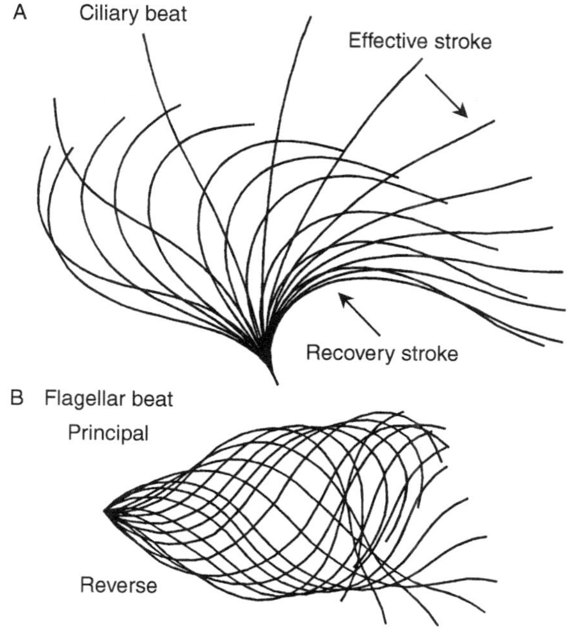

Figure 1.1 Ciliary and flagellar beating. (A) The typical beating pattern of a cilium is illustrated showing the biphasic alternation of a stiff effective stroke and a rolling recovery stroke. (B) The pattern of beating that is typically seen in flagella is illustrated. Bends initially produced near the base propagate toward the tip and alternate in the direction of bending. By generally accepted convention, the stronger curved bends are called the principal or P bends and the lesser curved bends are called the reverse or R bends. Both illustrations shown are output sequences from the geometric clutch computer model (A) simulating a 10 μm cilium with a fixed base and (B) simulating a 30 μm flagellum with a rotating base.

that work led to a series of experiments to determine what was required to sustain flagellar beating. This was owing to the fact that following rupture of the sperm head, the flagellum continued to beat for a while before slowing to a stop.

Another early observation was that pieces of cut bull sperm flagella immediately stopped beating following dissection, but a beat could almost always be restored if the flagellum was bent with a glass microprobe (Lindemann and Rikmenspoel, 1972a). The act of bending the flagellum restored the capacity for rhythmic beating in a repeatable way. This observation was clear evidence that the beating mechanism was responsive to mechanical input. A subsequent discovery showed that 4 mM ADP could also restore the capacity of rhythmic beating even without imposed bending (Lindemann and Rikmenspoel, 1972b).

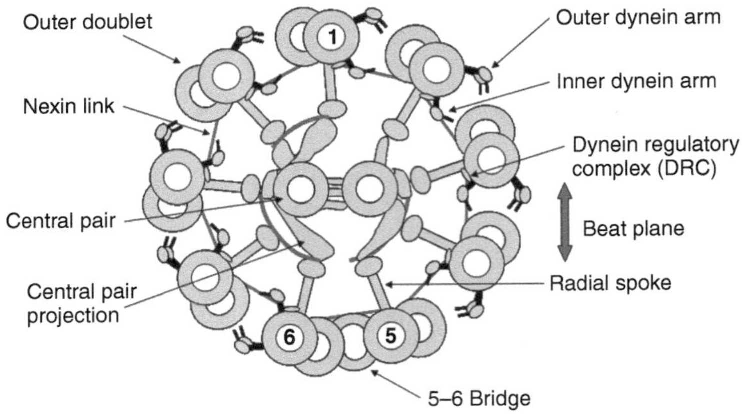

Figure 1.2 A schematic diagram of the eukaryotic flagellar/ciliary axoneme shown in cross section. Structures discussed in the chapter are labeled for the convenience of the reader. Reproduced from Lindemann (2007) with permission.

At about the same time, Summers and Gibbons (1971) demonstrated that fragmented sea urchin flagella would disintegrate in the presence of Mg-ATP, but only if they were initially treated with a protease (i.e., trypsin). This dramatic demonstration showed that the individual outer microtubule doublets can slide on each other by the action of the dynein arms and provided the first physical evidence that the doublets are held together by protein linkages that resist the sliding.

Two different methods of preparation can cause rat sperm to disintegrate by microtubule sliding. The sperm extrude doublets predominantly from either one side or the other side of the axoneme depending on which method is employed. The central pair plus doublets 3 and 8, and their associated outer dense fibers, frequently remain intact after both methods and could be seen in complete isolation forming a stable partition across the middle of the axoneme (Lindemann *et al.*, 1992). The existence of such a special structural partition in the axoneme was suggested by studies on compound cilia (Tamm and Tamm, 1981, 1984), and the new results suggested that it might be a more widespread feature of cilia and flagella.

The presence of such a partition across the axoneme should effect the bending of a cylinder of flexible filaments (the microtubules). To explore this concept, I constructed a wooden model using nine flexible basket weaving reeds arranged in a circle connected together by strands of silicone adhesive. The silicone strands are meant to be equivalent to the interdoublet linkages visualized in transmission electron microscopy (TEM) images (Linck, 1979; Warner, 1976; Witman *et al.*, 1978). The wooden model has a central partition formed by two centrally located reeds permanently linked via cotton string to the doublets corresponding to #3 and #8 in the axoneme. The nine outer wooden reed doublets are glued together at the base with an extra wooden dowel between each of the nine and reinforced at the base with string by crisscrossing the center opening. A similar basal anchoring system is seen at the basal bodies of real flagella and cilia (Pitelka, 1974; Ringo, 1967).

The wooden model confirmed that having a permanent partition makes the structure much stiffer in the plane parallel to the partition and more flexible in the plane perpendicular to the partition. This is expected and is likely part of the reason that most flagella tend to have a planar beat rather than a helical beat, a relationship first recognized by Gibbons (1961). Unexpectedly, the structure systematically collapsed and flattened when it was bent in the preferred bending direction, as shown in Fig. 1.3A.

Prior to this observation with the wooden model, little was published on the possible collapse of the axoneme when it is bent. Some of the TEM work of Warner (1978) seemed to show such a collapse, but TEM

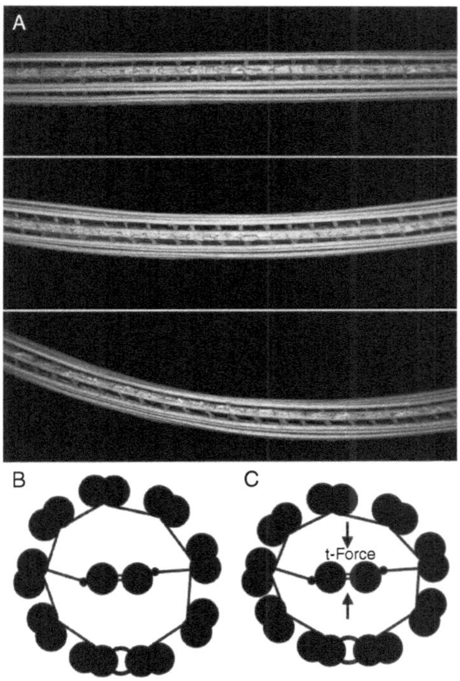

Figure 1.3 The effect of passive bending on a wooden model of the flagellar axoneme. In (A), the model is progressively bent and exhibits a collapse in the plane of bending. (B) Illustrates the spacing of the wooden reeds and interconnections when the model is straight. (C) Illustrates the change in configuration that results when the model is bent and the direction of the t-force which develops during bending. Reproduced from Lindemann (2004) with permission.

micrographs of flagella are tricky to interpret. Since they are thin slices through a cylindrical structure, the flattening can easily be attributed to the angle of the cut through the structure. Consequently, the evidence of distortion could be interpreted as a sectioning artifact, and little importance was given to this observation by most investigators.

After constructing the wooden model, we were faced with a physical dilemma. How could a real axoneme not suffer from the same effects of stress and strain as the wooden model? It became a problem in elementary physics. There are linkages between the doublets of a real flagellum that restrict free sliding. We know that because these linkages are cut by proteases. We also know they must be able to stretch, or the axoneme would be rigid and unable to bend. Consequently, the linkages must contribute tension and compression on the outer doublets when the structure bends. Basic physics dictates that a curved structure that is under tension will develop stress transverse to the curve. This is the principle that holds the road up under the arch of a suspension bridge; it works quite well. It is therefore an inescapable conclusion that there must be transverse forces (t-forces) acting across the axoneme when it is bent, and that these forces should act to distort the axoneme.

The recognition that transverse stress develops in the axoneme leads directly to a second, and possibly even more important, consideration. If passive bending generates tension on the doublets that distorts the axoneme, the active forces produced by the motor proteins must act similarly. In fact, when the dynein motors bend the axoneme, they do so by pulling on the doublets generating forces large enough to overcome the elastic resistance of the interdoublet linkages plus the external viscous drag on the flagellum. This requires that the tension on the doublets derived from the motor proteins must be larger than the tension which develops from stretching the elastic linkages. Once again, a curved flexible structure that is under tension must develop a transverse stress.

Figure 1.4 shows the direction of the transverse stress that develops when a bend is passively induced from the outside and when bending is caused by the active internal motors. Passive bending collapses the structure by creating a pinching force. This pushes the doublets closer together and is referred to as a positive t-force. This is what is observed with the wooden model. When the axoneme is actively bent by its own motor proteins, the transverse stress that develops produces an outwardly directed force couple that pries the doublets apart; for modeling purposes, this t-force is assigned a negative sign. This principle is the essence of the geometric clutch hypothesis.

From TEM micrographs, it appears that the outer doublets are positioned too far apart for the dynein arms to form connections. If the native condition of the doublet ring in the living cell is such that they need to move toward each other in order for the dyneins to form force bearing attachments, then things start to make sense. This is a possible explanation

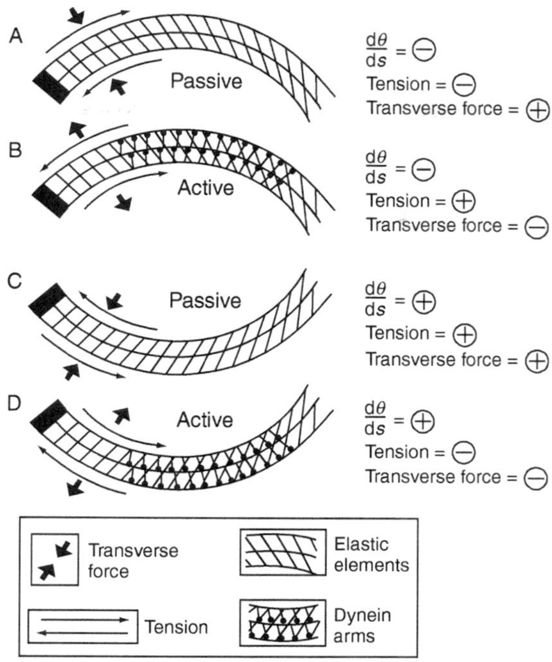

Figure 1.4 The t-force in passive and active bends. Panels (A)–(D) illustrate the direction of t-force that develops when the axoneme is bent passively (A and C) and actively (B and D) by the action of the dynein motors. Note that the direction of the resulting t-force couplet always compresses the diameter of the axoneme when the structure is passively bent and always distends the diameter of the axoneme when it is bent by the action of the dyneins. The mathematical convention used in the geometric clutch model is shown to the right of each illustration. Reprinted from Lindemann (1994a) with permission.

for the activation of the cutoff pieces of flagella when bent with a microprobe (Lindemann and Rikmenspoel, 1972a). The imposed bending partially collapses the axoneme and allows the dyneins to establish attachments and produce motive force. A similar mechanism can also account for switching the dyneins "off." Once dyneins engage, they actively bend the flagellum and generate a negative t-force which should be able to bring about their own disengagement by prying the doublets apart, as shown in Fig. 1.4.

This simple scheme was tested in a computer model to see if activating and deactivating the dyneins on the basis of the transverse stress can produce repetitive beating similar to that observed in flagella and cilia. The computer model confirmed that t-force can satisfactorily serve to regulate engagement and disengagement of the motors and produce sustained oscillations that resemble ciliary and flagellar beating (Lindemann, 1994a). However, making the attachment of the motors strictly dependent on t-force did not allow the model to initiate beating from a straight initial position, because in a straight flagellum there is no transverse stress.

A second version of the computer model utilized a stochastic probability of the dyneins attaching randomly when the t-force was zero and increased the probability further as the number of dynein attachments increased (Lindemann, 1994b). This created the potential for attachment cascades. The stochastic mechanism of attachment is quite reasonable. Dyneins have a flexible stem that attaches them to the doublet that they reside on, and thermal kinetic motions of the dynein head make it likely that a small number will bind to the adjacent doublet even when the interdoublet distance is at its maximum. The few dyneins that manage to attach will exert a spring tension drawing the doublets closer, since their equilibrium length is less than the interdoublet gap they are bridging, as illustrated in Fig. 1.5. This introduces the idea of dynein adhesion force. Attached dynein not only draws the doublets closer together but also resists prying them back apart. When stochastic behavior and the dynein adhesion effect were incorporated into the basic geometric clutch formulation, only then did the simulated cilia/flagella initiate beating from a straight starting position (Lindemann, 1994b).

These two concepts, regulation of both dynein attachment and detachment by the t-force developed from stress in the axoneme, and the dynein

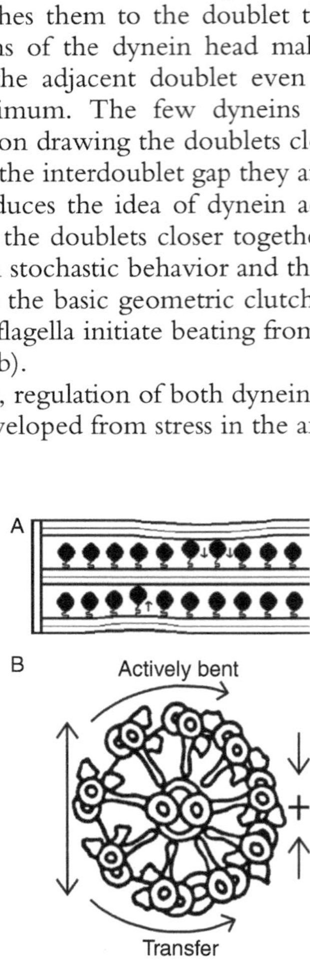

Figure 1.5 Dynein adhesion force. (A) Illustrates the relationship of the interdoublet spacing to the dynein length. The spacing of the doublets in the resting axoneme is greater than the dimension of dynein at rest. Random attachment of dyneins therefore results in a decrease in the interdoublet spacing. If sufficient attachment occurs, it can alter the spacing of the doublets in the intact axoneme as shown in (B). Reprinted from Lindemann (1994b) with permission.

adhesion force produced by the dynein crossbridges acting to draw the doublets together and resist detachment, are the two most essential parts of the geometric clutch hypothesis.

2. Experimental Support

There is a considerable experimental base of support for the geometric clutch hypothesis. Shortly after the publication of the two computer modeling studies (Lindemann, 1994a,b), a review article examined the experimental literature on cilia and flagella in the context of the new hypothesis (Lindemann and Kanous, 1997). Of the relevant points to emerge from that review, some are still noteworthy. The most conserved feature of all cilia and flagella capable of beating is that they have multiple doublet microtubules separated from each other by mechanical connections that maintain the spacing between them. The doublet size and spacing interval are actually more conserved than the basic 9 + 2 arrangement.

There are motile 9 + 0 axonemes, which lack a central apparatus, motile arrays of many doublets, and even flagella with other numbers of doublets (Dallai *et al.*, 2006; Ishijima *et al.*, 1988; Mencarelli, *et al.*, 2001; Phillips, 1974; Woolley, 1997). If they are motile, then the doublet spacing and interdoublet linkages are conserved. This is an important clue as to the nature of the basic beating mechanism and one that is often minimized. Therefore, to be a successful theory of flagellar beating, the proposed mechanism of oscillation must be sufficiently robust to explain coordination of a beat in axonemes lacking the central pair elements.

The first TEM studies of the ultrastructure of the axoneme showed that the spacing between the outer doublets is too large to be bridged by the dynein arms (Afzelius, 1959, 1961; Gibbons and Grimstone, 1960). This is a consistent feature in virtually all types of cilia and flagella when they are prepared from living material by the usual fixatives. In contrast, when flagella are in the rigid condition, called rigor waves, induced by rapidly removing ATP from reactivated detergent-extracted sperm cell models, the doublets are closer together and the dynein are observed bridging the doublets (Gibbons, 1975; Fig. 10A).

Figure 1.6 (Fig. 1 from Zanetti *et al.*, 1979) shows the dramatic difference that can be induced in the configuration of the axoneme as a result of variation in divalent cation concentration. In that study, flagella were first demembranated with a detergent and then fixed in the presence low or high Mg^{2+} or Ca^{2+}. A high divalent cation condition increases the formation of dynein crossbridges, and decreases the interdoublet distance as well as the flagellar diameter (Warner, 1978; Zanetti *et al.*, 1979). Apparently, the attachment of the dyneins is capable of pulling the doublets closer and holding them together.

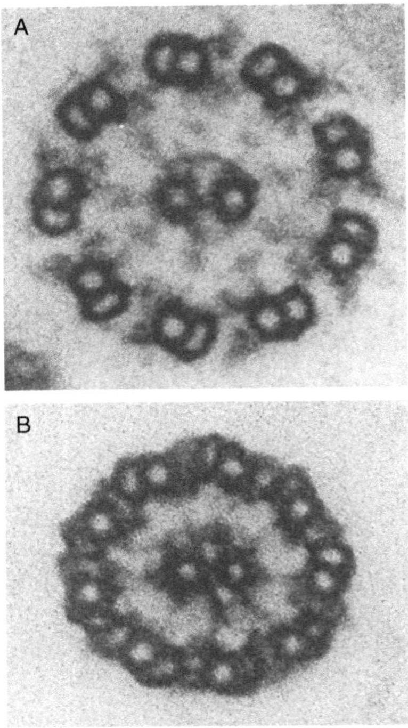

Figure 1.6 Axoneme configurations. This figure, reprinted from Zanetti *et al.* (1979), shows the extreme change in the spacing of the outer doublets and spokes that result when all of the dynein arms are bridged to their adjacent doublet. Both axonemes displayed are cross sections of isolated *Tetrahymena* cilia, demembranated and fixed in 40 mM HEPES buffer (pH 7.4). The axoneme in (B) was in the added presence of 6 mM $MgSO_4$ which resulted in all dynein arms forming bridges to the B subfibers. Reproduced with permission. Magnification $\times$ 210,000.

These observations support the idea that when the dyneins grab onto their binding sites, they hold on to the neighboring doublets with considerable tenacity, enough to alter the spacing of the doublets. Since they can create enough tension to pull the doublets together, likewise they should also resist imposed stress acting to pulling the doublets apart. These observations support the concept of dynein adhesion force.

Given these basic facts about the axoneme, it is reasonable to expect that the imposed stress within a bent axoneme would assist in engaging the dyneins by pushing some of the doublets closer together. There are many experimental studies that confirm this point. The bending of an isolated fragment of flagellum can activate beating (Lindemann and Rikmenspoel, 1972a). Intact flagella that are immotile (Ishikawa and Shingyoji, 2007; Okuno and Hiramoto, 1976) or are immotile due to a deleterious mutation (Hayashibe *et al.*, 1997; Omoto *et al.*, 1996) can sometimes be activated by bending. In addition, it was shown that bending an axoneme partially digested with elastase can activate interdoublet sliding (Hayashi and Shingyoji, 2008; Morita and Shingyoji, 2004).

The t-force that develops in the beat cycle is dependent on the tension experienced by the doublets. If the basal anchor (i.e., the basal body, or in the case of mammalian sperm the connecting piece) is removed, the tension on the doublets does not accumulate at the basal end of the flagellum. In the geometric clutch view of things, this reduces both the torque that bends the axoneme and the t-force available for switching. It was postulated that the basal anchor plays a role in beat initiation and bend propagation direction (Lindemann, 2007; Lindemann and Kanous, 1995). As noted earlier, this can explain the loss of spontaneous beating when bull sperm are dissected (Lindemann and Rikmenspoel, 1972a). Fujimura and Okuno (2006) confirmed the requirement of a fixed end for spontaneous beating in sea urchin sperm.

It is reasonable to expect that, once the dyneins do attach, they will resist their own detachment because they have the property of adhesion and can exert some force to hold the doublets together. This property would allow the dyneins to remain attached long enough to act as motors and bend the flagellum. When the stress that the motors impose on the doublets becomes large enough, it is equally reasonable to propose that it might overcome the adhesion of the dyneins and terminate their motor action. This is the fundamental basis of the beating mechanism proposed in the geometric clutch hypothesis.

The proposed mechanism is simple enough to be intuitive, but it is defined well enough that it can be tested. To be of utility, a hypothesis must have predictive value and be testable. To this end, a number of published studies tested various predictions of the model. If switching depends on the t-force, and the t-force is determined by the product of tension on the doublets and local curvature, then devising situations where this product is sufficiently reduced should interrupt the switching and create an arrest of the beat. This was done using reactivated bull sperm and blocking the motion to prevent curvature development or clipping the flagellum to reduce t-force development. Both techniques resulted in the predicted arrest behavior and the results coincided with the prediction of the geometric clutch computer model (Holcomb-Wygle *et al.*, 1999; Schmitz *et al.*, 2000).

There is documentation that Ni^{2+} disables the dyneins on one side of the axoneme, while leaving the opposite set functional (Kanous *et al.*, 1993; Lindemann *et al.*, 1980). An extensive study of Ni^{2+}-inhibited reactivated bull sperm showed that dyneins on the unaffected side can be activated to produce a beat if a microprobe is used to bend the flagellum and provide the bending that the disabled dyneins would normally contribute (Lindemann *et al.*, 1995). This result is also predicted by the geometric computer model.

In an effort to recruit the efforts of other investigators, predictions of the model and ways that it could be tested experimentally were published (Lindemann, 2007; Lindemann and Kanous, 1995; Lindemann and Lesich, 2010). Many of the laboratories that compared their experimental results with the underlying assumptions of the geometric clutch hypothesis are cited in this review.

3. The Underlying Assumptions

The geometric clutch hypothesis depends on a number of key assumptions that are amenable to experimental testing. These assumptions are as follows:

1. There must be elastic linkages that resist microtubule sliding and create stress (t-force) between the microtubules when the axoneme is bent. This is necessary because the hypothesis requires that passive bending of the flagellum create a t-force that acts to push doublets together in the plane of bending and this positive t-force is instrumental in the activation of dynein.
2. There must be distortion of the axoneme that accompanies the waves of bending during the beat cycle. This is essential to the hypothesis because it is the interdoublet spacing that determines the activation and deactivation of the dynein motors.
3. Transverse stress, in the range present in the flagellar axoneme, must be able to detach working dyneins from their binding sites. This is crucial, as the hypothesis contends that switching the dyneins off in the beat cycle is regulated by the accumulation of stress (negative t-force).
4. The binding affinity of dynein for its attachment to the B subtubule of the adjacent doublets should change the switch point of the beat cycle and alter the curvature that develops during the beat cycle. This is a direct consequence of the switching event being a balance of dynein adhesion and t-force in the geometric clutch hypothesis.

Direct experimental observations that are cogent to each of these theoretical assumptions of the geometric clutch hypothesis are examined next.

4. Support for the Four Basic Assumptions

4.1. Are there elastic linkages that resist sliding?

The first evidence for elastic linkages between the outer doublets that resist sliding came from the work of Summers and Gibbons (1971). They showed that interdoublet sliding would not occur even in fragmented sea urchin sperm flagella unless proteolytic digestion with trypsin was employed. This implied that a protein connection(s) between the doublets restricted the free sliding of the doublets unless the links are destroyed. Brokaw (1971, 1972a, b) showed that the sliding doublet model of flagellar bending could produce life-like motility patterns in a computer model if a linear elastic resistance was acting between the doublets. Yagi and Kamiya (1995) used force-calibrated glass microneedles to directly measure the longitudinal resistance to shear in *Chlamydomonas* flagella lacking the central pair. From this they estimated the elastic constant for the interdoublet linkages.

These results establish the existence of elastic interdoublet linkages, but a combination of reports negated the concept of permanent elastic connections. Bozkurt and Woolley's (1993) morphological study of the configuration of the nexin links in relationship to interdoublet sliding showed that the links did not stretch but maintained their morphology after sliding occurred. This suggested the interdoublet connections are movable linkages. Kamiya's lab (Minoura *et al.*, 1999) subsequently published a more extensive study of the interdoublet elasticity that found a more complex behavior which also seemed to suggest the interdoublet linkages could step.

A more recent study (Lindemann *et al.*, 2005) examined the mechanical behavior of rat sperm when the dynein motors were disabled by high concentrations of sodium metavanadate, a condition that is reported to put dynein into the unbound state (Sale and Gibbons, 1979). When flagella are bent with a glass microprobe, the part of the flagellum beyond the probe contact point exhibits a bend in the opposite direction of the imposed bend (a "counterbend"), as seen in Fig. 1.7 (Lindemann *et al.*, 2005). When released from the probe, the bend and the counterbend both relax and the flagellum returns to a nearly straight configuration (Fig. 1.7A). This counterbend behavior can only be explained if there are permanent elastic links between the doublets that return to their original length when the flagellum is released.

The large sperm of rodents have a number of accessory structures that are not present in more typical 9 + 2 flagella. A subsequent study by Pelle *et al.* (2009) on sea urchin sperm is an important verification that counterbend behavior is also present in a more conventional 9 + 2 flagellum. From that investigation, using a computer model of a passive sea urchin flagellum created by Charles Brokaw, it was determined that the interdoublet elasticity of sea urchin sperm is quite large and corresponds to a linear elastic model of interdoublet elastic resistance.

How can this apparent discrepancy in the behavior of elastic linkages be explained? One possibility is that there is a moveable structure in addition to the protein connections that provide the shear resistance. This is implicated in a Minoura *et al.* (1999) study that showed that the elastic resistance had both a stepwise and a cumulative component. It is also suggested in the most recent morphological studies of the interdoublet linkages. Cryoelectron

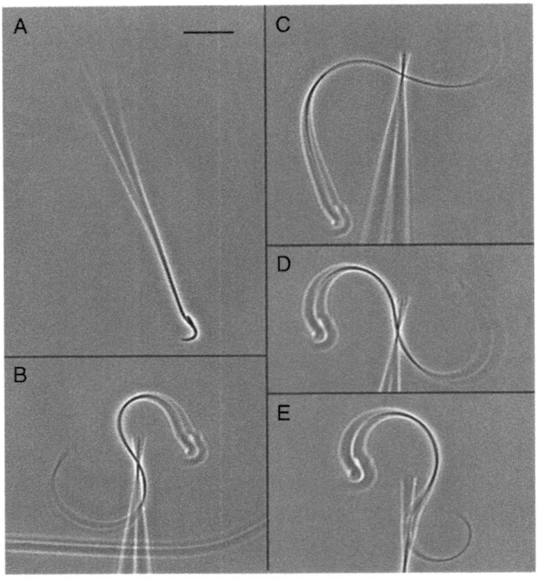

Figure 1.7 The counterbend response of rat sperm. (A) Image of a Triton X-100 demembranated rat sperm inhibited with 50 μM sodium vanadate in the presence of 0.1 mM ATP. This treatment disables the dynein arms. When the flagellum is bent with a microprobe, the portion of the flagellum beyond the probe exhibits a bend in the opposite direction that is proportional to the amount of induced shear (B–E). Counterbend behavior can only be explained if elastic resistance to interdoublet shear is present in the axoneme. Bar = 20 μm. Reproduced from Lindemann *et al.* (2005) with permission.

tomography examination found that the nexin links are a huge complex of proteins that share identity with the dynein regulatory complex (DRC; Heuser *et al.*, 2009; Nicastro *et al.*, 2006). These are most likely the band-like linkages that Bozkurt and Woolley (1993) saw in their micrographs, as they are bulky and ribbon like. These may be the elements that do not appear to stretch, but appear to translocate in their micrographs.

Conversely, another research study (Baron *et al.*, 2007) identified specific proteins that are functionally required to hold the ring of nine doublets together in a *Trypanosome* flagellum. These proteins are orthologs of known *Chlamydomonas* proteins, but are not among the proteins of the DRC that are part of the structure identified by Heuser *et al.* (2009). The case for interdoublet elastic resistance is convincing, but the identity of the linkage that is responsible for the resistance is still uncertain. Evidence suggests that there is more than one interdoublet linker. The specific proteins that provide the interdoublet spacing and elastic resistance are just now emerging.

4.2. Does the real axoneme show distortion during beating?

The working mechanism of the geometric clutch hypothesis is integrally tied to the idea that as the dynein engage and disengage during the operation of the flagellum, the doublets move closer together and further apart. Further, the action of t-force on the elastic elements between the doublets should, in principle, cause the entire axoneme to distort as the t-force becomes large in an active bend. An analysis of the magnitude of the t-force in real flagella is presented in detail in Lindemann (2003). One of the most problematic weaknesses in support of the hypothesis has been the dearth of hard evidence that there is any distortion of the axoneme accompanying flagellar beating. Gibbons (1975) specifically looked for such distortion in flagella rapidly locked into rigor so that bending waves appeared to be preserved. There was little evidence of distortion found. Flagella have been extensively examined by EM and therefore, the lack of evidence for structural distortion is a legitimate concern. This deficiency was pointed out by other investigators (Wirschell *et al.*, 2007).

While the evidence for distortion of the axoneme during the beat is sparse, it is not completely lacking. Gibbons and Gibbons (1973) performed an extensive study of the ultrastructure of reactivated sperm, both with and without the outer arms present. The micrographs of control sperm show specific outer doublets with apparently completed dynein bridges. They stated in their analysis of their data, "The average spacing between tubules appears to be rather less when a bridge is present than when it is not, suggesting that formation of a crossbridge is associated with a moving together of the affected doublet tubules." This is direct support for the idea that the active doublets, in the regions where dyneins are engaged, move closer to each other.

Part of the problem of observing the relative movements of the doublets during the beat cycle is that structural distortion during the beat is a product of stress acting on an elastic solid. If the stress is terminated, then the elastic solid will recoil to its own mechanical equilibrium. Most EM fixation methods are relatively slow and stop motility before the specimen is sufficiently stabilized to prevent elastic recoil of the flagellum. Consequently, stresses are allowed time to dissipate before the specimen is mechanically stable. The critical observation to determine if the fixed specimen captured the stress and distortion present during the beat is whether or not the waveform of the beat is also preserved. If the forces acting to bend the flagellum are locked in place by the fixation, then the flagellum should retain its shape.

Mitchell (2003) published a study designed to look at the orientation of the central pair relative to the bending cycle of the *Chlamydomonas* flagellum. A fixation protocol successfully preserved the waveform of the beating flagella. The flagella were sectioned parallel to the surface of the slide onto which the *Chlamydomonas* were adhered. This resulted in a sampling of flagella that not only preserved their native waveform, but were also sectioned longitudinally. Micrographs were selected that showed the central pair in its entirety and included both a bent and a straight portion of the flagellum. This insured that the sample displayed the widest dimension of the flagellum as the section included the central pair. The micrographs showed an important finding; there was considerable change in the diameter of the axoneme between the straight and curved regions of the bending waves.

A second study documented the extent of the axoneme distortion and compared it to predictions of the geometric clutch computer model (Lindemann and Mitchell, 2007). Figure 1.8 is reproduced from that report and shows a side by side comparison of the axoneme from bent and straight regions of the same flagellum. There was an average diameter distortion of approximately 40 nm and this yielded an estimate of the Young's modulus for the flagellum of approximately 0.02 MPa.

A 40 nm distortion of the axoneme diameter is quite large and corresponds to a 24% change in the overall diameter. This is sufficient repositioning of the doublets to alter the probability of dynein crossbridges forming. Unfortunately, while this is very encouraging and fits well with the geometric clutch model, it does not tell us how the distortion is distributed around the ring of nine. As discussed in the paper, it does not exclude the possibility that the diameter change could result from distortion of the ring of nine into an oval with no net change in the interdoublet

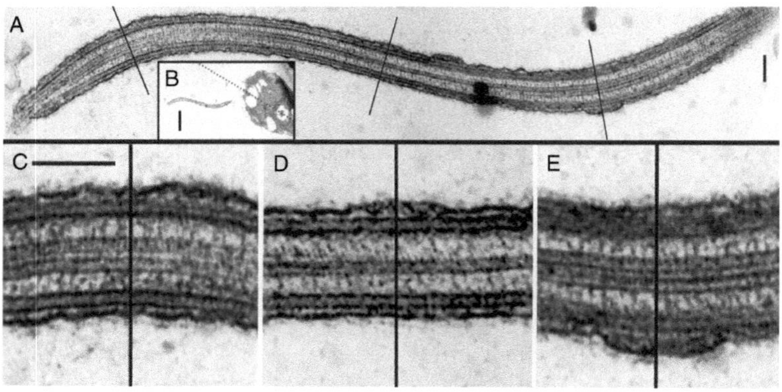

Figure 1.8 Diameter distortion during flagellar beating in *Chlamydomonas*. (A) The location of the positions where diameter measurements were taken from an electron micrograph of a *Chlamydomonas* that was rapidly fixed while beating. (B) Image of the relative position of the flagellum to the whole organism. (C–E) The three axoneme diameter measurement positions side by side for visual comparison. Bars = A, C–E, 0.2 μm; B, 1 μm. Reprinted from Lindemann and Mitchell (2007) with permission.

spacing. Thus far, it is the most direct evidence that the axoneme distorts during the beat, and by a significant amount.

So how does one account for the contradictory evidence from the earlier studies (Gibbons, 1975; Gibbons and Gibbons, 1974)? There is evidence given in these reports that the rigor waves did trap some distortion. The observed rigor waves were not entirely stable after fixation. When fixed with 2% glutaraldehyde for 75 min at pH 8.0, they often exhibited a torsional shift in the plane of each preserved bend relative to its proximal and distal neighbors. If it is assumed that at pH 8.0 the rigor bridges are only formed from dynein that are bound to their adjacent doublet at the instant when rigor was induced, then there should also be stretches of dynein that are not in rigor because they were unbound at the time of fixation. This is supported by the TEM of the fixed sperm in the pH 8.0 condition. These free dyneins would be the very ones detached by t-force just prior to induction of rigor and might remain free through the fixation process. This would mean that the rigor axonemes consist of rigid areas where rigor bridges are fixed in place, alternated with areas where the doublets are not bound together. If the free doublets were pushed apart from each other by t-force, as a consequence of the switching process as envisioned in the geometric clutch mechanism, these unbound doublets would be under tension and compression. The only way to relieve some of that stress would be to twist the axoneme in the regions between the rigor segments. Coincidentally, this is what is observed in these samples and it might be the way to reconcile the disparate findings.

4.3. Can t-force actually detach working dyneins?

Perhaps the most important element of the geometric clutch hypothesis is the contention that the accumulation of tension created by the motor action of the dyneins themselves is sufficient to serve as the switch mechanism to turn the motors "off" at a critical switch point. This principle is illustrated in Fig. 1.9.

In order for this to work as an effective switching mechanism, the t-force developed in a beating flagellum must be of the same order as the maximum force that the stalk attachment of the dynein heavy chain can hold. The attachment of the dynein stalk to the B subtubule of the adjacent doublet is also the transmission point to allow the motor protein to pull on the adjacent doublet for the purpose of outer doublet sliding translocation. Therefore, it is reasonable to expect that if this linkage is to be terminated by the t-force, then the t-force would have to be about as large as the maximum force that the dynein motors can deliver. This issue was addressed in Lindemann (2003) and it was determined that the t-force per micron of flagellum at the switch point of beat reversal is nearly equal to the maximum dynein force per micron of flagella. Furthermore, the relationship was true

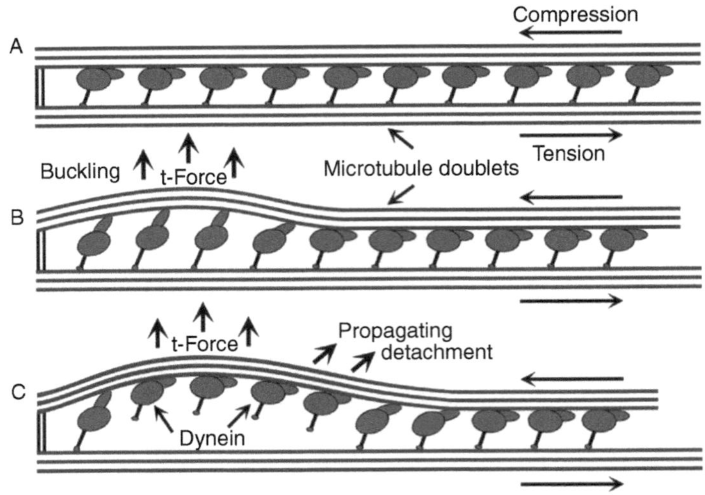

Figure 1.9 An illustration of the mechanism proposed by Brokaw (2009) to explain the results of the two doublet experiment of Aoyama and Kamiya (2005). (A) Two doublets associate by dynein attachment and dynein motor activity results in the compression of the upper doublet and tension on the lower doublet. In (B), the doublet under compression undergoes buckling, which creates a local curvature and t-force. In (C), the t-force overcomes the adhesion of the dyneins and produces a spreading area of dynein detachment. Reprinted from Lindemann (2009) with permission.

for both sea urchin sperm and bull sperm, which exhibit substantial differences in curvature development, flagellar size, and stiffness.

While the coincidence of the similarity between the dynein stalling force and the t-force at the switch point is suggestive, it is not proof of a t-force-driven switching mechanism. An experiment by Kamiya and Okagaki (1986) is a more convincing proof of the switching principle. In their experiment, two individual doublets from a frayed *Chlamydomonas* axoneme produced a repetitive cycle of attachment, bending, and separation. It is a direct demonstration of the principle that the dyneins can pull away from their binding sites when bending, and thus the resultant t-force reaches a critical value. This crucial experiment was cited as important support for the t-force switching mechanism in the original formulation of the geometric clutch hypothesis (Lindemann, 1994a,b). However, a subsequent report cast doubt on this interpretation.

Aoyama and Kamiya (2005) published a more extensive follow-up study of the frayed axoneme experiment. They concluded that there were at least three different patterns of association between pairs of adjacent doublets and, most significantly, dynein release often occurred without the development of substantial curvature. They interpreted this result as precluding a t-force-based switching mechanism. However, Brokaw (2009) conducted an extensive independent analysis of data from the two doublet experiment. He found that it was possible to quite accurately simulate the experimental results based on a detachment mechanism which depends on the t-force. He showed that the doublet that is compressed by the action of the dyneins experiences bucking and the local buckling generates sufficient curvature to create a t-force large enough to overcome the adhesion of the dyneins. This mechanism is illustrated in Fig. 1.9.

In summary, there is experimental support for the contention that the t-force that develops during flagellar beating is sufficient to overcome the adhesion of the dynein to their binding sites on the B subtubule. There is also strong experimental and theoretical foundation to contend that the dyneins are able to terminate their own action by a t-force-based release mechanism.

4.4. Does the microtubule-binding affinity of dynein regulate the switch point?

In the geometric clutch interpretation of the beat cycle, the switching event that terminates the action of active dyneins is the balance of t-force and dynein adhesion. When the t-force exceeds the dynein adhesion force, this initiates a release of the dynein from their binding sites on the B subtubule of the adjacent doublet. We have seen solid evidence that the development of t-force can separate adjacent doublets and terminate the action of dynein. The next question centers on the interaction of the dynein stalk with tubulin. If the geometric clutch interpretation is correct, then factors that change the microtubule-binding affinity of the dynein stalk must result in changes in the switch point of the beat cycle.

Are there factors that alter the microtubule-binding affinity of the dynein heavy chain? There is considerable evidence that ADP plays a regulatory role in this regard. The first reports, published by Omoto and collaborators (Frey *et al.*, 1997; Kinoshita *et al.*, 1995; Omoto *et al.*, 1996), reported that ADP could initiate motility in certain paralyzed mutants of demembranated *Chlamydomonas* that were immotile with ATP alone. Based on this evidence, they proposed that ADP plays a role in regulating dynein function.

This view is also supported by the observation that ADP facilitates more complete microtubule sliding in isolated axonemes (Kinoshita *et al.*, 1995) and increases rates of microtubule sliding on isolated inner arm dyneins (Yagi, 2000). ADP appears to act directly on the dynein heavy chain (Shiroguchi and Toyoshima, 2001) and its mode of action is to increase the duration or intensity of the high affinity binding state of the dynein (Kon *et al.*, 2004; Silvanovich *et al.*, 2003). The high affinity state is believed to be the force producing phase of the dynein crossbridge cycle and it is the ADP-bound state that is the force producing intermediate (Tani and Kamimura, 1999).

Molecular dissection of the dynein nucleotide-binding domains showed that in addition to the principal ATP hydrolysis site at AAA1, binding of nucleotide at the other sites also effects the functioning of the dynein heavy chain (Kon *et al.*, 2004; Numata *et al.*, 2008). The nucleotide-binding site at AAA3 is also an ATP hydrolysis site (Reck-Peterson and Vale, 2004; Takahashi *et al.*, 2004). Specifically, the binding of nucleotide at the P-loop of AAA3 alters the microtubule-binding affinity at the stalk (Cho *et al.*, 2008; Kon *et al.*, 2004; Silvanovich *et al.*, 2003). High concentrations of free ADP would be expected to increase the residency time of ADP at the regulatory sites on the dynein head and increase the duration of the high affinity (force producing) phase of the crossbridge cycle (Johnson, 1985). A summary of this proposed mechanism of nucleotide regulation was described by Inoue and Shingyoji (2007) and is summarized in Fig. 1.10.

The preponderance of evidence is that the microtubule-binding affinity of the dynein stalk is increased by ADP residence at a regulatory site.

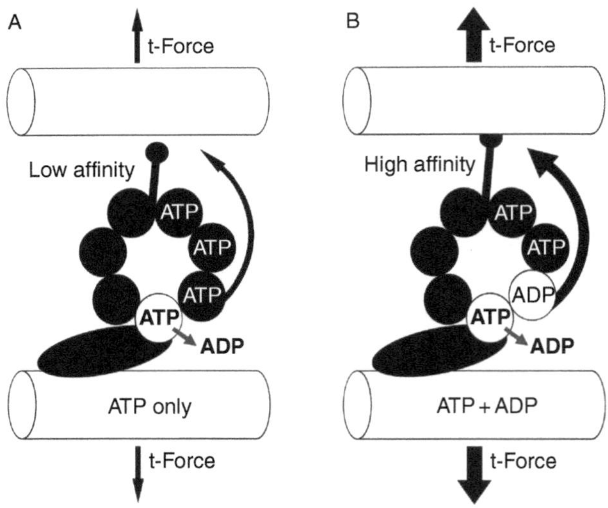

Figure 1.10 The connection between nucleotide regulation of dynein microtubule-binding affinity and dynein adhesion. Based on experimental evidence, it was proposed that ADP binding to a regulatory site on the dynein head governs the microtubule-binding affinity of the dynein stalk (Inoue and Shingyoji, 2007). In the current conception of the power stroke of dynein, the globular AAA domain (the dynein head) rotates on the linker region of the N-terminal domain (Burgess *et al.*, 2003; Roberts *et al.*, 2009). As the globular head rotates, force is transmitted to the adjacent doublet via the dynein stalk. The dynein stalk is the structure that attaches to the B subtubule of the adjacent doublet during flagellar beating. Therefore, when ADP is present in high concentration, it requires a greater t-force to pull dynein from its binding sites which forces switching to occur at a higher t-force (and curvature). Reprinted from Lesich *et al.* (2008) with permission.

What effect does ADP have on the beat cycle? If the geometric clutch mechanism of dynein switching is correct, then increasing the microtubule-binding affinity of dynein attachment to the B subtubule should alter the balance point between dynein adhesion and t-force, such that a greater t-force is required to bring about dynein detachment. Since t-force develops as a product of the curvature and tension on the outer doublets, a greater curvature should be developed at the switch point that triggers dynein detachment.

The documented effect of ADP on demembranated bull sperm models reactivated with 0.1 mM ATP (Lesich *et al.*, 2008) found the expected result. There is a significant increase in the curvature, and hence the t-force, at the point of beat direction reversal when 1 or 4 mM ADP is present, as seen in Fig. 1.11. In that report, the experimental result was simulated by the geometric clutch computer model by varying the single modeling parameter that corresponds to the dynein adhesion, as shown in Fig. 1.12.

As a final consideration, there are currently at least two competing views of the mechanism of force production by the dynein heavy chain. One view, based on the change in configuration of the stem and AAA motor domain in isolated dynein, suggests that the rotation of the dynein head on the linker region of the stem is the mechanism of force production (Burgess *et al.*, 2003; Roberts *et al.*, 2009). The other view, based on *in situ* imaging of the inner and outer dynein arms, suggests that the dynein head does not rotate but that it shifts laterally along the A subtubule to produce sliding (King, 2010; Movassagh *et al.*, 2010; Ueno *et al.*, 2008). In either case, there is no disagreement that the force for microtubule sliding is transmitted

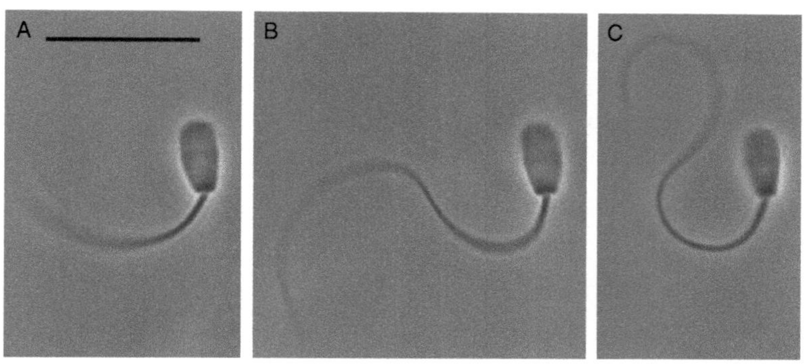

Figure 1.11 The effect of ADP on the flagellar switch point in bull sperm. Bull sperm demembranated with Triton X-100 and reactivated with 0.1 mM Mg-ATP and varying concentrations of ADP (A) no ADP, (B) 1 mM ADP, (C) 4 mM ADP. Cells are attached by their heads to the glass coverslip bottom of an imaging chamber. Images show the sperm at the point of beat reversal where the P bend is at maximum curvature and the flagellum changes direction; this is the switch point of the beat cycle. Bar = 20 μm. Reprinted from Lesich *et al.* (2008) with permission.

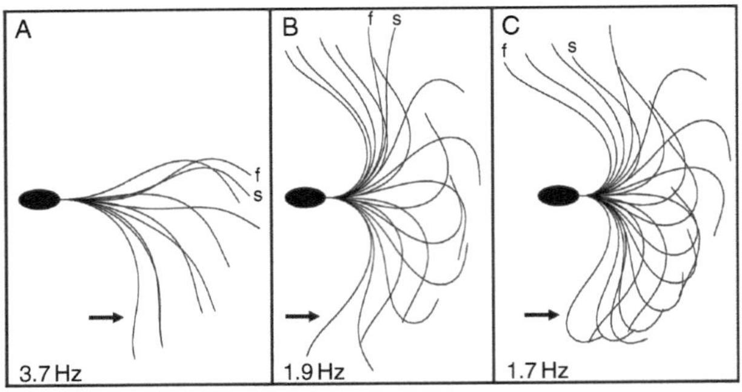

Figure 1.12 The effect of varying dynein adhesion in the geometric clutch computer model of a bull sperm. The geometric clutch computer model for bull sperm flagella (Lindemann, 1996) was used to explore the effects of dynein adhesion on the beat cycle. The model outputs show the resulting change in the simulated beat cycle caused by increasing the scaling factor which increases the dynein adhesion in the model by 15% between (A) and (B) and by an additional 5% between (B) and (C). All other modeling parameters where held constant. Arrows point to the frame that is the switch point for reversal of beat direction. As shown in Fig. 1.11 for reactivated bull sperm, there is an increase in the curvature of the P bend at the switch point. The simulation also shows a reduction in the beat frequency as dynein adhesion is increased; this is identical to the effect of ADP on reactivated sperm. Reprinted from Lesich *et al.* (2008) with permission.

between the adjacent tubules by the stalk attachment to the B subtubule of the adjacent doublet. In both putative mechanisms, there must be an alternation of a high affinity binding state and a low affinity state at the stalk microtubule-binding site. Without a high and low affinity state at this binding site, it is not possible to alternately transmit force through the stalk attachment during the power stroke and then move the stalk to the next binding site with minimal resistance. Consequently, both views of the dynein motor mechanism are subject to the same dynamic relationship between microtubule-binding affinity and interdoublet adhesion.

5. Discussion

Figure 1.13 summarizes events in the flagellar beat cycle as they occur in the geometric clutch computer model (reprinted from Lindemann, 1994b). The left and right columns of the figure represent views of the doublets on opposite sides of the same axoneme.

In panel 1, the side of the axoneme with the slightly higher probability of spontaneous dynein engagement initiates a cascade of attachment. This cascade is possible because the random attachment of a few dynein

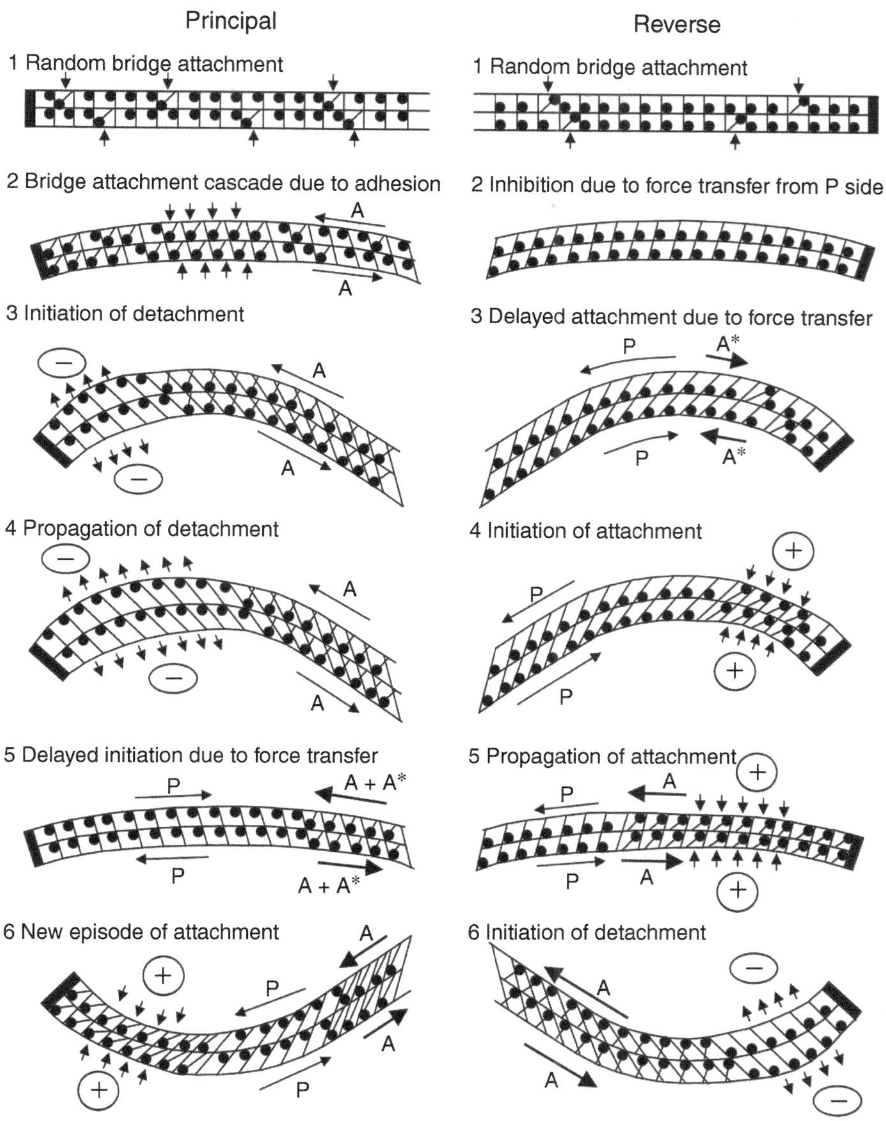

Figure 1.13 Events in the beat cycle of a flagellum or cilium according to the geometric clutch computer model. The figure illustrates the events on the two opposing sides of the axoneme as the geometric clutch computer model completes a beat cycle from a straight starting position. Time advances from top to bottom. A full description of the events in the cycle is given in the text. Reprinted from Lindemann (1994b) with permission.

contributes adhesion force that pulls the doublets together and increases the probability of further attachments. When a length of dyneins have attached, as in panel 2, they generate motive force to bend the flagellum and to inhibit the dyneins on the other side of the axoneme by pulling those doublets apart (a consequence of the interdoublet elastic elements). As the active bend reaches a critical curvature, the tension and compression on the active doublets create a negative, outwardly directed t-force that pulls the doublets apart, as is shown in panel 3. The separation of the doublets starts near the flagellar base, where the t-force is greatest, and spreads distally as in panel 4. On the opposite, and inactive, side of the axoneme, the imposed bending acts to stretch the interdoublet elastic linkages and create a positive t-force that pushes the doublets closer together and increases the probability of dyneins engaging on that side. In panels 4 and 5, the engagement of dyneins, on what was the inactive side, starts a cascade of attachment which shifts the dominance to this side of the axoneme and initiates the reverse bend of the beat cycle. As depicted in panel 6, once the reverse bend is formed, the process repeats and beating continues into a new cycle.

In the geometric clutch mechanism, coordination of dynein action is dependent on variations in the doublet spacing introduced by strain between the doublets pairs of the axoneme. Consequently, in this coordination scheme, even a single pair of doublets should be able to engage and disengage the dynein motors to establish a rudimentary beat cycle, as seen experimentally in the study of Aoyama and Kamiya (2005). For this reason, the same mechanism is also compatible with beating of unusual axonemal configurations such as 9 + 0 flagella.

Without the mechanical restraining influence of the central pair to define a preferred beat plane, the geometric clutch mechanism would be expected to produce a helical beat. This is because in the geometric clutch view each doublet pair is a potential oscillator. In the absence of a preferred bending plane, each doublet pair would contribute equally to the generation of a bending wave, and since they are arranged in a circle, this would result in a helical beat. Observations on real 9 + 0 flagella appear to confirm that they do beat with a helical motion (Ishijima *et al.*, 1988; Woolley, 1997).

When viewed looking toward the base of the flagellum, the dynein motors always project from the doublet they are on in a clockwise direction around the axoneme. Therefore, the vector direction of the torque contributed by each doublet pair, if they act in succession around the ring of nine would also rotate clockwise. This would explain the consistent chirality of the helical beat component of cilia and flagella.

Most cilia and flagella can initiate a beat from a straight position as in the scheme described above. Certain conditions, like high ATP in the absence of ADP, will yield axonemes that jitter or twitch without developing a full beat. In those instances, bending the flagellum will often start coordinated beating (Lindemann and Rikmenspoel, 1971, 1972a). This is consistent with the idea that passive bending aids in dynein attachment and activation. Such bend-induced activation requires interdoublet elastic elements. The evidence for the existence of interdoublet elastic elements was presented. The initiation of a spontaneous cascade of attachment without the aid of imposed bending depends on the dynein adhesion. There is evidence that ADP facilitates the adhesion of dynein by increasing the high affinity state of the dyneins. Predictably, this would also facilitate the initiation of beating, and it does. In mammalian sperm (Lindemann and Rikmenspoel, 1972b), sea urchin sperm (Ishikawa and Shingyoji, 2007), and in *Chlamydomonas* (Frey *et al.*, 1997; Omoto *et al.*, 1996), ADP will sometimes initiate spontaneous beating when the axoneme is compromised by damage or by mutation. This fits the geometric clutch model, as ADP enhances dynein adhesion and facilitates a cascade of attachment.

In the current conception of the dynein power stroke, when ATP is hydrolyzed at the P-loop of AAA1, the AAA domain head of the dynein molecule undergoes a rotation on the linker region of the N-terminal domain of the dynein heavy chain (Burgess *et al.*, 2003; Roberts *et al.*, 2009). The force generated by the power stroke is transmitted to the adjacent doublet through the stalk of the dynein head which has a microtubule-binding site at its tip (see Fig. 1.10). Consequently, both the adhesion of the dynein to the adjacent doublet and the transfer of force for interdoublet sliding are transmitted through this single binding site. Hence, there is a functional relationship between the dynein motive force and the dynein adhesion force.

The off switch for the dyneins in the model is created by the accumulation of t-force as a result of active bending. In this report, we reviewed the evidence that supports this switching mechanism. The magnitude of the t-force is sufficient in an intact flagellum to account for switching and for dynein release in the two doublet experiments. It is also consistent with the force that axonemal dynein has been measured to generate (Sakakibara *et al.*, 1999; Schmitz *et al.*, 2000; Shingyoji *et al.*, 1998). Furthermore, ADP has the predicted effect of increasing the curvature and t-force at the point of beat reversal.

Other factors may also affect the balance of dynein adhesion and t-force in a beating flagellum. Possibilities include the regulation of the waveform of the flagellar beat by Ca^{2+}, and the glutamylation of tubulin. Calcium is interesting from the standpoint that all cilia and flagella have a fundamental response to Ca^{2+} that alters the waveform of the beat in a way that modulates the principal/reverse bend symmetry. Since waveform is, to a large degree, dictated by the switching thresholds, it is likely that the Ca^{2+} regulatory pathway differentially alters the switching threshold of the two sets of dynein that bend the flagellum in opposite directions.

One possibility is that the nucleotide regulation of dynein microtubule-binding affinity is Ca^{2+} ion dependent. An even more attractive possibility is that Ca^{2+} works through certain interdoublet attachments, specifically those identified by Heuser *et al.* (2009) as being identical to the DRC. It is known that the DRC complex contains centrin (LeDizet and Piperno, 1995; Piperno *et al.*, 1990, 1992, 1994), a known Ca^{2+}-responsive protein, making it likely that these linkages change their configuration when Ca^{2+} is bound. Their role may be to alter the resting interdoublet spacing and bias the probability of dynein attachment.

The DRC is also known to contain protein kinases that can inactivate certain dynein inner arms; specifically I1 was identified (Wirschell *et al.*, 2007). I1, also called dynein f, was shown to be specially suited to hang on to microtubules (Kotani *et al.*, 2007). This dynein may be especially important for contributing the majority share of dynein adhesion needed for switching. A reduction or increase in the activity of this particular subset of inner arms would have a very profound effect on the switch point.

The recent interest in glutamylation of tubulin as a regulatory factor (Kubo *et al.*, 2010; Mitchell, 2010; Suryavanshi *et al.*, 2010) is noteworthy in that tubulin glutamylation may exert its effects by changing the dynamics of the dynein–tubulin bridge and hence regulate the dynein adhesion by an alternate mechanism. This is an independent factor that could change the switch point balance. The evidence points to this factor being most important for inner arm function. Consequently, there should be profound effects of glutamylation on the switch point, as the inner arms are known to have the greatest influence on the waveform of the beat (Brokaw, 1999; Brokaw and Kamiya, 1987; Lindemann, 2002).

The distortion of the axoneme is of crucial importance to the verification of the geometric clutch hypothesis. While it remains to be established that the doublets move closer when dyneins are active and further apart when the dyneins are inactive, there is some evidence that this may be the case. Sakakibara *et al.* (2004) showed that there is a change in axonemal diameter which accompanies dynein activity in immobilized axonemes of sea urchin sperm. Lindemann and Mitchell (2007) showed a diameter change corresponding to bent and straight regions of *Chlamydomonas* flagella rapidly fixed during beating. Clearly, additional support and documentation will be necessary to establish that the observed diameter changes are related to the engagement and disengagement of dyneins. Fortunately, new techniques, such as cryoelectron tomography, may be capable of observing spatial relationships in unfixed, vitrified material produced from live, beating cilia and flagella. It may soon be possible to confirm or reject this important and necessary element of the hypothesis.

In conclusion, there is now concrete experimental evidence to support each of the four major assumptions that underlie the beating mechanism proposed in the geometric clutch hypothesis. While additional experimental verification is still required, the specific issues that need to be explored are well defined and amenable to experimental investigation.

ACKNOWLEDGMENT

I thank Kathleen Lesich for her assistance in the preparation and editing of the figures and chapter.

REFERENCES

Afzelius, B. (1959). Electron microscopy of the sperm tail; results obtained with a new fixative. *J. Biophys. Biochem. Cytol.* **5,** 269–278.

Afzelius, B. A. (1961). The fine structure of the cilia from ctenophore swimming-plates. *J. Biophys. Biochem. Cytol.* **9,** 383–394.

Aoyama, S., and Kamiya, R. (2005). Cyclical interactions between two outer doublet microtubules in split flagellar axonemes. *Biophys. J.* **89,** 3261–3268.

Baron, D. M., Ralston, K. S., Kabututu, Z. P., and Hill, K. L. (2007). Functional genomics in *Trypanosoma brucei* identifies evolutionarily conserved components of motile flagella. *J. Cell Sci.* **120,** 478–491.

Bozkurt, H. H., and Woolley, D. M. (1993). Morphology of nexin links in relation to interdoublet sliding in the sperm flagellum. *Cell Motil. Cytoskeleton* **24,** 109–118.

Brokaw, C. J. (1971). Bend propagation by a sliding filament model for flagella. *J. Exp. Biol.* **55,** 289–304.

Brokaw, C. J. (1972a). Computer simulation of flagellar movement. I. Demonstration of stable bend propagation and bend initiation by the sliding filament model. *Biophys. J.* **12,** 564–586.

Brokaw, C. J. (1972b). Flagellar movement: A sliding filament model. *Science* **178,** 455–462.

Brokaw, C. J. (1999). Computer simulation of flagellar movement: VII. Conventional but functionally different cross-bridge models for inner and outer arm dyneins can explain the effects of outer arm dynein removal. *Cell Motil. Cytoskeleton* **42,** 134–148.

Brokaw, C. J. (2009). Simulation of cyclic dynein-driven sliding, splitting, and reassociation in an outer doublet pair. *Biophys. J.* **97,** 2939–2947.

Brokaw, C. J., and Kamiya, R. (1987). Bending patterns of *Chlamydomonas* flagella: IV. Mutants with defects in inner and outer dynein arms indicate differences in dynein arm function. *Cell Motil. Cytoskeleton* **8,** 68–75.

Burgess, S. A., Walker, M., Sakakibara, H., Knight, P. J., and Oiwa, K. (2003). Dynein structure and power stroke. *Nature* **421,** 715–718.

Cho, C., Reck-Peterson, S. L., and Vale, R. D. (2008). Regulatory ATPase sites of cytoplasmic dynein affect processivity and force generation. *J. Biol. Chem.* **283,** 25839–25845.

Dallai, R., Lupetti, P., and Mencarelli, C. (2006). Unusual axonemes of hexapod spermatozoa. *Int. Rev. Cytol.* **254,** 45–99.

Frey, E., Brokaw, C. J., and Omoto, C. K. (1997). Reactivation at low ATP distinguishes among classes of paralyzed flagella mutants. *Cell Motil. Cytoskeleton* **38,** 91–99.

Fujimura, M., and Okuno, M. (2006). Requirement of the fixed end for spontaneous beating in flagella. *J. Exp. Biol.* **209,** 1336–1343.

Gibbons, I. R. (1961). The relationship between the fine structure and direction of beat in gill cilia of a lamellibranch mollusc. *J. Biophys. Biochem. Cytol.* **11,** 179–205.

Gibbons, I. R. (1975). The molecular basis of flagellar motility in sea urchin spermatozoa. *In* "Molecules and Cell Movement," (S. Inou and R. E. Stephens, eds.), pp. 207–232. Raven Press, New York.

Gibbons, B., and Gibbons, I. R. (1973). The effect of partial extraction of dynein arms on the movement of reactivated sea-urchin sperm. *J. Cell Sci.* **13,** 337–357.

Gibbons, B., and Gibbons, I. R. (1974). Properties of flagellar "rigor waves" formed by abrupt removal of adenosine triphosphate from actively swimming sea urchin sperm. *J. Cell Biol.* **63,** 970–985.
Gibbons, I. R., and Grimstone, A. V. (1960). On flagellar structure in certain flagellates. *J. Biophys. Biochem. Cytol.* **7,** 697–715.
Hayashi, S., and Shingyoji, C. (2008). Mechanism of flagellar oscillation-bending-induced switching of dynein activity in elastase-treated axonemes of sea urchin sperm. *J. Cell Sci.* **121,** 2833–2843.
Hayashibe, K., Shingyoji, C., and Kamiya, R. (1997). Induction of temporary beating in paralyzed flagella of *Chlamydomonas* mutants by application of external force. *Cell Motil. Cytoskeleton* **37,** 232–239.
Heuser, T., Raytchev, M., Krell, J., Porter, M. E., and Nicastro, D. (2009). The dynein regulatory complex is the nexin link and a major regulatory node in cilia and flagella. *J. Cell Biol.* **187,** 921–933.
Holcomb-Wygle, D. L., Schmitz, K. A., and Lindemann, C. B. (1999). Flagellar arrest behavior predicted by the Geometric Clutch model is confirmed experimentally by micromanipulation experiments on reactivated bull sperm. *Cell Motil. Cytoskeleton* **44,** 177–189.
Inoue, Y., and Shingyoji, C. (2007). The roles of noncatalytic ATP binding and ADP binding in the regulation of dynein motile activity in flagella. *Cell Motil. Cytoskeleton* **64,** 690–704.
Ishijima, S., Sekiguchi, K., and Hiramoto, Y. (1988). Comparative study of the beat patterns of American and Asian horseshoe crab sperm: Evidence for a role of the central pair complex in forming planar waveforms in flagella. *Cell Motil. Cytoskeleton* **9,** 264–270.
Ishikawa, R., and Shingyoji, C. (2007). Induction of beating by imposed bending or mechanical pulse in demembranated, motionless sea urchin sperm flagella at very low ATP concentrations. *Cell Struct. Funct.* **32,** 17–27.
Johnson, K. A. (1985). Pathway of the microtubule-dynein ATPase and the structure of dynein: A comparison with actomyosin. *Ann. Rev. Biophys. Chem.* **14,** 161–188.
Kamiya, R., and Okagaki, T. (1986). Cyclical bending of two outer-doublet microtubules in frayed axonemes of *Chlamydomonas. Cell Motil. Cytoskeleton* **6,** 580–585.
Kanous, K. S., Casey, C., and Lindemann, C. B. (1993). Inhibition of microtubule sliding by Ni^{2+} and Cd^{2+}: Evidence for a differential response of certain microtubule pairs within the bovine sperm axoneme. *Cell Motil. Cytoskeleton* **26,** 66–76.
King, S. M. (2010). Axonemal dyneins winch the cilium. *Nat. Struct. Mol. Biol.* **17,** 673–674.
Kinoshita, S., Miki-Noumura, T., and Omoto, C. K. (1995). Regulatory role of nucleotides in axonemal function. *Cell Motil. Cytoskeleton* **32,** 46–54.
Kon, T., Nishiura, M., Ohkura, R., Toyoshima, Y. Y., and Sutoh, K. (2004). Distinct functions of nucleotide-binding/hydrolysis sites in the four AAA modules of cytoplasmic dynein. *Biochemistry* **43,** 11266–11274.
Kotani, N., Sakakibara, H., Burgess, S. A., Kojima, H., and Oiwa, K. (2007). Mechanical properties of inner-arm Dynein-f (Dynein i1) studied with in vitro motility assays. *Biophys. J.* **93,** 886–894.
Kubo, T., Yanagisawa, H. A., Yagi, T., Hirono, M., and Kamiya, R. (2010). Tubulin polyglutamylation regulates axonemal motility by modulating activities of inner-arm dyneins. *Curr. Biol.* **20,** 441–445.
LeDizet, M., and Piperno, G. (1995). The light chain p28 associates with a subset of inner dynein arm heavy chains in *Chlamydomonas* axonemes. *Mol. Biol. Cell* **6,** 697–711.
Lesich, K. A., Pelle, D. W., and Lindemann, C. B. (2008). Insights into the mechanism of ADP action on flagellar motility derived from studies on bull sperm. *Biophys. J.* **95,** 472–482.

Linck, R. W. (1979). Advances in the ultrastructural analysis of the sperm flagellar axoneme. *In* "The Spermatozoon. Maturation, Motility, Surface Properties and Comparative Aspects," (D. W. Fawcett and J. M. Bedford, eds.), pp. 99–115. Urban & Schwarzenberg, Inc., Baltimore.

Lindemann, C. B. (1994a). A "Geometric Clutch" hypothesis to explain oscillations of the axoneme of cilia and flagella. *J. Theor. Biol.* **168,** 175–189.

Lindemann, C. B. (1994b). A model of flagellar and ciliary functioning which uses the forces transverse to the axoneme as the regulator of dynein activation. *Cell Motil. Cytoskeleton* **29,** 141–154.

Lindemann, C. B. (1996). Functional significance of the outer dense fibers of mammalian sperm examined by computer simulations with the geometric clutch model. *Cell Motil. Cytoskeleton* **34,** 258–270.

Lindemann, C. B. (2002). Geometric Clutch model version 3: The role of the inner and outer arm dyneins in the ciliary beat. *Cell Motil. Cytoskeleton* **52,** 242–254.

Lindemann, C. B. (2003). Structural-functional relationships of the dynein, spokes, and central-pair projections predicted from an analysis of the forces acting within a flagellum. *Biophys. J.* **84,** 4115–4126.

Lindemann, C. B. (2004). Testing the geometric clutch hypothesis. *Biol. Cell* **96,** 681–690.

Lindemann, C. B. (2007). The geometric clutch as a working hypothesis for future research on cilia and flagella. *Ann. N. Y. Acad. Sci.* **1101,** 477–493.

Lindemann, C. B. (2009). Heart of the beat (the flagellar beat, that is). *Biophys. J.* **97,** 2865–2866.

Lindemann, C. B., and Kanous, K. S. (1995). "Geometric clutch" hypothesis of axonemal function: Key issues and testable predictions. *Cell Motil. Cytoskeleton* **31,** 1–8.

Lindemann, C. B., and Kanous, K. S. (1997). A model for flagellar motility. *Int. Rev. Cytol.* **173,** 1–72.

Lindemann, C. B., and Lesich, K. A. (2010). Flagellar and ciliary beating: The proven and the possible. *J. Cell Sci.* **123,** 519–528.

Lindemann, C. B., and Mitchell, D. R. (2007). Evidence for axonemal distortion during the flagellar beat of *Chlamydomonas*. *Cell Motil. Cytoskeleton* **64,** 580–589.

Lindemann, C., and Rikmenspoel, R. (1971). Intracellular potentials in bull spermatozoa. *J. Physiol.* **219,** 127–138.

Lindemann, C. B., and Rikmenspoel, R. (1972a). Sperm flagella: Autonomous oscillations of the contractile system. *Science* **175,** 337–338.

Lindemann, C. B., and Rikmenspoel, R. (1972b). Sperm flagellar motion maintained by ADP. *Exp. Cell Res.* **73,** 255–259.

Lindemann, C. B., Fentie, I., and Rikmenspoel, R. (1980). A selective effect of Ni2+ on wave initiation in bull sperm flagella. *J. Cell Biol.* **87,** 420–426.

Lindemann, C. B., Orlando, A., and Kanous, K. S. (1992). The flagellar beat of rat sperm is organized by the interaction of two functionally distinct populations of dynein bridges with a stable central axonemal partition. *J. Cell Sci.* **102,** 249–260.

Lindemann, C. B., Walker, J. M., and Kanous, K. S. (1995). Ni^{2+} inhibition induces asymmetry in axonemal functioning and bend initiation of bull sperm. *Cell Motil. Cytoskeleton* **30,** 8–16.

Lindemann, C. B., Macauley, L. J., and Lesich, K. A. (2005). The counterbend phenomenon in dynein disabled rat sperm flagella and what it reveals about the interdoublet elasticity. *Biophys. J.* **89,** 1165–1174.

Mencarelli, C., Lupetti, P., Rosetto, M., Mercati, D., Heuser, J. E., and Dallai, R. (2001). Molecular structure of dynein and motility of a giant sperm axoneme provided with only the outer dynein arm. *Cell Motil. Cytoskeleton* **50,** 129–146.

Minoura, I., Yagi, T., and Kamiya, R. (1999). Direct measurement of the inter-doublet elasticity in flagellar axonemes. *Cell Struct. Funct.* **24,** 27–33.

Mitchell, D. R. (2003). Orientation of the central pair complex during flagellar bend formation in *Chlamydomonas*. *Cell Motil. Cytoskeleton* **56,** 120–129.
Mitchell, D. R. (2010). Polyglutamylation: The GLU that makes microtubules sticky. *Curr. Biol.* **20,** R234–R236.
Morita, Y., and Shingyoji, C. (2004). Effects of imposed bending on microtubule sliding in sperm flagella. *Curr. Biol.* **14,** 2113–2118.
Movassagh, T., Bui, K. H., Sakakibara, H., Oiwa, K., and Ishikawa, T. (2010). Nucleotide-induced global conformational changes of flagellar dynein arms revealed by *in situ* analysis. *Nat. Struct. Mol. Biol.* **6,** 673–674.
Nicastro, D., Schwartz, C., Pierson, J., Gaudette, R., Porter, M. E., and McIntosh, J. R. (2006). The molecular architecture of axonemes revealed by cryoelectron tomography. *Science* **313,** 944–948.
Numata, N., Kon, T., Shima, T., Imamula, K., Mogami, T., Ohkura, R., and Sutoh, K. (2008). Molecular mechanism of force generation by dynein, a molecular motor belonging to the AAA+ family. *Biochem. Soc. Trans.* **36,** 131–135.
Okuno, M., and Hiramoto, Y. (1976). Mechanical stimulation of starfish sperm flagella. *J. Exp. Biol.* **65,** 401–413.
Omoto, C. K., Yagi, T., Kurimoto, E., and Kamiya, R. (1996). Ability of paralyzed flagella mutants of *Chlamydomonas* to move. *Cell Motil. Cytoskeleton* **33,** 88–94.
Pelle, D. W., Brokaw, C. J., Lesich, K. A., and Lindemann, C. B. (2009). Mechanical properties of the passive sea urchin sperm flagellum. *Cell Motil. Cytoskeleton* **66,** 721–735.
Phillips, D. M. (1974). Structural variants in invertebrate sperm flagella and their relationship to motility. *In* "Cilia and Flagella," (M. A. Sleigh, ed.), pp. 379–402. Academic Press, London.
Piperno, G., Ramanis, Z., Smith, E. F., and Sale, W. S. (1990). Three distinct inner dynein arms in *Chlamydomonas* flagella: Molecular composition and location in the axoneme. *J. Cell Biol.* **110,** 379–389.
Piperno, G., Mead, K., and Shestak, W. (1992). The inner dynein arms I2 interact with a "dynein regulatory complex" in *Chlamydomonas* flagella. *J. Cell Biol.* **118,** 1455–1463.
Piperno, G., Mead, K., LeDizet, M., and Moscatelli, A. (1994). Mutations in the "dynein regulatory complex" alter the ATP-insensitive binding sites for inner arm dyneins in *Chlamydomonas* axonemes. *J. Cell Biol.* **125,** 1109–1117.
Pitelka, D. R. (1974). Basal bodies and root structures. *In* "Cilia and Flagella," (M. A. Sleigh, ed.), pp. 437–469. Academic Press, London.
Reck-Peterson, S. L., and Vale, R. D. (2004). Molecular dissection of the roles of nucleotide binding and hydrolysis in dynein's AAA domains in *Saccharomyces cerevisiae*. *Proc. Natl. Acad. Sci. USA* **101,** 1491–1495.
Ringo, D. L. (1967). Flagellar motion and fine structure of the flagellar apparatus in *Chlamydomonas*. *J. Cell Biol.* **33,** 543–571.
Roberts, A. J., Numata, N., Walker, M. L., Kato, Y. S., Malkova, B., Kon, T., Ohkura, R., Arisaka, F., Knight, P. J., Sutoh, K., and Burgess, S. A. (2009). AAA+ ring and linker swing mechanism in the dynein motor. *Cell* **136,** 485–495.
Sakakibara, H., Kojima, H., Sakai, Y., Katayama, E., and Oiwa, K. (1999). Inner-arm dynein c of *Chlamydomonas* flagella is a single-headed processive motor. *Nature* **400,** 586–590.
Sakakibara, H. M., Kunioka, Y., Yamada, T., and Kamimura, S. (2004). Diameter oscillation of axonemes in sea-urchin sperm flagella. *Biophys. J.* **86,** 346–352.
Sale, W. S., and Gibbons, I. R. (1979). Study of the mechanism of vanadate inhibition of the dynein cross-bridge cycle in sea urchin sperm flagella. *J. Cell Biol.* **82,** 291–298.
Schmitz, K. A., Holcomb-Wygle, D. L., Oberski, D. J., and Lindemann, C. B. (2000). Measurement of the force produced by an intact bull sperm flagellum in isometric arrest and estimation of the dynein stall force. *Biophys. J.* **79,** 468–478.

Shingyoji, C., Higuchi, H., Yoshimura, M., Katayama, E., and Yanagida, T. (1998). Dynein arms are oscillating force generators. *Nature* **393,** 711–714.

Shiroguchi, K., and Toyoshima, Y. Y. (2001). Regulation of monomeric dynein activity by ATP and ADP concentrations. *Cell Motil. Cytoskeleton* **49,** 189–199.

Silvanovich, A., Li, M. G., Serr, M., Mische, S., and Hays, T. S. (2003). The third P-loop domain in cytoplasmic dynein heavy chain is essential for dynein motor function and ATP-sensitive microtubule binding. *Mol. Biol. Cell* **14,** 1355–1365.

Sleigh, M. A. (1962). The Biology of Cilia and Flagella. Pergamon Press, Ltd., Oxford.

Summers, K. E., and Gibbons, I. R. (1971). Adenosine triphosphate-induced sliding of tubules in trypsin-treated flagella of sea-urchin sperm. *Proc. Natl. Acad. Sci. USA* **68,** 3092–3096.

Suryavanshi, S., Edde, B., Fox, L. A., Guerrero, S., Hard, R., Hennessey, T., Kabi, A., Malison, D., Pennock, D., Sale, W. S., Wloga, D., and Gaertig, J. (2010). Tubulin glutamylation regulates ciliary motility by altering inner dynein arm activity. *Curr. Biol.* **20,** 435–440.

Takahashi, Y., Edamatsu, M., and Toyoshima, Y. Y. (2004). Multiple ATP-hydrolyzing sites that potentially function in cytoplasmic dynein. *Proc. Natl. Acad. Sci. USA* **101,** 12865–12869.

Tamm, S. L., and Tamm, S. (1981). Ciliary reversal without rotation of axonemal structures in ctenophore comb plates. *J. Cell Biol.* **89,** 495–509.

Tamm, S. L., and Tamm, S. (1984). Alternate patterns of doublet microtubule sliding in ATP-disintegrated macrocilia of the ctenophore *Beroë*. *J. Cell Biol.* **99,** 1364–1371.

Tani, T., and Kamimura, S. (1999). Dynein-ADP as a force-generating intermediate revealed by a rapid reactivation of flagellar axoneme. *Biophys. J.* **77,** 1518–1527.

Ueno, H., Yasunaga, T., Shingyoji, C., and Hirose, K. (2008). Dynein pulls microtubules without rotating its stalk. *Proc. Natl. Acad. Sci. USA* **105,** 19702–19707.

Warner, F. D. (1976). Ciliary inter-microtubule bridges. *J. Cell Sci.* **20,** 101–114.

Warner, F. D. (1978). Cation-induced attachment of ciliary dynein cross-bridges. *J. Cell Biol.* **77,** R19–R26.

Wirschell, M., Hendrickson, T., and Sale, W. S. (2007). Keeping an eye on I1: I1 dynein as a model for flagellar dynein assembly and regulation. *Cell Motil. Cytoskeleton* **64,** 569–579.

Witman, G. B., Plummer, J., and Sander, G. (1978). *Chlamydomonas* flagellar mutants lacking radial spokes and central tubules. Structure, composition, and function of specific axonemal components. *J. Cell Biol.* **76,** 729–747.

Woolley, D. M. (1997). Studies on the eel sperm flagellum. The structure of the inner dynein arm complex. *J. Cell Sci.* **110,** 85–94.

Yagi, T. (2000). ADP-dependent microtubule translocation by flagellar inner-arm dyneins. *Cell Struct. Funct.* **25,** 263–267.

Yagi, T., and Kamiya, R. (1995). Novel mode of hyper-oscillation in the paralyzed axoneme of a *Chlamydomonas* mutant lacking the central-pair microtubules. *Cell Motil. Cytoskeleton* **31,** 207–214.

Zanetti, N. C., Mitchell, D. R., and Warner, F. D. (1979). Effects of divalent cations on dynein cross bridging and ciliary microtubule sliding. *J. Cell Biol.* **80,** 573–588.

CHAPTER TWO

From Cilia Hydrodynamics to Zebrafish Embryonic Development

Willy Supatto* *and* Julien Vermot†

Contents

* Laboratory for Optics and Biosciences, Ecole Polytechnique, Centre National de Recherche Scientifique (CNRS) UMR 7645, and Institut National de Santé et de Recherche Médicale (INSERM) U696, Palaiseau, France

† Institut de Génétique et de Biologie Moléculaire et Cellulaire (IGBMC), Institut National de Santé et de Recherche Médicale (INSERM) U964, Centre National de Recherche Scientifique (CNRS) UMR 1704, and Université de Strasbourg, Illkirch, France

Current Topics in Developmental Biology, Volume 95
ISSN 0070-2153, DOI: 10.1016/B978-0-12-385065-2.00002-5

Abstract

Embryonic development involves the cellular integration of chemical and physical stimuli. A key physical input is the mechanical stress generated during embryonic morphogenesis. This process necessitates tensile forces at the tissue scale such as during axis elongation and budding, as well as at the cellular scale when cells migrate and contract. Furthermore, cells can generate forces using motile cilia to produce flow. Cilia-driven flows are critical throughout embryonic development but little is known about the diversity of the forces they exert and the role of the mechanical stresses they generate. In this chapter, through an examination of zebrafish development, we highlight what is known about the role of hydrodynamics mediated by beating cilia and examine the physical features of flow fields from the modeling and experimental perspectives. We review imaging strategies to visualize and quantify beating cilia and the flow they generate *in vivo*. Finally, we describe the function of hydrodynamics during left–right embryonic patterning and inner ear development. Ideally, continued progress in these areas will help to address a key conceptual problem in developmental biology, which is to understand the interplay between environmental constraints and genetic control during morphogenesis.

1. Introduction

Cilia are tail-like organelles that protrude out of nearly all vertebrate cells. Their roles in moving fluid along epithelium, such as the respiratory tract or the fallopian tubes, are well known in humans (Baker and Beales, 2009). Yet, this past decade identified many more functions for this organelle. Overall, one can define two types of cilia: motile cilia involved in moving fluids and primary cilia (most often immotile) that can sense chemicals and/or mechanical inputs. This dual activity seems conserved in most vertebrates (Shah *et al.*, 2009), and it is thought that primary cilia and motile cilia were once the same structure that specialized toward motility, sensitivity, or both (Jekely and Arendt, 2006).

Cilia structure is variable and usually defines their function. In eukaryotic cells, cilia are made of microtubules and classified according to their internal molecular arrangement and their ability to move. Usually, cells assemble only one cilium but a number of specialized cells can assemble up to 300 cilia. While motile cilia are commonly involved in cell motility, such as sperm, their prominent role in the embryo is to generate fluid flow. Moving fluids can participate in numerous processes and cilia activities turn out to be involved in diverse developmental functions: from generating frictional forces in the left–right organizer (McGrath *et al.*, 2003) to act as a mixer in the olfactory pit (Castleman *et al.*, 2009) or to attract particles in the inner ear (Colantonio *et al.*, 2009). Interestingly, the rules dictating these features strictly depend on fundamental fluid dynamics that were articulated 160 years ago by Stokes (1851). This work, initially applied to aerodynamics and hydrodynamics, became slowly incorporated into microfluidics (Stone *et al.*, 2004), bioengineering (Vilfan *et al.*, 2010; White and Grosh, 2005; Yoganathan *et al.*, 2004), microbiology (Purcell, 1997; Short *et al.*, 2006; Solari *et al.*, 2006), physiology (Purcell, 1977), and, more recently, developmental biology (Cartwright *et al.*, 2009).

The diversity of cilia motility and fluid mechanics in the embryo is starting to be studied *in vivo*. The number of flows that can be generated by such simple structure is limited, but the combination of cilia beating pattern and particular topology of the environment can participate to build up complex flows. Importantly, specific flow shapes are critical for embryonic development. The understanding of fluid mechanics principles and the use of modeling are essential to address the emergence of complexity through simple hydrodynamic interactions. In addition, the constant progress of live imaging has greatly changed the views of cilia flows. Several examples are paving the way for describing the complexity of cilia-driven advections and their roles in organizing flow in zebrafish embryos.

In this chapter, we focus on what is known about cilia-driven flows during zebrafish embryogenesis, with an emphasis on fluid dynamics and its relevance in the emergence of higher order aspects of morphogenesis. We first review the roles of cilia motility during zebrafish embryonic development. Next, we consider basic rules explaining the fluid dynamic at work at the cilium scale. We start with the simplest possible hydrodynamic rules, working up for basic geometry. We next address how modeling and experimental approaches are driving the investigation of flow *in vivo* and conclude with the examination of fluid dynamics into organs whose development relies on fluid dynamics, the otolithic morphogenesis and left–right axis specification in zebrafish.

2. Motile Cilia and Zebrafish Development

Zebrafish embryo contains numerous cavities and tubes. As in many chordates, most of them are extensively ciliated. For example, the nervous system contains a ciliated epithelial cell layer that delimits the ventral canal of the spinal cord in mouse, xenopus, amphioxus, and zebrafish (Caspary *et al.*, 2007; Dale *et al.*, 1987; Kramer-Zucker *et al.*, 2005; Nakao and Ishizawa, 1984; Fig. 2.1). Other cavities such as the left–right organizer, also called the Kupffer's vesicle (KV) in zebrafish, the developing kidneys, and the inner ear also contain cilia (Kramer-Zucker *et al.*, 2005; Fig. 2.1). Cilia are usually visualized *in situ* using antibodies directed against the acetylated tubulin (Essner *et al.*, 2002; Fig. 2.2), but motile cilia can also

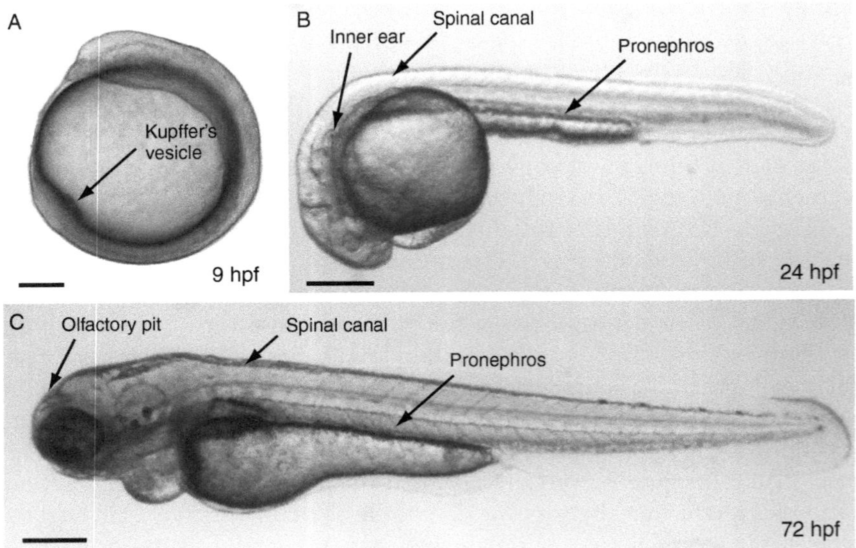

Figure 2.1 Ciliated organs in the zebrafish embryo. Side views of 10 hpf (A), 24 hpf (B), and 72 hpf (C) showing reported areas containing motile cilia (black arrows). Scale bar: 150 μm.

be located in zebrafish embryo by looking at the expression of cilia specific genes. A good example is *foxj1*, a transcription factor that controls the expression of genes involved in the formation of motile cilia and which is important for embryonic development (Stubbs *et al.*, 2008; Yu *et al.*, 2008). Its expression starts at gastrulation in the forerunner cells, which constitute the precursors of the KV early on during embryogenesis, and remains strong in the vesicle itself. Its expression is also detected in the ventral floor plate of the spinal cord, the inner ear, the pronephros, the developing kidneys (Aamar and Dawid, 2008), and in the olfactory pit, where motile cilia are actively mixing the environment (Castleman *et al.*, 2009). In many vertebrates, cilia motility has been shown to be critical for the development and function of these organs. In zebrafish, the consequences of cilia immobility have been studied in detail using an array of cilia mutants (Drummond, 2009). The most common phenotypes are the presence of kidney cysts, left–right defects, inner ear defects, a curly tail, and hydrocephaly. It is to note that other transcription factors have been shown to control cilia assembly through the activation of *foxj1* expression, in particular, the rfx family which are differentially expressed in developing tissues. *rfx* genes are expressed in diverse developing organs, such as in the islet cell lineage of the pancreas in vertebrates (Smith *et al.*, 2010b; Soyer *et al.*, 2010) or sensory neurons in *Caenorhabditis elegans* (Perkins *et al.*, 1986) and *Drosophila* (Vandaele *et al.*,

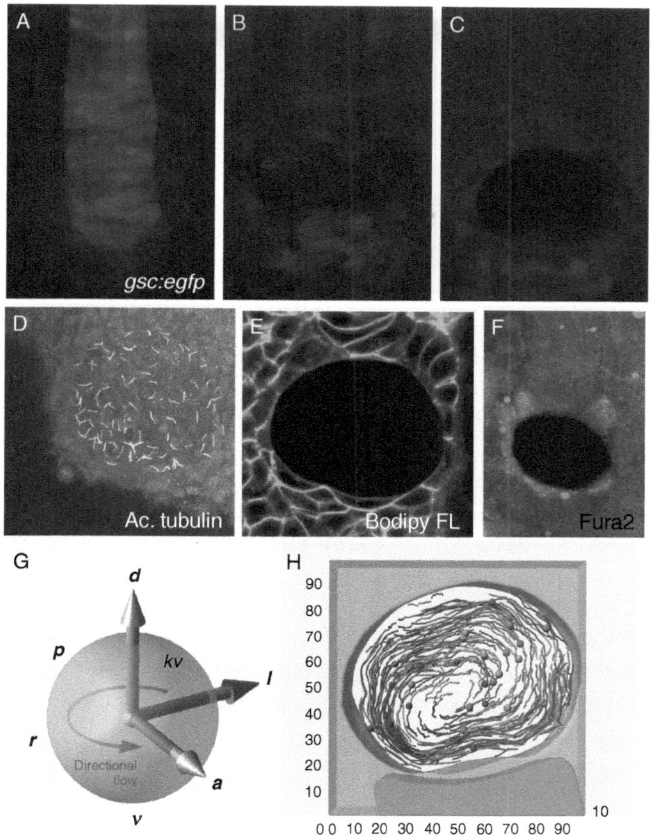

Figure 2.2 Anatomy of the Kupffer's vesicle. (A and B) Visualization and localization of the Kupffer's vesicle with the gsc:GFP line. Confocal sections at increasing depth showing the localization of the Kupffer's vesicle just below the midline. (D) Labeling using antibodies directed against acetylated tubulin reveals cilia localization in the Kupffer's vesicle. (E) Cell membrane labeling using Bodipy FL staining underlines the spherical shape of the vesicle. (F) Calcium indicator (Fura2) shows asymmetric activation. This asymmetric activation is essential for the left–right embryonic patterning. (G) Scheme representing the Kupffer's vesicle in 3D and the direction of the cilia-driven flow. (H) Particle tracking in the Kupffer's vesicle demonstrates the circularity of the flow *in vivo*. The *Tg(gsc:eGFP)* line is a kind gift from the Houart lab (King's College London, UK). Acetylated tubulin labeling was performed as in Colantonio *et al.* (2009), Bodipy FL labeling as in Hove *et al.* (2003). Panel (H) was obtained as described in Supatto *et al.* (2008). (See Color Insert.)

2001). Whether, these cells display motile cilia remains to be properly demonstrated

The developmental origins of these phenotypes are unclear. The presence of kidney cyst is not always due to cilia motility defects and has been correlated with abnormal cellular rearrangement (Sullivan-Brown *et al.*, 2008). Explanations for the curly tail phenotype are lacking, but it is tempting to speculate that this phenotype is related with the abnormal extracellular matrix formation seen in polycystin mutants, as cilia function and the polycystins are strongly interdependent (Mangos *et al.*, 2010). In contrast, hydrocephaly seems specific to cilia motility defects. It correlates with the absence of the spinal cord flow but experimental evidence is lacking to explain how flow can contribute to brain cavity morphogenesis. Left–right defects and otolithic defects in the inner ear are better understood and were clearly linked with fluid flow defects (Colantonio *et al.*, 2009; Essner *et al.*, 2005; Kramer-Zucker *et al.*, 2005). The KV is a transient cavity visible from 8 to 11 h after fertilization underneath the notochord whose role is to break the intrinsic embryonic axis of symmetry (Essner *et al.*, 2005; Kramer-Zucker *et al.*, 2005; Fig. 2.2). Experimental evidence clearly shows the presence of a directional flow (Fig. 2.2) which is critical to establish the left-sided expression of genes involved in providing left identity, such as *southpaw* and *pitx2* (Kramer-Zucker *et al.*, 2005).

Overall, a key issue in addressing the roles of motile cilia during embryogenesis has been to deal with the difficulties in bridging theoretical hydrodynamics with *in vivo* responses. In the following sections, we try to reconcile the theory of flow with what is known *in vivo* and to integrate this information into more complex structures such as the inner ear and the left–right organizer of zebrafish.

3. Structure of Motile Cilia in the Developing Zebrafish

A key parameter dictating cilia-mediated hydrodynamics is the type of beat they generate (Fig. 2.3). The beat pattern seems related with the internal organization of the cilia but many unresolved questions remain concerning the correlation of structure and cilia beat in different developing organs. Motile cilia are composed of microtubules and are classified according to their microtubule organization into two groups: 9 + 0 and 9 + 2 (Fig. 2.4). The axoneme of 9 + 2 motile cilia is composed of nine peripheral microtubule doublets and two central microtubules (the central pair). The motility depends on the presence of dynein arms that are attached to the microtubules, the dynein regulatory complex (DRC) and by radial spokes (Lindemann and Lesich, 2010). Interestingly, vertebrates 9 + 2 cilia can significantly bend during its motion with effective and recovery strokes, while 9 + 0 cilia have an almost perfect circular motion (Nonaka *et al.*, 1998, 2005). The structural basis of this difference is still open for interpretation and the relationship between protein arrangement within

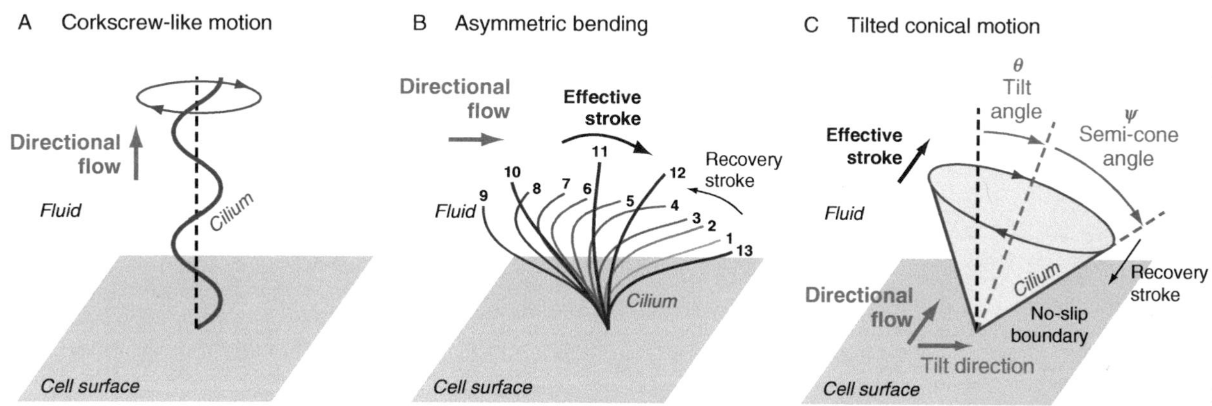

Figure 2.3 Pumping flow with motile cilia at low Reynolds numbers: three types of spatially asymmetric beating patterns observed experimentally: (A) helical motion or corkscrew-like motion which pump fluids along the cilium, (B) asymmetric motion which is related to an asymmetric bending of the cilium during its movement, (C) cylindrical rotation with a tilted cilium. In (B) and (C), the effective stroke corresponds to the cilium momentum where fluid is moved efficiently in the direction of motion, whereas poor transport occurs during the cilia recovery phase.

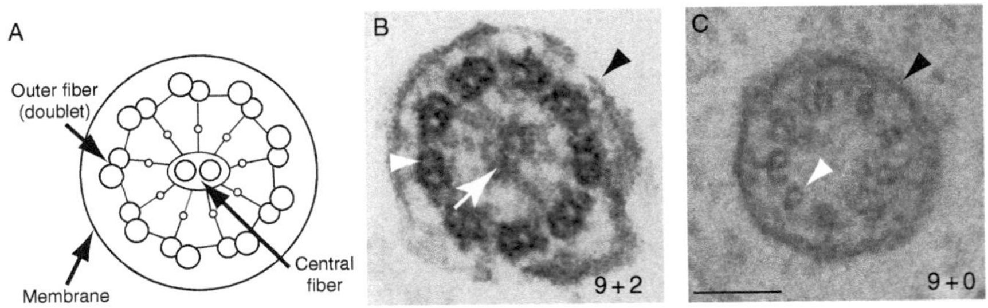

Figure 2.4 Ultrastructure of cilia: schematic drawing depicting a transverse section through a 9 + 2 cilium (for more information about the internal structure of the cilium see Lindeman in this book issue Lindemann, 2011) (A) and 9 + 2 and 9 + 0 cilia at 48 hpf (B and C) observed in the zebrafish inner ear using TEM (black arrowhead points to the membrane, white arrowhead to the outer fiber and the white arrow to the central pair). The intrinsic organization of the cilia can be seen: 9 doublets of microtubules surround the central pair in 9 + 2 cilia. No central pair is seen in 9 + 0 cilia. Scale bar: 100 nm. Electron microscopy imaging was performed as described in Pisam *et al.* (2002).

cilia and its three-dimensional (3D) motion remains to be better established in vertebrates. In protists, the presence of the central pair of microtubules is critical in shaping the overall 3D motion of the beating cilium. The so-called central pair hypothesis constitutes an attractive view of the structural basis of cilia beat: the primary role of the associated dynein motors of the nine outer doublet microtubules power the microtubule sliding that ultimately results in flagellar bending whereas the central pair acts as a coordinator of the doublet sliding through the radial spokes and the DRC (Lindemann and Lesich, 2010). Several lines of evidence also suggest that the presence of the central pair helps in establishing more powerful planar effective strokes and that the central spokes act as a stress transducer (Smith and Yang, 2004). In humans and zebrafish, mutants of the radial spokes heads affect cilia motion (Castleman *et al.*, 2009) and the DRC is critical for proper motility in zebrafish (Colantonio *et al.*, 2009). Importantly, abnormal cilia motion found in human patients affected with primary cilia dyskinesia can be predicted from ultrastructural defects observed by transmission electron microscopy (Chilvers *et al.*, 2003). Nevertheless, while most of the motile cilia are of 9 + 2 type in zebrafish embryo (Sarmah *et al.*, 2007), they are not all displaying a typical waveform motion. For example, the spinal canal cilia have been shown to display circular motion (Essner *et al.*, 2005), even though 9 + 2 cilia populate this structure. Importantly, the chirality of the ciliary structure originates from microtubule and dynein organization (Afzelius, 1999) and is thought to determine the rotation direction of beating cilia (Hilfinger and Julicher, 2008).

Zebrafish constitutes a model of choice to address the functions of motor protein because of its amenability to genetics and morpholino knockdown.

It helped for the identification of the FGF signaling pathway as one of the few signaling cascade involved in the control of cilia size and biogenesis (Neugebauer *et al.*, 2009). Interestingly, many FGF roles during morphogenesis correlate with phenotypes observed in mutants of motile cilia. Considering the cellular organization of cilia, the Notch signaling pathway has been shown to affect the multiciliated versus monociliated fate in zebrafish pronephros (Liu *et al.*, 2007). It is to note that notch also controls cilia length in the KV (Lopes *et al.*, 2010). In terms of motility, little is known about potential factors that could control the direction and frequency of beating cilia in the zebrafish embryo. It has been proposed that the inositol kinase (Ipk1 or ippk) is controlling cilia length and frequency (Sarmah *et al.*, 2007). Surprisingly, Ipk1 is localized in the centrosomes and basal bodies, suggesting that its role in controlling cilia activity is not acting through the central pair, DRC, or the radial spokes.

Overall, most of these structural features will influence the fluid flow generated by cilia. Importantly, understanding the links between cilium beating pattern and the flow pattern generated by cilia requires understanding the specificity of fluid dynamics at this scale. In Section 3, we will discuss hydrodynamics from the fluid mechanics point of view.

4. The Physics Side of Cilia-Mediated Flow: From Modeling to Experiment

Paralleling the studies of cilia structure and function during embryogenesis is a body of work that aims at understanding the basis of the hydrodynamics generated by cilia. In this part, we review what is known about the theory of cilia-mediated flow and its relevance when compared with *in vivo* observations.

4.1. The basics of hydrodynamics at the cilium scale: Stokes flow or fluid dynamics at low Reynolds number

Essentially, the main function of motile cilia is to generate fluid flow at the micrometer scale. However, fluid dynamics are governed by laws that are not trivial at this scale and the resulting fluid flow presents features that are not intuitive when used to human scale. In order to understand the fundamental principles governing cilia-driven flow and behind its modeling and simulation, it helps to go back to the basics of fluid mechanics theory and introduce the Reynolds number (Re). This dimensionless number characterizes the nature of a fluid flow and the relative contribution of inertia and viscous dissipation. In practice, flows with the same Re will display the same properties. For an object of typical length L moving at typical velocity U, in a fluid of dynamic viscosity η and density ρ, the *Re* is defined as $Re = UL\rho/\mu$. It also reads $Re = UL/\upsilon$ using the kinematic viscosity $\upsilon = \mu/\rho$. The cilia-driven flow involved in zebrafish development exhibits characteristic scales, $L < 100$ μm and $U < 100$ μm s^{-1} (Table 2.1). The resulting *Re* using the kinetic viscosity of water ($\upsilon \approx 10^6$ μm^2 s^{-1}) is $Re < 10^{-2}$. As a consequence, due to the small length and velocity scales, the flow generated by beating cilia is characterized by a low *Re* ($Re \ll 1$): it is governed by Stokes equations and is referred to as creeping flow or Stokes flow. More generally, fluid dynamics involved in most microscopic biological systems (Purcell, 1977) and in microfluidic devices (Stone *et al.*, 2004) works at low *Re*. At the human scale, a swimmer ($L \approx 1$ m and $U \approx 1$ m s^{-1}) should swim in a fluid that is 10^8 times more viscous than water to experience such fluid behavior.

From a modeling perspective, the low value of the *Re* enables the simplification of the Navier–Stokes flow equation, the general model governing fluid dynamics. As described in Fig. 2.5, each term of the Navier–Stokes flow equation scales as a force per unit volume and *Re* estimates the relative contribution of the inertial and viscous forces in this equation ($\mathbf{f}_{\text{inertia}}$ and $\mathbf{f}_{\text{viscosity}}$ terms in Fig. 2.5, respectively). If $Re \ll 1$, the inertial forces can be neglected compared to the viscous forces and the Navier–Stokes flow equation can be approximated by the linear Stokes flow equation (see Fig. 2.5 for details). This equation is the starting point of any modeling of cilia-driven flows (see Cartwright *et al.*, 2004; Smith *et al.*, 2007; Vilfan and Julicher, 2006 for instance). The linear nature of this equation simplifies its solving and allows applying the superposition theorem: for instance, the flow generated by an ensemble of cilia can be approximated as the sum of the flow induced by each single cilium. In addition, the linearity and the time reversibility of Stokes flow equation provides fundamental properties of the flow that can be generated by motile cilia.

4.2. Directional flow

Once cilia beat in their fluidic environment, how can they generate a directional flow? Whereas the main role of motile cilia is thought to be efficient fluid pumping, the first challenge resulting from the Stokes flow equation is the difficulty to obtain a net flow. The absence of inertia in such a flow results in a velocity that is simply proportional to the force applied to the fluid: in other words, it means that if a cilium stops beating and applying a force on the surrounding fluid, the flow velocity drops instantaneously. More importantly, the time reversibility of the Stokes flow equation means that time asymmetry is not sufficient to create a net flow (reversing time does not affect the equation). This property has been described by Purcell as the scallop theorem (Purcell, 1977): in the absence of inertia at low *Re*, the time reversible movement of a scallop would not be sufficient to achieve

Table 2.1 Cilia properties in various embryonic models

Species	Organ	Structure	Mono- or multiciliated cells	Frequency (Hz)	Length (μm)	Rotation direction (view from the cilium tip)	θ (deg)	ψ (deg)	Directional flow velocity (μm s^{-1})
Zebrafish	Left–right organizer (Kupffer's vesicle)	9 + 2[a]	Mono[a]	29.7 ± 0.3[b] 26.2 ± 1.6[a]	3.3 ± 1.1[a] 3.7 ± 0.8[c]	Clockwise[b,d,e] Counterclockwise[a]	30[f]		10–50[f]
	Inner ear	9 + 2 9 + 0 (this study)	Mono[g]	34 ± 6[g]	5.9 ± 0.2[g]		~90[h]		~10–50[h]
	Central canal spinal cord	9 + 0[a]	Mono[a]	12.3 ± 3.4[a]	2.1 ± 0.7[a]				0.45 ± 0.03[a]
	Pronephric ducts	9 + 2[a]	Mono and multi[a]	20.0 ± 3.2[a]	8.8 ± 2[a]	Clockwise[a]			
Mouse	Left–right organizer (node)	9 + 0[d]	Mono[d]	10.7 ± 2.8[i]	5.5[i]	Clockwise[i]	35–40[i]	40–50[i]	4 (leftward)[i] −2 (rightward)[i] 50[j]
	Brain ventricles		Multi[k]	12–17[k]	8–10[k]	Wave pattern[k]			
Medaka	Left–right organizer (Kupffer's vesicle)	9 + 2[l]	Mono[l]	42.7 ± 2.6[i]	5[i]	Clockwise[i]	35–40[i]	40–50[i]	7.4 ± 3.6[i]
Xenopous	Gastrocoel roof plate			20–25[m]	3–5[m]	Clockwise[m]			3.5 (leftward)[m]

Cilia are characterized by their internal microtubule structure, by their number at the cell surface (mono- or multiciliated), their beating frequency, length, direction, and rotation direction. ψ and θ give a good indication about the ability of cilia to generate a directional flow. When divergent informations are reported in the literature, several indications are present in the same box. For example, both clockwise and counterclockwise rotation directions have been reported in zebrafish.

[a] Kramer-Zucker *et al.* (2005).
[b] Okabe *et al.* (2008).
[c] Lopes *et al.* (2010).
[d] Nonaka *et al.* (1998).
[e] Okada *et al.* (1999).
[f] Supatto *et al.* (2008).
[g] Colantonio *et al.* (2009).
[h] Wu *et al.* (2011).
[i] Okada *et al.* (2005).
[j] Nonaka *et al.* (2002).
[k] Hirota *et al.* (2010).
[l] Kobayashi *et al.* (2010).
[m] Schweickert *et al.* (2007).

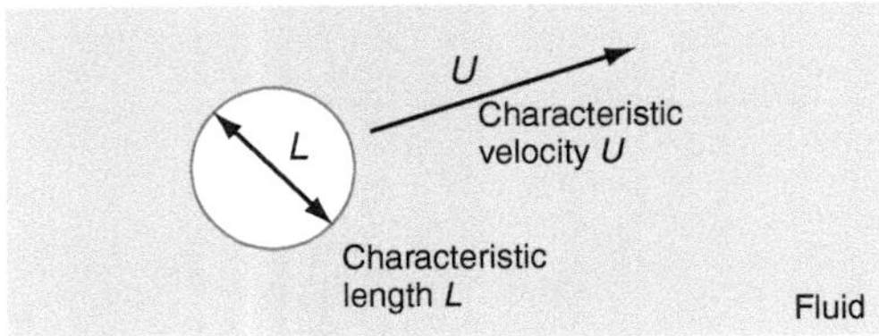

Navier–Stokes flow equation:

$$\rho\frac{\partial \mathbf{u}}{\partial t} + \rho(\mathbf{u}.\nabla)\mathbf{u} = -\nabla p + \mu\nabla^2\mathbf{u} + \mathbf{f}_{\text{ext}}$$

$$\mathbf{f}_{\text{inertia}} = \mathbf{f}_{\text{pressure}} + \mathbf{f}_{\text{viscosity}} + \mathbf{f}_{\text{ext}}$$

scales with...

$\nabla = \begin{pmatrix} \frac{\partial}{\partial x} \\ \frac{\partial}{\partial y} \\ \frac{\partial}{\partial z} \end{pmatrix}$	$\frac{1}{L}$
Time t	$\frac{L}{U}$
$\mathbf{f}_{\text{viscosity}} = \mu\nabla^2\mathbf{u}$	$\mu\frac{U}{L^2}$
$\mathbf{f}_{\text{inertia}} = \rho\frac{\partial \mathbf{u}}{\partial t} + \rho(\mathbf{u}.\nabla)\mathbf{u}$	$\rho\frac{U^2}{L}$
$\frac{\mathbf{f}_{\text{inertia}}}{\mathbf{f}_{\text{viscosity}}}$	$\frac{\rho\frac{U^2}{L}}{\mu\frac{U}{L^2}} = \frac{\rho U L}{\mu} = \frac{U L}{\nu} = Re$ Reynolds number

If $Re \ll 1$,

Stokes flow equation:

$$0 = -\nabla p + \mu\nabla^2\mathbf{u} + \mathbf{f}_{\text{ext}}$$

$$\cancel{\mathbf{f}_{\text{inertia}}} = \mathbf{f}_{\text{pressure}} + \mathbf{f}_{\text{viscosity}} + \mathbf{f}_{\text{ext}}$$

Figure 2.5 Fluid dynamics at low Reynolds number. The velocity field **u** of a fluid can be generally described by the Navier–Stokes flow equation. Each term of this equation scales as a force per unit volume: $\mathbf{f}_{\text{inertia}}$ includes the time-dependent ($\rho\partial\mathbf{u}/\partial t$) and the nonlinear ($\rho u \cdot \nabla \mathbf{u}$) intertial components (blue), $\mathbf{f}_{\text{pressure}}$ is the force generated by a pressure gradient (green), $\mathbf{f}_{\text{viscosity}}$ is the viscous dissipation term (orange), and $\mathbf{f}_{\text{ext}}$ corresponds to external forces applied to the liquid (such as force generated by a motile cilium). The Reynolds number (*Re*) compares $\mathbf{f}_{\text{inertia}}$ with $\mathbf{f}_{\text{viscosity}}$ to check if one of these terms can be neglected in the Navier–Stokes equation. $Re = UL\rho/\mu$ is obtained by scaling each term using characteristic velocity U and length L of the system. $Re \ll 1$ means that viscous forces dominate inertia. In this case, the inertial term can be neglected to obtain the Stokes flow equation. (For interpretation of the references to color in this figure legend, the reader is referred to the Web version of this chapter.)

propulsion, the scallop would simply move back and forth, and the motion pattern would remain the same whether slow or fast, whether forward or backward in time. Similarly, the proposition that a beating cilium could generate a net flow simply by changing its angular velocity while rotating is incorrect (Raya and Belmonte, 2006). In order to produce a directional flow at low *Re*, a beating cilium needs an asymmetry is space or shape.

Models predict that a directional flow cannot be produced by either a stiff cilium rotating in free space (Buceta *et al.*, 2005) or a distribution of cilia arranged on a spatially asymmetric pattern (Cartwright *et al.*, 2004). Presumably due to the cilia ultrastructure, length, and/or orientation, three types of spatially asymmetric cilia beating patterns have been proposed theoretically and observed experimentally in developing embryos (Fig. 2.3):

(i) *The corkscrew-like motion* (Fig. 2.3A): This motion pattern is well known and is used by flagella to propel bacteria (Purcell, 1997), but is not common in cell epithelium. However, such cilia motion pattern has been observed during zebrafish kidney development (Kramer-Zucker *et al.*, 2005). In this case, the direction of the net flow generated is parallel to the rotation axis of the cilium. The direction depends on the rotation direction and the helix orientation: if they are opposite, the flow moves from the base to the tip of the cilium (e.g., right-handed helix and clockwise rotation).

(ii) *The asymmetric bending* (Fig. 2.3B): In this case, the cilium changes its shape while rotating: during one beating cycle, its bending is not the same on one way compared to the other, producing an effective and a recovery stroke. This beating pattern looks like the breast stroke in swimming. The flow direction is perpendicular to the main axis of the cilium (rotation axis) and depends on the bending asymmetry (relative direction of effective/recovery strokes). Flow direction does not directly depend on the cilium direction of rotation, but on the orientation of the effective stokes. Such beating pattern has been observed in adult or embryonic epithelium in mouse (Hirota *et al.*, 2010; Sanderson and Sleigh, 1981) and possibly in other species, such as in the left–right organizer in xenopus (Schweickert *et al.*, 2007). In the case of asymmetric bending, it has been shown that the fluid pumping efficiency can be increased when coupled cilia are densely distributed on a surface and synchronize their cycle to form metachronal waves (Gueron and Levit-Gurevich, 1999).

(iii) *Tilted conical motion* (Fig. 2.3C): Directional flow generated by cilia exhibiting symmetrical circular motion, without asymmetric shape as in the two previous cases, has been observed experimentally in mouse, rabbit, and fish left–right organizer (Okada *et al.*, 2005). Recent models proposed that the spatial symmetry can be broken without bending of the cilium if the rotation axis of the cilium is not perpendicular to the cell membrane, but tilted with an angle θ toward the normal of this surface (Fig. 2.3C; Cartwright *et al.*, 2004; Smith *et al.*, 2007; Vilfan and Julicher, 2006). In this case, the cilium interaction with the cell surface plays a critical role: applying the no-slip boundary condition (the velocity of the fluid at the cell surface must match the velocity of the surface itself), the rotation forms an effective stroke when the cilium is far from the surface and a recovery stroke close to the surface (Fig. 2.3C). As the effective and recovery strokes are defined by the interaction between cilium and cell surface, the direction of the net flow generated depends on the cilium tilt and on the cell surface. The directional flow is perpendicular to the direction of the cilium tilt and parallel to the surface. The orientation of the flow depends on both the angular tilt orientation and the cilia rotation sense. The model predictions of a cilium tilt have been experimentally confirmed *in vivo* in mouse (Okada *et al.*, 2005) and in zebrafish (Supatto *et al.*, 2008; Fig. 2.6). Interestingly, among the three asymmetric beating patterns generated by biological cilia, the tilted conical motion is the easiest to reproduce artificially as it does not require controlling the bending of the cilium. For this reason, the generation of directional flow at low *Re* in microfluidic devices can be obtained using stiff artificial cilia beating with tilted conical motion (Nonaka *et al.*, 2005; Shields *et al.*, 2010; Vilfan *et al.*, 2010).

4.3. Mixing with chaotic advection

The second important consequence of Stokes flow properties is that mixing is hard at low *Re*. The absence of inertial and nonlinear terms in the Stokes flow equation prevents the occurrence of turbulent behaviors. Without turbulence, flow mixing relies on molecular diffusion and is usually slow. In this case, it is known that the only efficient way to create mixing is by using chaotic advection (Stone *et al.*, 2004). Chaotic advection enables molecular diffusion to take place on efficient time scales.

In general, the flow generated by motile cilia can help increasing the relative contribution of advection compared to diffusion in the transport of particles in a fluid. In fluid mechanics, this relative contribution is quantified by the Peclet number: $Pe = UL/D$, with U and L the characteristic length and velocity scales (as used in *Re*) and D the particle diffusion coefficient. $Pe \gg 1$ means that the advection dominates the particle transport. The mixing efficiency thus increases with *Pe* (Stone *et al.*, 2004).

In the case of chaotic advection, although the flow is laminar and dominated by viscous forces, fluid particle trajectories are chaotic, in the sense of being sensitive to initial conditions. In practice, it can be demonstrated experimentally or in simulations by following the trajectories of several particles starting at close positions in space: if trapped into a chaotic

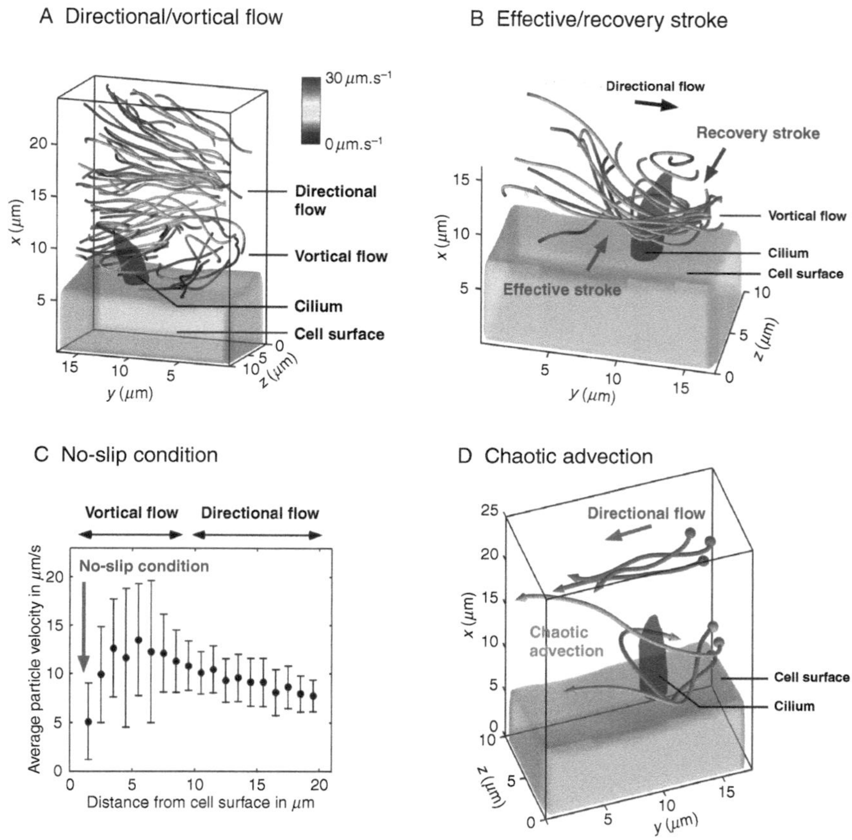

Figure 2.6 Predicted flow features from simulations are experimentally validated in the zebrafish's left–right organizer. (A) The flow is vortical close to the cilium and the cell membrane and directional far away. (B) Effective and recovery strokes are identified in the flow surrounding a beating cilium with faster velocity on one side and slower velocity on the other. (C) Experimental verification of the no-slip condition: the average flow velocity drop to zero close to the cell membrane. (D) The chaotic nature of the flow surrounding a beating cilium is confirmed experimentally by following particles starting nearby in the vicinity of the cilium and exhibiting drastically different trajectories (green), whereas particle trapped in the directional flow (red) stay close together. Experimental data were obtained using the same procedure as described in Supatto *et al.* (2008). Velocity field display and analysis were done using custom Matlab (The MathWorks) scripts. (See Color Insert.)

flow, these trajectories cannot be predicted and the distance between these particles drastically increase in time. Importantly, chaotic advection generated around the envelope of motile cilia has been predicted by simulations (Smith *et al.*, 2007) and demonstrated experimentally in living zebrafish embryos (Supatto *et al.*, 2008; Fig. 2.6D). Whether this chaotic behavior is actively used or avoided during embryonic development is interesting and remains unclear. Recent studies suggest that chaotic advection near beating cilia could be an active mechanism involved in the otolith formation during zebrafish inner ear development (Wu *et al.*, 2011) and (Riley *et al.*, 1997). Efficient fluid mixing using chaotic advection has been recently applied in microfluidics with artificial motile cilia (Fahrni *et al.*, 2009; Shields *et al.*, 2010).

4.4. The left–right organizer: Interplay between models and experiments

Efforts have been made to address the hydrodynamics involved in the left–right symmetry breaking during early vertebrate development. Several key predictions have been made through modeling of the leftward directional flow generated in left–right organizers. This system illustrates well the interplay between fluid mechanics, modeling, and experimental investigation of cilia-driven flow in embryos. Whereas most of the experimental data and models have been obtained in mouse, this system allows interesting comparisons with the zebrafish left–right organizer (KV).

In recent years, several models and simulations have been developed based on mouse node experimental data. In these theoretical studies, the cilium is modeled either as an infinitesimal sphere rotating in its place (Cartwright *et al.*, 2004), a small sphere moving on a fixed trajectory in the vicinity of a planar surface (Vilfan and Julicher, 2006) or a slender body (Smith *et al.*, 2007). Simple time averaged modeling allows explaining the generation of a directional flow with tilted cilia (Cartwright *et al.*, 2004), whereas more complex modeling considering the time-dependent dynamics of cilia allows capturing detailed features, such as the chaotic advection occurring in the vortical flow surrounding cilia (Smith *et al.*, 2007). Whatever the complexity of the model used, it helps our understanding of the biological process by explaining new observations, raising specific predictions, and suggesting the next set of experiments. In this section, we will review the main flow features predicted by modeling and simulations based on the mouse node system, describe the validation of these features in the zebrafish KV compared to the mouse node, and discuss the open questions raised by each point.

The origin of the spatial asymmetry permitting the generation of a directional flow within the left–right organizer was the first question assessed by theorists. The proposition that cilia were beating with a tilted conical motion (Fig. 2.3C) was first suggested by Cartwright *et al.* (2004) and further investigated by demonstrating the importance of the cell surface and the no-slip boundary condition to obtain efficient and recovery strokes (Smith *et al.*, 2007; Vilfan and Julicher, 2006). The posterior tilt of cilia has been first observed in mouse (Okada *et al.*, 2005). This topology promotes a directional flow toward the left direction with cilia rotating clockwise. Among the asymmetric cilia beating patterns, it is interesting to mention that only the tilted conical motion allows breaking the left–right symmetry without preexisting left–right asymmetry at the scale of the cilium (Fig. 2.6) as the directional flow in this particular case is perpendicular to the tilt direction, this pattern allows translating an anterior–posterior or dorsal–ventral asymmetry into a left–right asymmetry. This is not true for cork-screw-like motion or asymmetric bending: a net flow in the left–right direction would require having a preexisting left–right asymmetry at the level of the cilium. In zebrafish, the flow circulating around the dorsal–ventral axis within the KV (Fig. 2.2) cannot be due to a posterior tilt, but due to a dorsal tilt. Importantly, this dorsal tilt has been observed *in vivo* (Supatto *et al.*, 2008). Nevertheless, cilia with posterior tilt have been observed in the dorsal roof of the KV (Kramer-Zucker *et al.*, 2005; Okabe *et al.*, 2008). Such topology cannot produce the observed flow around the dorsal–ventral axis. It is noted that the beating pattern of KV's cilia is different from the mouse nodal cilia since they do not share the same ultrastructure (9 + 2 instead of 9 + 0 in mouse; Kramer-Zucker *et al.*, 2005; Nonaka *et al.*, 1998). Overall, a statistical analysis of the distribution of cilia orientations within the entire KV would greatly help to solve this issue.

A key aspect of cilia-driven flow is the balance between effective/recovery strokes and the role of the no-slip conditions. In viscous fluids, the fluid will have zero velocity relative to a solid boundary. In case of cilia tilt, a fraction of the circular motion occurs while being closer to the membrane where moving fluid is more difficult. As a consequence, the induced flow magnitude is smaller than when circular motion occurs far from the cell membrane, producing a net flow (Smith *et al.*, 2007; Vilfan and Julicher, 2006). The no-slip condition has been demonstrated experimentally in zebrafish KV where the average flow velocity drops to 0 close to the cell surface (Fig. 2.6C) as well as in engineered cilia arrays (Shields *et al.*, 2010). As a result, the flow velocity is stronger on the side of the effective stroke, compared to the recovery stroke as shown experimentally (Fig. 2.6B). The optimal directional flow obtained for a tilt angle $\theta \approx 35°$ and a semi-cone angle $\psi \approx 55°$ (Fig. 2.6), with $\theta + \psi \approx 90°$ (Smith *et al.*, 2008, 2010a). Such angles have been reported in mouse, rabbit, and medaka embryos (Okada *et al.*, 2005). The far field directional flow velocity scale with $1/r^2$, with r the distance from the cilium (Cartwright *et al.*, 2004; Smith *et al.*, 2010a; Vilfan and Julicher, 2006). In fish, $\theta \approx 35°$ (Supatto *et al.*, 2008) or $\theta \approx 45°$ (Kramer-Zucker *et al.*, 2005) have been observed. We show in Fig. 2.7C that this angle can be even higher suggesting that θ presents strong fluctuations *in vivo*. The crucial role of this angle raises the question of its origin *in vivo*. In mouse, cilia tilt seems dynamic and

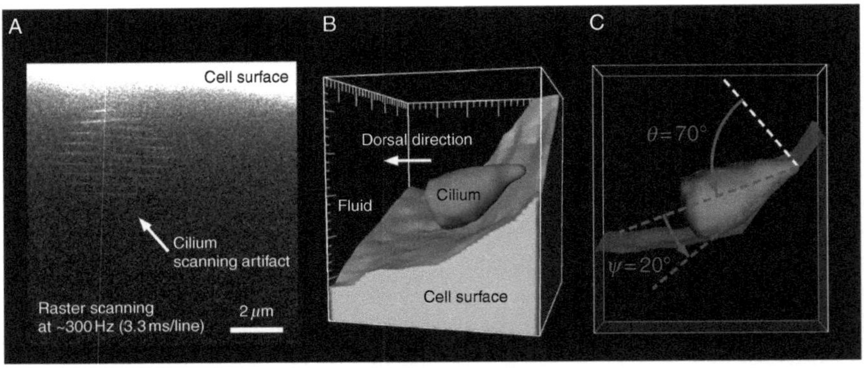

Figure 2.7 Characterizing 3D cilia features using standard laser scanning microscopy and scanning artifact. Cilia labeled with Bodipy TR are imaged in the zebrafish Kupffer's vesicle with two-photon laser scanning microscopy (A). The raster scanning is adjusted to be slow enough (3.3 ms per horizontal line) to capture several beats during the 2D image acquisition: in this case, stripped pattern is observed (arrows in A) corresponding to a slow scanning artifact. The distance between each strip corresponds to one period of cilium rotation: here, 10 lines correspond to $\sim$33 ms period and $\sim$30 Hz of beating frequency. Recording this pattern at different z-positions allows to manually trace the contour of each stripped pattern and reconstruct the cilium envelop in 3D (B). The strong tilt direction toward the dorsal direction appears clearly. The tilt angle θ and the semi-cone angle ψ can be measured (C): here, the tilt angle is specifically strong with $\theta \sim 70°$ and $\psi \sim 20°$. Scale in (B) is 2 μm per tick. Experimental data were obtained using zebrafish embryo preparation and labeling as described in Supatto *et al.* (2008). Two-photon excited fluorescence imaging was done at 820 nm wavelength using a Chameleon Ultra laser (Coherent), a Zeiss LSM510 microscope, and a 40×/NA 1.1 objective lens (Zeiss) on six somite-stage embryo. Image display and analysis were performed using ImageJ (http://rsb.info.nih.gov/ij/) and Imaris (Bitplane).

dependent on the planar cell polarity pathway (Marshall and Kintner, 2008). Most importantly, a clear coupling between cilia and directional flow has been shown in ependymal cilia (Guirao *et al.*, 2010). In mouse, it seems that the shape of the cell membrane and the planar polarity also plays a role in controlling cilia tilt (Hashimoto *et al.*, 2010).

Due to the low *Re* of the system, the influence of cilia vanishes quickly in space. However, cilia motion causes unsteady behavior in its vicinity: mathematical modeling clearly shows that a vortical flow (unsteady) is generated close to the cilium and a directional flow (steady) far from the cilium (Smith *et al.*, 2007). This chaotic nature of the vortical flow is seen in artificial setups (Shields *et al.*, 2010) and in zebrafish KV (Supatto *et al.*, 2008; Fig. 2.6D). So far, the relative contribution of directional and chaotic advection in cilia-driven flow is not known. A simple back-of-the-envelope calculation in the KV indicates that the chaotic advection occurs in more than 50% of the total volume of the KV assuming the chaotic behavior at the periphery of the KV reaches up to 10 μm from the cell membrane in vesicle of 80 μm diameter (as shown experimentally in Supatto *et al.*, 2008). Mixing is thus prominent in the KV, meaning that most of the fluid and factors secreted in the vesicle would not travel through a linear path but will rather have a complex, erratic motion when close to the cilium.

More sophisticated models are now taking into account the geometry of the left–right organizer and the spatial distribution and the density of cilia in mice. Such models show interesting features, such as backward flow due to back pressure gradient in the enclosed nodal cavity (Cartwright *et al.*, 2007; Smith *et al.*, 2010a). A circular flow is thus generated in the mouse node, very much like in fish. Yet, the backward motion observed in mouse spins around the anterior–posterior axis contrary to the dorsoventral axis in fish. Furthermore, there are cilia distributed all around the vesicle in fish as opposed to mice. Thus, while many features are common between mice and fish, each species remain distinct and might have developed different ways to break the embryonic symmetry using the same basic ciliary machinery.

5. Experimental Investigation of Cilia-Driven Fluid Flow in Developing Embryos

5.1. Challenges and experimental models

The experimental investigation of cilia-driven flow and its function during embryonic development necessitates measuring parameters characterizing motile cilia and fluid flow *in vivo*. As seen earlier, key information is held in the fluid flow velocity field as well as in the cilia positioning, beating pattern and orientation. Collecting this information in embryos is difficult because of the following challenges:

(i) *Multiple time scales are required*: cilia and flow dynamics are amongst the fastest processes occurring during embryonic morphogenesis and require high temporal resolution (up to ~10 ms time resolution). At the same time, the long-term roles of cilia-mediated flow, such as activating signal transduction pathways and the resulting cellular processes, can be slow (hours or days).

(ii) *Multiple spatial scales are required*: cilia act at the subcellular scale (μm) and the flow-generated acts at the entire organ scale (mm).

(iii) Specific experimental imaging strategies that do not compromise normal biology have to be developed. Ciliated structures are usually located in cavities that are difficult to access experimentally. Moreover, they experience strong and large scale morphogenetic movements and are often located deep inside light scattering tissues.

The zebrafish embryo offers several advantages for visualizing fluid flows *in vivo*. First, the embryo is transparent and has very little autofluorescence and scattering, allowing an optimal use of optical imaging. Second, embryos are resistant to phototoxicity and are easy to culture permitting the use of long-term light imaging and the collection of *in vivo* data at high resolution. Of note, recent zebrafish transgenic lines expressing fluorescent proteins specifically in cilia or in ciliated embryonic areas have been generated (Borovina *et al.*, 2010; Tian *et al.*, 2009). This facilitates the localization of the region of interest greatly and will certainly accelerate cilia imaging developments. Nevertheless, imaging cilia-driven flows have been successfully implemented in several vertebrate species. In particular, several approaches have been developed in mice, fish, frog, and rabbits to look at the left–right organizer flow through the tracking of fluorescent beads in the environment of the cilia (Okada and Hirokawa, 2009; Okada *et al.*, 2005; Schweickert *et al.*, 2007). Tissue explants can also be an excellent approach when one can extract out and cultivate the ciliated cells without affecting their activity. Brain slices, and the primary culture of ependymal cells of the forebrain cavities or from the trachea, allow to image both flow and cilia activity (Guirao *et al.*, 2010; Hirota *et al.*, 2010; Zahm *et al.*, 1990). However, it is difficult to know whether the boundary conditions in the explants are intact, especially when looking at the far field flows.

5.2. Mapping microscopic flow field in living embryos

In the field of fluid mechanics, the experimental investigation of fluid dynamics usually relies on seeding the flow with tracer particles, imaging the seeded flow and measuring the velocity field using image processing (Raffel *et al.*, 1998). Particle image velocimetry (PIV) and particle tracking velocimetry (PTV) are the most typical techniques to quantify flow velocity. PIV is based on image cross-correlation to estimate the displacement of patterns between successive images in a statistical manner (Raffel *et al.*, 1998). PIV is routinely used in fluid mechanics labs and has found recent applications in developmental biology to map blood flow dynamics during heart development (Hove *et al.*, 2003) or to quantify the morphogenetic movements shaping embryos (Supatto *et al.*, 2005; Zamir *et al.*, 2006). PTV is an alternative approach and relies on particle segmentation in each image followed by the tracking of these objects between successive images. In fluid mechanics, these two techniques are related to the Eulerian and the Lagrangian descriptions of the fluid flow: using PIV, the motion of the fluid is obtained at fixed positions in space distributed on a regular grid (Eulerian description such as in Hirota *et al.*, 2010), whereas using PTV, the velocity is measured along trajectories of specific particles trapped into the flow (Lagrangian description such as in Fig. 2.6). The choice between these two descriptions usually depends on practical considerations: for instance, by following the unpredictable trajectories of particles, the Lagrangian description is well suited for demonstrating chaotic advection in the zebrafish KV (Fig. 2.6D).

From an experimental perspective, PTV presents several advantages compared to PIV: it allows higher spatial resolution and manual or automated particle tracking that can be performed in 2D or 3D, whereas PIV usually works only for 2D analyses and cannot be performed manually. To be accurate, PIV requires a high density of tracer particles, whereas PTV works at low density. Commercial and free open source packages exist to perform PIV or PTV (see MatPIV http://folk.uio.no/jks/matpiv/, or the ImageJ plugin http://www.mosaic.ethz.ch/Downloads/ParticleTracker, for instance).

Both PIV and PTV have been used to investigate cilia-driven flow during embryonic development, such as in the left–right organizer (PIV in mouse (Hirota *et al.*, 2010), PTV in xenopus (Schweickert *et al.*, 2007), or zebrafish (Supatto *et al.*, 2008)). Seeding the flow with tracer particles remains a critical step for applying such strategy in developmental biology. In mice, flow in the left–right organizer has been imaged in cultured embryos and the velocity of the steady flow has been measured by tracking the motion of beads (see Table 2.1 and Okada and Hirokawa, 2009). In zebrafish, such an experiment necessitates an injection of beads within the KV cavity using a needle. In practice, this step remains invasive when one wants to look at fine hydrodynamics. An alternative approach has been recently developed to address the flow field with much milder side effects: the flow is seeded with microscopic fluorescently labeled cell debris generated by targeting a single cell with femtosecond subcellular ablation as described in Fig. 2.8 (Supatto *et al.*, 2008).

The experimental mapping of the flow field using PIV or PTV allows quantifying basic features of the flow such as the average speed of the directional flow generated by beating cilia (Table 2.1). It also permits further characterization of the flow and the experimental validation of predictions from models and simulations, as well as feeding further modeling with precise experimental data. The investigation of the velocity field generated within the zebrafish KV using PTV is presented in Fig. 2.6. This study illustrates how experimental investigation of the fluid flow can validate the flow features predicted by simulations (see Section 3). The flow features are measured within the whole vesicle (Fig. 2.8), as well as in the vicinity of single cilia. As predicted by simulations, two types of flows are observed: a laminar flow, which is directional and breaks to left–right symmetry, and a vortical flow surrounding the cilium. The chaotic nature predicted by simulations can be demonstrated experimentally using such a quantitative approach (Fig. 2.6D).

5.3. Measuring cilia features *in vivo*

Investigating the role of cilia-driven flow in embryonic development requires the experimental characterization of cilia *in vivo*: the key features includes cilia spatial position and density, beating pattern, rotation speed and

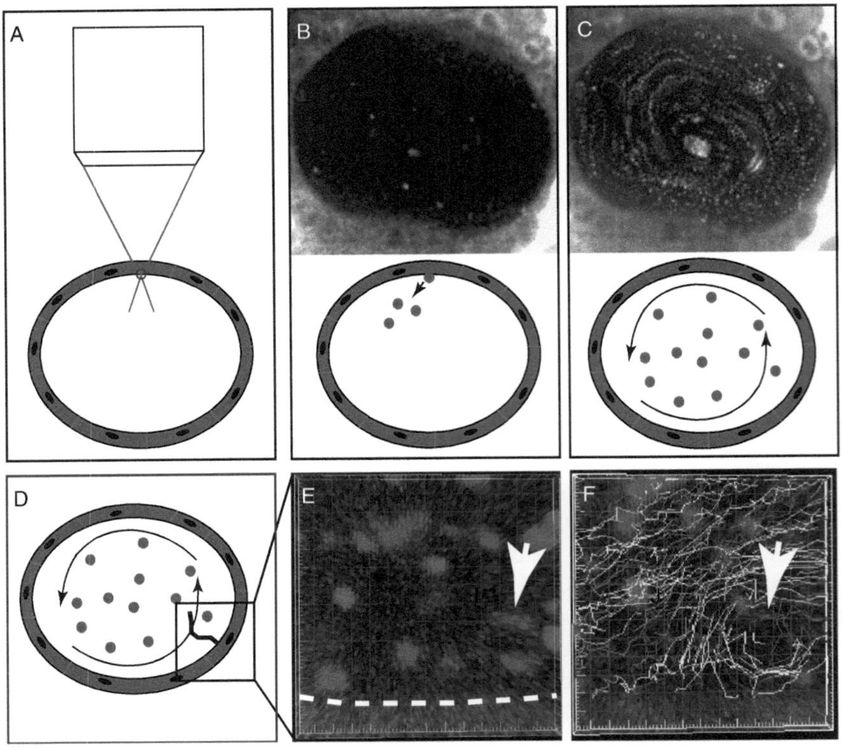

Figure 2.8 Mapping the cilia-driven flow within the zebrafish Kupffer's vesicle (KV) using femtosecond laser ablation, fast imaging and 3D-particle tracking. The KV cells are labeled with Bodipy TR and a single cell lining the cavity is targeted with tightly focused femtosecond laser pulses at 820 nm wavelength (A). A subcellular femtosecond laser ablation is performed to generate fluorescent microdebris seeding the flow (B). The nonlinear effect used in this technique allows a high spatial confinement and a low invasiveness, thus permitting the targeting of subcellular regions deep inside the embryo. Fast confocal 3D imaging at 4 *z*-stack per second capture the movement of the tracer particles: in (C), time-lapse images are superimposed to show the circular trajectories of the flow around the dorsal–ventral axis of the embryo. Cilia close to the KV surface (D) can be identified by following particles trapped in a vortical flow (arrows in E and F) and mapping the flow using 3D-particle tracking (D). Experimental data were obtained using the same procedure as described in Supatto *et al.* (2008). Image display and analysis were performed using ImageJ (http://rsb.info.nih.gov/ij/) and Imaris (Bitplane). (See Color Insert.)

direction, amplitude, tilt, or bending. However, these parameters cannot be measured experimentally using direct imaging of cilia using conventional technologies, such as confocal microscopy. Since cilia are moving extremely fast (Table 2.1) and in 3D, capturing cilia shape and motion requires both 3D resolution and time resolution that are at the limit of the state-of-the-art imaging technology. We can easily estimate the ideal time resolution required to acquire 3D-time lapses capturing the full rotation movement of a cilium beating at 30 Hz: with 10 images per z-stack and 10 z-stacks per cilium rotation, it would require imaging at 3000 frame per second (fps) and acquiring z-stacks at 300 Hz with sufficient signal, which can be challenging when cilia are located deep in light scattering tissues. High-speed imaging of cilia using bright field microscopy without fluorescent labeling has been reported (acquisitions at 500 fps in Okada and Hirokawa, 2009). However, this technique lacks 3D resolution and depth penetration: it is only adapted for 2D arrays of cilia close to the embryo surface such as in the mouse node and does not provide measurements of every cilia features that would require 3D spatial reconstruction. Recent developments in the fast fluorescence microscopy field allow high-speed fluorescence imaging (Vermot *et al.*, 2008). 2D imaging of cilia with commercial fast confocal microscopy up to 900 fps has been recently reported (Hirota *et al.*, 2010), corresponding in this case to typically 50 frames per cilium beat, which is more than enough to capture the movement. However, the ideal z-stack acquisition frequency required to capture the full 3D movements of cilia is currently unreachable through commercial imaging setups. To circumvent this limitation, experimentalists have developed tricks to obtain the required information without fully capturing the 3D movements of beating cilia *in vivo*. We report in the following section several of these techniques used to study cilia-driven flow in the left–right organizer.

To demonstrate cilia beat with a tilted conical motion in the mouse left–right organizer (Fig. 2.3C), high-speed 2D bright field imaging has been used to estimate cilia position, tilt angle θ, semi-cone angle ψ, and tilt orientation (Nonaka *et al.*, 2005; Okada *et al.*, 1999, 2005). To circumvent the lack of 3D resolution, the 2D traces of cilia tips were analyzed and compared with the 2D projection of an expected tilted cone. It allowed estimating the tilt angle and orientation, as described in Fig. 2.9. This ruse works only if we assume the cilia are positioned on a 2D flat epithelium perpendicular to the optical axis of the microscope. In the case of a curved epithelium, such as in the zebrafish KV, it cannot give accurate measurement of cilia orientation.

An alternative approach to characterize cilia features without actual 3D imaging of them has been recently reported in zebrafish (Supatto *et al.*, 2008). From 3D particle trajectories trapped into the vortical flow surrounding a cilium (Figs. 2.6A and 2.8F), several features can be indirectly characterized: cilium position in space (Fig. 2.8F), tilt direction and angle θ, rotation direction (clockwise/anticlockwise), and effective and recovery strokes (Fig. 2.6B). Importantly, the particles trapped in the vicinity of a cilium are moving much slower than the cilium itself in Stokes flow (Smith *et al.*, 2010a). For this reason, mapping the flow surrounding a cilium allows obtaining 3D data with lower time resolution than required for the direct 3D imaging of cilia (Supatto *et al.*, 2008). However, this indirect approach

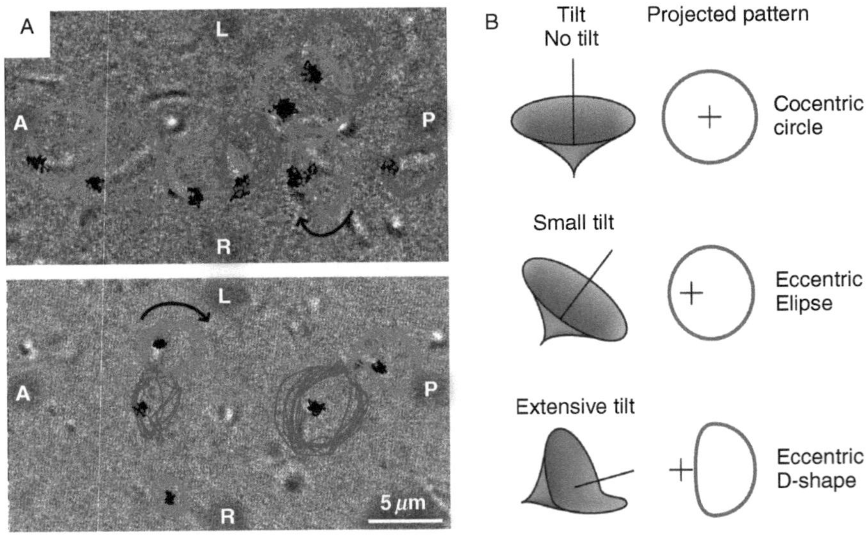

Figure 2.9 *In vivo* experimental characterization of cilia beating pattern in the mouse node without labeling or 3D resolution using DIC microscopy (figure reproduced from Nonaka *et al.*, 2005). (A) Trace of node cilia in enhanced DIC images after background subtraction. The black spots represent the positions of cilia roots, and tip trajectories are in blue, green, and orange. (B) Relationship between observed trip trajectories as projected pattern (circle, ellipse, or D-shape, corresponding to green, blue, and orange trajectories in (A), respectively) and expected beating patterns in 3D. The arrows show the clockwise rotation of cilia. A, P, L, and R refer to anterior, posterior, left, and right sides of the node, respectively. (For interpretation of the references to color in this figure legend, the reader is referred to the Web version of this chapter.)

fails at characterizing properties such as cilium length, shape, rotation speed, or semi-cone angle ψ and requires an extensive analysis for characterizing each investigated cilium.

Interestingly, experimental artifacts due to a lack of time resolution can be used to extract useful information and measure cilia features *in vivo*. Hadjantonakis *et al.* (2008) reported a scanning artifact using slow acquisition speed with standard laser scanning microscopy allows biologists to detect the presence of cilia in the mouse node. In this case, the laser scanning is slow enough to capture several traces of cilium signal during the 2D acquisition of an image and produces a stripped pattern as shown in the zebrafish KV in Fig. 2.7A. Recording this pattern at different z-positions allows reconstructing the cilium envelope in 3D (Fig. 2.7B) and to measure the tilt orientation and angle θ or the semi-cone angle ψ (Fig. 2.7C). In addition, the distance between the 2D stripes provide a measure of cilium beating frequency (Fig. 2.7A). This approach does not permit measuring the direction of rotation (clockwise/anticlockwise).

Bringing together these imaging approaches allow characterizing cilia features in living embryos without requiring the full 3D capture of cilia beats. Even if these technical ruses work better at low cilia density, they open access to simultaneous measurement of several cilia and permit to follow processes at multiple spatial scales. For instance, flow mapping allows simultaneously following the flow within an entire organ and close to individual cilia (Supatto *et al.*, 2008).

5.4. Next experimental challenges: Flow/cilia manipulation and functional imaging

While flow and cilia features can be measured experimentally in living embryos using imaging, the full investigation of the role of cilia during embryonic development brings other experimental challenges. For instance, to fully understand the biophysics of cilia and the interplay between mechanical properties and biological function, it can be critical to manipulate cilia properties (density, clockwise/anticlockwise rotation, speed, tilt angle and direction, amplitude, direction, etc.) or the external flow. Pioneering experimental studies reported the effect of manipulating external flow on a ciliated epithelium (Guirao *et al.*, 2010) and on the mouse embryo node (Nonaka *et al.*, 2005) or the application of external forces on cilia using magnetic manipulations (Hill *et al.*, 2010). However, such manipulation is extremely difficult to perform in live embryos, especially when ciliated structures are internal. Working on explants or cultured epithelia allows direct access to cilia and facilitates measurements that are challenging to perform *in vivo,* such as through magnetic manipulation (Hill *et al.*, 2010). Alternative strategies using optical tweezers have been successfully used in the zebrafish inner ear (Riley *et al.*, 1997; Wu *et al.*, 2011) and should open access to other organs as well. Coupling structural imaging of cilia, flow mapping and *in vivo* functional imaging of the biological response such as calcium signaling, gene expression dynamics, or cellular processes is clearly the next challenge to meet. The constant progress in optics, molecular imaging and engineering will certainly help to reach this milestone.

6. Role of Cilia-Driven Flow in the Kupffer's Vesicle in Zebrafish

The KV is a transient cavity visible from 8 to 11 h after fertilization underneath the notochord (Fig. 2.2). The epithelium of the KV is made of 9 + 2 monociliated cells that generate a counterclockwise fluid flow (Fig. 2.2G–H). This flow triggers asymmetric calcium response on the left side of the cavity (Fig. 2.2F; Francescatto *et al.*, 2010; Sarmah *et al.*, 2005) and is involved in establishing and maintaining the left–right asymmetry of the body axis (Essner *et al.*, 2005; Kramer-Zucker *et al.*, 2005). Two hypotheses coexist to explain the role of flow in this process: either it biases the transport of biochemical signals toward the left side of the KV or it generates a physical stimulus which is read differently according to the direction of the flow. So far, none of the two hypotheses has been ruled out. The asymmetric calcium response is dependent on two channel receptors, the ryanodine receptor *ryr3* and *pkd2* (TRPP2) that are expressed by the KV cells (Francescatto *et al.*, 2010). *pkd2* is a mechanosensory receptor whose function is tightly associated with primary cilia, thus it has been proposed that flow is sensed by cilia through *pkd2* which in turn triggers calcium flux. However, a direct link between these three events is difficult to establish and theoretical work based on measures performed in the mouse suggests that the shear forces are too low for cilia to sense flow directionality (Cartwright *et al.*, 2008). Even though the presence of sensory, immotile cilia is still a matter of debate in the fish (Borovina *et al.*, 2010; Okabe *et al.*, 2008), it is essential to quantify the flow forces generated within the left–right organizer to address this question. Typically, the flow map within the KV displays different velocities according to the *A–P* axis. It is not surprising as cilia are enriched in the anterior pole of the vesicle (Okabe *et al.*, 2008). This reflects the linear relationship postulated by the Stokes flow where the activity of cilia adds to each other showing that directional flow is obtained because of a dorsal cilia tilt. Overall, questions still remain whether flow, morphogen, and physical influences coexist in the process of left–right specification in vertebrates. It is certain that the theory of hydrodynamics can lead us toward novel hypotheses, which will greatly benefit the developmental biology field. Nevertheless, the biological mechanisms at work during this process remain poorly understood in the zebrafish, especially when considering that many observations in the mouse node such as the presence of vesicular release in the vesicle and the presence of sensory cilia are yet lacking in fish. The anatomy and the genetics have greatly diverged between fish and mice, and it is possible that slightly different mechanisms operate in these two species.

7. Role of Cilia-Driven Flow in the Developing Inner Ear in Zebrafish

Besides the well-known left–right flow, cilia-driven flows are necessary in multiple developing organs (Cartwright *et al.*, 2009). Recent characterization of inner ear flow uncovered alternative roles to fluid motion during embryogenesis. In zebrafish, as in humans and other vertebrates, balance is mediated by mechanical sensors in the inner ear. These sensors

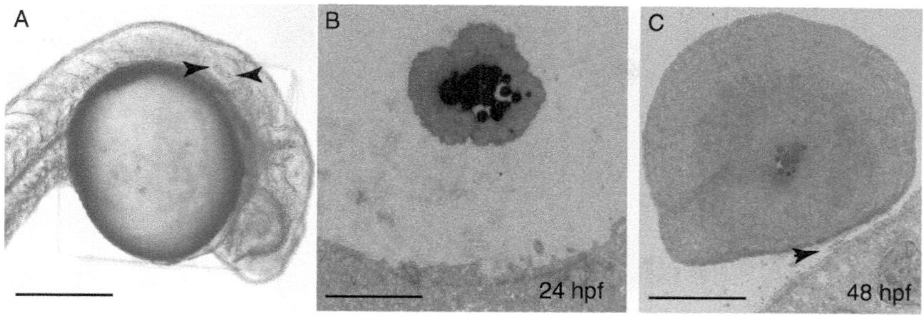

Figure 2.10 Otolithic biomineralization in zebrafish. (A) Side view of a zebrafish at 20 hpf showing the inner ear and the two otolithic aggregates. Otoliths first appear after 18 h postfertilization in the otic cavity. (B and C) Transverse section through the otolith at 22 hpf (B) and 48 hpf (C) showing an electron dense inner part (nucleus) and radial biomineralization starting around the nucleus viewed by transmission electron microscopy (TEM). Radial biomineralization is prominent at that stage. Sensory cilia bundles of the hair cells attached to the otolith at 48 hpf (arrow heads). Scale bar 100 μm in (A) and 5 μm in (B and C). Electron microscopy imaging was performed as described in Pisam *et al.*, 2002).

consist of biomineralized composite crystals, called otoliths (ear-stones), situated atop cilia bundles on the surface of epithelial cells (Fig. 2.10). They provide an inertial mass that facilitates deflection of cilia bundles in response to vibration, gravity, and linear acceleration essential for hearing and balance through specialized cells called hair cells (Fig. 2.10; Yu *et al.*, 2011). In zebrafish, the growth phase of otolith development constitutes a key step during the biomineralization process. Otoliths form at the top of tether cilia, located at the anterior and posterior poles of the inner ear from smaller dense clusters referred to as "spherules," which are self-aggregating particles secreted from the apical portions of the epithelial cells lining the inner ear cavity (Fig. 2.10). At the site of cilium–otolith attachment are clusters of motile cilia inducing local flow. Electron microscopy analysis shows that otolith growth starts as a nucleus of spherules aggregating at the top of a tether cilium (Fig. 2.10; Pisam *et al.*, 2002). At 30 hpf, a mineralized ovoid otolith is visible (Pisam *et al.*, 2002; Sollner *et al.*, 2003) and concentric arrays of spherule deposition are seen at the periphery of the nascent otolith (Fig. 2.10; Pisam *et al.*, 2002). Cilia-driven flow is required for proper otolith formation, but the nature of the flow dynamics at work in the inner ear and the precise contribution of local flow at sites of biomineralization are now the focus of recent investigations. On the basis of high-speed video microscopy of cilia motility and quantitative analysis of precursor particle movements in wild-type and immotile cilia mutant embryos, a cilium-dependent hydrodynamic mechanism for otolith biogenesis has been proposed (Fig. 2.11). In this model, motility of tether cilia at the poles of the otic vesicle establishes a vortex that attracts otolith precursors

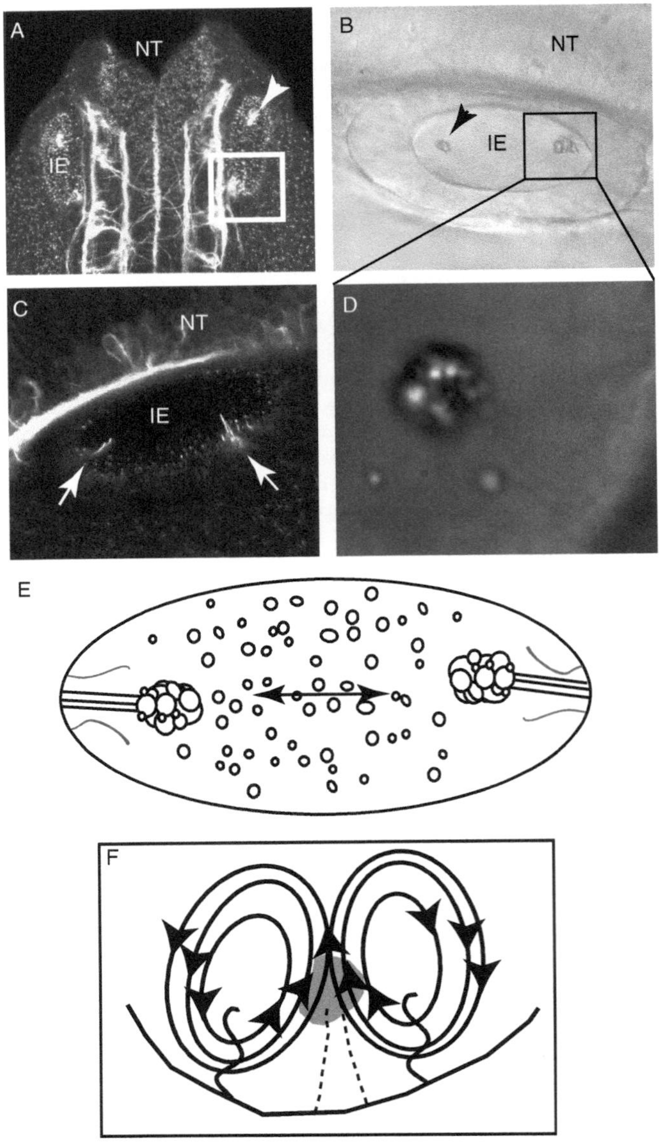

Figure 2.11 Cilia-mediated flow and otolith biogenesis. (A) Dorsal view of embryo labeled with anti-acetylated tubulin antibody showing the neural tube (NT) and axonal projections as well as primary cilia throughout the embryos. (B) Side view of the inner ear visualized through bright field microscopy showing otolith (arrowhead) within the inner ear (IE). (C) Cilia labeling in the inner ear using anti-acetylated tubulin antibody reveals two types of cilia: long tether and short cilia. Immunohistochemistry was performed as described in Colantonio *et al.* (2009). (D) Otolith view with bright field microscopy as described in Colantonio *et al.* (2009), showing spherules next to it. (E and F) Model for motile cilia function during otolith biogenesis.

(Fig. 2.11E and F), thereby biasing the distribution of precursor particles towards the two patches of tether cilia. Such a mechanism would favor preferential otolith seeding at the poles of the otic vesicle. At the otic vesicle poles, tether cilia motility further serves to disperse precursor particles locally. This model, while attractive, needs some clarifications in terms of precise hydrodynamics, most importantly in terms of the hydrodynamic basis of spherule attraction. The hydrodynamic features in the inner ear are particularly interesting because, as opposed to other known cilia-mediated flows, there is no directional flow. It is the activity of isolated clusters of beating cilia that dictates a local flow, which is not able to take over the whole cavity (Wu *et al.*, 2011). It clearly shows that the topology, such as the localization of cilia motility in key areas of the inner ear, is dictating the hydrodynamics in the organ. Importantly, the role of mixing in this system is critical for dictating the shape of the otolith demonstrating that cilia can combine different function through generation of different flow regime (Wu *et al.*, 2011).

8. Conclusion

Throughout the years of cilia research conducted in the past, fluid mechanics and modeling greatly helped experimenters in identifying proper questions and in developing appropriate methods to address them. Today, many questions remain unanswered, which will undoubtedly drive cilia investigations in the coming years. For example, the crucial role of the tilting angle in the genesis of directional flow raises important unknown questions: what are the structural origins of the tilt? How is it regulated and through which dynamics? Last, how does the coupling between cilia and flow influence each other so that the system can reach a steady state? Furthermore, a lot remains to be said about the roles of chaotic advections in cilia-driven flows. We showed that this is a prominent effect in the KV and in the inner ear, but it is certainly true in other ciliated organs and species. The continuous effort to combine experimental approaches through live imaging, modeling, and forward and reverse genetics in the zebrafish are expected to provide us answers to some of these interesting questions in the near future.

The study of the role of cilia-driven flow in embryonic development is an exciting example of interdisciplinary investigation. As reviewed here, the fruitful interplay between fluid mechanics, mechanical modeling, microscopy, image processing, experimental micromanipulation, and developmental genetics permits gaining important knowledge and promises future exciting discoveries in the field. A central driving force for the understanding of the biology of cilia-flow during embryonic development is the continuous necessity for integrating knowledge and experimental data originating from various disciplines (fluid mechanics, modeling, imaging and image processing, genetics, etc.). Another very recent trend is the fact that the biology of cilia structure is feeding biomimetic research and finds of applications in microfluidic devices inspired by biology (Shields *et al.*, 2010; Vilfan *et al.*, 2010). There is a lot to bet that many other biological systems will benefit from such an integrated way of approaching science.

ACKNOWLEDGMENTS

We thank D. Wu and L. Thornton for thoughtful comments on the chapter. We are grateful to N. Messaddeq and J. L. Weickert for help with the electron microscopy and the Fraser laboratory at Caltech for sharing reagents. We also thank the IGBMC, Institut de Génétique et de Biologie Moléculaire et Cellulaire for assistance, the Caltech Biological Imaging Center and the ImagoSeine imaging facility at Institut Jacques Monod for sharing equipment. J. V. is supported by the Human Frontier Science Program (HFSP), INSERM, and the Fondation pour la Recherche Medical (FRM). J. V. and W. S. are supported by Marie Curie International Reintegration Grants within the 7th European Community Framework Programme.

REFERENCES

Aamar, E., and Dawid, I. B. (2008). Isolation and expression analysis of foxj1 and foxj1.2 in zebrafish embryos. *Int. J. Dev. Biol.* **52,** 985–991.

Afzelius, B. A. (1999). Asymmetry of cilia and of mice and men. *Int. J. Dev. Biol.* **43,** 283–286.

Baker, K., and Beales, P. L. (2009). Making sense of cilia in disease: The human ciliopathies. *Am. J. Med. Genet. C Semin. Med. Genet.* **151C,** 281–295.

Borovina, A., Superina, S., Voskas, D., and Ciruna, B. (2010). Vangl2 directs the posterior tilting and asymmetric localization of motile primary cilia. *Nat. Cell Biol.* **12,** 407–412.

Buceta, J., Ibanes, M., Rasskin-Gutman, D., Okada, Y., Hirokawa, N., and Izpisua-Belmonte, J. C. (2005). Nodal cilia dynamics and the specification of the left/right axis in early vertebrate embryo development. *Biophys. J.* **89,** 2199–2209.

Cartwright, J. H. E., Piro, O., and Tuval, I. (2004). Fluid-dynamical basis of the embryonic development of left-right asymmetry in vertebrates. *Proc. Natl. Acad. Sci. USA* **101,** 7234–7239.

Cartwright, J. H. E., Piro, N., Piro, O., and Tuval, I. (2007). Embryonic nodal flow and the dynamics of nodal vesicular parcels. *J. R. Soc. Interface* **4,** 49–55.

Cartwright, J. H., Piro, N., Piro, O., and Tuval, I. (2008). Fluid dynamics of nodal flow and left-right patterning in development. *Dev. Dyn.* **237,** 3477–3490.

Cartwright, J. H., Piro, O., and Tuval, I. (2009). Fluid dynamics in developmental biology: Moving fluids that shape ontogeny. *HFSP J.* **3,** 77–93.

Caspary, T., Larkins, C. E., and Anderson, K. V. (2007). The graded response to Sonic Hedgehog depends on cilia architecture. *Dev. Cell* **12,** 767–778.

Castleman, V. H., Romio, L., Chodhari, R., Hirst, R. A., de Castro, S. C., Parker, K. A., Ybot-Gonzalez, P., Emes, R. D., Wilson, S. W., Wallis, C., Johnson, C. A., Herrera, R. J., *et al.* (2009). Mutations in radial spoke head protein genes RSPH9 and RSPH4A cause primary ciliary dyskinesia with central-microtubular-pair abnormalities. *Am. J. Hum. Genet.* **84,** 197–209.

Chilvers, M. A., Rutman, A., and O'Callaghan, C. (2003). Ciliary beat pattern is associated with specific ultrastructural defects in primary ciliary dyskinesia. *J. Allergy Clin. Immunol.* **112,** 518–524.

Colantonio, J. R., Vermot, J., Wu, D., Langenbacher, A. D., Fraser, S., Chen, J. N., and Hill, K. L. (2009). The dynein regulatory complex is required for ciliary motility and otolith biogenesis in the inner ear. *Nature* **457,** 205–209.

Dale, N., Roberts, A., Ottersen, O. P., and Storm-Mathisen, J. (1987). The morphology and distribution of 'Kolmer-Agduhr cells', a class of cerebrospinal-fluid-contacting neurons revealed in the frog embryo spinal cord by GABA immunocytochemistry. *Proc. R. Soc. Lond. B Biol. Sci.* **232,** 193–203.

Drummond, I. (2009). Studying cilia in zebrafish. *Methods Cell Biol.* **93,** 197–217.

Essner, J. J., Vogan, K. J., Wagner, M. K., Tabin, C. J., Yost, H. J., and Brueckner, M. (2002). Conserved function for embryonic nodal cilia. *Nature* **418,** 37–38.

Essner, J. J., Amack, J. D., Nyholm, M. K., Harris, E. B., and Yost, H. J. (2005). Kupffer's vesicle is a ciliated organ of asymmetry in the zebrafish embryo that initiates left-right development of the brain, heart and gut. *Development* **132,** 1247–1260.

Fahrni, F., Prins, M. W. J., and van Ijzendoorn, L. J. (2009). Micro-fluidic actuation using magnetic artificial cilia. *Lab Chip* **9,** 3413–3421.

Francescatto, L., Rothschild, S. C., Myers, A. L., and Tombes, R. M. (2010). The activation of membrane targeted CaMK-II in the zebrafish Kupffer's vesicle is required for left-right asymmetry. *Development* **137,** 2753–2762.

Gueron, S., and Levit-Gurevich, K. (1999). Energetic considerations of ciliary beating and the advantage of metachronal coordination. *Proc. Natl. Acad. Sci. USA* **96,** 12240–12245.

Guirao, B., Meunier, A., Mortaud, S., Aguilar, A., Corsi, J. M., Strehl, L., Hirota, Y., Desoeuvre, A., Boutin, C., Han, Y. G., Mirzadeh, Z., Cremer, H., *et al.* (2010). Coupling between hydrodynamic forces and planar cell polarity orients mammalian motile cilia. *Nat. Cell Biol.* **12,** 341–350.

Hadjantonakis, A. K., Pisano, E., and Papaioannou, V. E. (2008). Tbx6 regulates left/right patterning in mouse embryos through effects on nodal cilia and perinodal signaling. *PLoS ONE* **3,** e2511.

Hashimoto, M., Shinohara, K., Wang, J., Ikeuchi, S., Yoshiba, S., Meno, C., Nonaka, S., Takada, S., Hatta, K., Wynshaw-Boris, A., and Hamada, H. (2010). Planar polarization of node cells determines the rotational axis of node cilia. *Nat. Cell Biol.* **12,** 170–176.

Hilfinger, A., and Julicher, F. (2008). The chirality of ciliary beats. *Phys. Biol.* **5**. Article number is 016003.

Hill, D. B., Swaminathan, V., Estes, A., Cribb, J., O'Brien, E. T., Davis, C. W., and Superfine, R. (2010). Force generation and dynamics of individual cilia under external loading. *Biophys. J.* **98,** 57–66.

Hirota, Y., Meunier, A., Huang, S., Shimozawa, T., Yamada, O., Kida, Y. S., Inoue, M., Ito, T., Kato, H., Sakaguchi, M., Sunabori, T., Nakaya, M. A., *et al.* (2010). Planar polarity of multiciliated ependymal cells involves the anterior migration of basal bodies regulated by non-muscle myosin II. *Development* **137,** 3037–3046.

Hove, J. R., Köster, R. W., Forouhar, A. S., Acevedo-Bolton, G., Fraser, S. E., and Gharib, M. (2003). Intracardiac fluid forces are an essential epigenetic factor for embryonic cardiogenesis. *Nature* **421,** 172–177.

Jekely, G., and Arendt, D. (2006). Evolution of intraflagellar transport from coated vesicles and autogenous origin of the eukaryotic cilium. *Bioessays* **28,** 191–198.

Kobayashi, D., Iijima, N., Hagiwara, H., Kamura, K., Takeda, H., and Yokoyama, T. (2010). Characterization of the medaka (*Oryzias latipes*) primary ciliary dyskinesia mutant, jaodori: Redundant and distinct roles of dynein axonemal intermediate chain 2 (dnai2) in motile cilia. *Dev. Biol.* **347,** 62–70.

Kramer-Zucker, A. G., Olale, F., Haycraft, C. J., Yoder, B. K., Schier, A. F., and Drummond, I. A. (2005). Cilia-driven fluid flow in the zebrafish pronephros, brain and Kupffer's vesicle is required for normal organogenesis. *Development* **132,** 1907–1921.

Lindemann, C. B. (2011). Experimental evidence for the geometric clutch hypothesis. *Curr. Top. Dev. Biol.* **95,** 1–31.

Lindemann, C. B., and Lesich, K. A. (2010). Flagellar and ciliary beating: The proven and the possible. *J. Cell Sci.* **123,** 519–528.

Liu, Y., Pathak, N., Kramer-Zucker, A., and Drummond, I. A. (2007). Notch signaling controls the differentiation of transporting epithelia and multiciliated cells in the zebrafish pronephros. *Development* **134,** 1111–1122.

Lopes, S. S., Lourenço, R., Pacheco, L., Moreno, N., Kreiling, J., and Saúde, L. (2010). Notch signalling regulates left-right asymmetry through ciliary length control. *Development* **137,** 3625–3632.

Mangos, S., Lam, P. Y., Zhao, A., Liu, Y., Mudumana, S., Vasilyev, A., Liu, A., and Drummond, I. A. (2010). The ADPKD genes pkd1a/b and pkd2 regulate extracellular matrix formation. *Dis. Model. Mech.* **3,** 354–365.

Marshall, W. F., and Kintner, C. (2008). Cilia orientation and the fluid mechanics of development. *Curr. Opin. Cell Biol.* **20,** 48–52.

McGrath, J., Somlo, S., Makova, S., Tian, X., and Brueckner, M. (2003). Two populations of node monocilia initiate left-right asymmetry in the mouse. *Cell* **114,** 61–73.

Nakao, T., and Ishizawa, A. (1984). Light- and electron-microscopic observations of the tail bud of the larval lamprey (Lampetra japonica), with special reference to neural tube formation. *Am. J. Anat.* **170,** 55–71.

Neugebauer, J. M., Amack, J. D., Peterson, A. G., Bisgrove, B. W., and Yost, H. J. (2009). FGF signalling during embryo development regulates cilia length in diverse epithelia. *Nature* **458,** 651–654.

Nonaka, S., Tanaka, Y., Okada, Y., Takeda, S., Harada, A., Kanai, Y., Kido, M., and Hirokawa, N. (1998). Randomization of left-right asymmetry due to loss of nodal cilia generating leftward flow of extraembryonic fluid in mice lacking KIF3B motor protein. *Cell* **95,** 829–837.

Nonaka, S., Shiratori, H., Saijoh, Y., and Hamada, H. (2002). Determination of left-right patterning of the mouse embryo by artificial nodal flow. *Nature* **418,** 96–99.

Nonaka, S., Yoshiba, S., Watanabe, D., Ikeuchi, S., Goto, T., Marshall, W. F., and Hamada, H. (2005). De novo formation of left-right asymmetry by posterior tilt of nodal cilia. *PLoS Biol.* **3,** e268.

Okabe, N., Xu, B., and Burdine, R. D. (2008). Fluid dynamics in zebrafish Kupffer's vesicle. *Dev. Dyn.* **237,** 3602–3612.

Okada, Y., and Hirokawa, N. (2009). Observation of nodal cilia movement and measurement of nodal flow. *Methods Cell Biol.* **91,** 265–285.

Okada, Y., Nonaka, S., Tanaka, Y., Saijoh, Y., Hamada, H., and Hirokawa, N. (1999). Abnormal nodal flow precedes situs inversus in iv and inv mice. *Mol. Cell* **4,** 459–468.

Okada, Y., Takeda, S., Tanaka, Y., Belmonte, J. C., and Hirokawa, N. (2005). Mechanism of nodal flow: A conserved symmetry breaking event in left-right axis determination. *Cell* **121,** 633–644.

Perkins, L. A., Hedgecock, E. M., Thomson, J. N., and Culotti, J. G. (1986). Mutant sensory cilia in the nematode *Caenorhabditis elegans*. *Dev. Biol.* **117,** 456–487.

Pisam, M., Jammet, C., and Laurent, D. (2002). First steps of otolith formation of the zebrafish: Role of glycogen? *Cell Tissue Res.* **310,** 163–168.

Purcell, E. M. (1977). Life at low Reynolds number. *Am. J. Phys.* **45,** 3–11.

Purcell, E. M. (1997). The efficiency of propulsion by a rotating flagellum. *Proc. Natl. Acad. Sci. USA* **94,** 11307–11311.

Raffel, M., Willert, C., and Kompenhans, J. (1998). Particle Image Velocimetry: A Practical Guide. Springer-Verlag, Berlin.
Raya, A., and Belmonte, J. C. I. (2006). Left-right asymmetry in the vertebrate embryo: From early information to higher-level integration. *Nat. Rev. Genet.* **7,** 283–293.
Riley, B. B., Zhu, C., Janetopoulos, C., and Aufderheide, K. J. (1997). A critical period of ear development controlled by distinct populations of ciliated cells in the zebrafish. *Dev. Biol.* **191,** 191–201.
Sanderson, M. J., and Sleigh, M. A. (1981). Ciliary activity of cultured rabbit tracheal epithelium—Beat pattern and metachrony. *J. Cell Sci.* **47,** 331–347.
Sarmah, B., Latimer, A. J., Appel, B., and Wente, S. R. (2005). Inositol polyphosphates regulate zebrafish left-right asymmetry. *Dev. Cell* **9,** 133–145.
Sarmah, B., Winfrey, V. P., Olson, G. E., Appel, B., and Wente, S. R. (2007). A role for the inositol kinase Ipk1 in ciliary beating and length maintenance. *Proc. Natl. Acad. Sci. USA* **104,** 19843–19848.
Schweickert, A., Weber, T., Beyer, T., Vick, P., Bogusch, S., Feistel, K., and Blum, M. (2007). Cilia-driven leftward flow determines laterality in Xenopus. *Curr. Biol.* **17,** 60–66.
Shah, A. S., Ben-Shahar, Y., Moninger, T. O., Kline, J. N., and Welsh, M. J. (2009). Motile cilia of human airway epithelia are chemosensory. *Science* **325,** 1131–1134.
Shields, A. R., Fiser, B. L., Evans, B. A., Falvo, M. R., Washburn, S., and Superfine, R. (2010). From the Cover: Biomimetic cilia arrays generate simultaneous pumping and mixing regimes. *Proc. Natl. Acad. Sci. USA* **107,** 15670–15675.
Short, M. B., Solari, C. A., Ganguly, S., Powers, T. R., Kessler, J. O., and Goldstein, R. E. (2006). Flows driven by flagella of multicellular organisms enhance long-range molecular transport. *Proc. Natl. Acad. Sci. USA* **103,** 8315–8319.
Smith, E. F., and Yang, P. (2004). The radial spokes and central apparatus: Mechano-chemical transducers that regulate flagellar motility. *Cell Motil. Cytoskeleton* **57,** 8–17.
Smith, D. J., Gaffney, E. A., and Blake, J. R. (2007). Discrete cilia modeling with singularity distributions: Application to the embryonic node and the airway surface liquid. *Bull. Math. Biol.* **69,** 1477–1510.
Smith, D. J., Blake, J. R., and Gaffney, E. A. (2008). Fluid mechanics of nodal flow due to embryonic primary cilia. *J. R. Soc. Interface* **5,** 567–573.
Smith, D., Smith, A., and Blake, J. (2010a). Mathematical embryology: The fluid mechanics of nodal cilia. *J. Eng. Math.* 1–25, doi:10.1007/s10665-010-9383-y.
Smith, S. B., Qu, H. Q., Taleb, N., Kishimoto, N. Y., Scheel, D. W., Lu, Y., Patch, A. M., Grabs, R., Wang, J. H., Lynn, F. C., Miyatsuka, T., Mitchell, J., *et al.* (2010b). Rfx6 directs islet formation and insulin production in mice and humans. *Nature* **463,** 775–780.
Solari, C. A., Ganguly, S., Kessler, J. O., Michod, R. E., and Goldstein, R. E. (2006). Multicellularity and the functional interdependence of motility and molecular transport. *Proc. Natl. Acad. Sci. USA* **103,** 1353–1358.
Sollner, C., Burghammer, M., Busch-Nentwich, E., Berger, J., Schwarz, H., Riekel, C., and Nicolson, T. (2003). Control of crystal size and lattice formation by starmaker in otolith biomineralization. *Science* **302,** 282–286.
Soyer, J., Flasse, L., Raffelsberger, W., Beucher, A., Orvain, C., Peers, B., Ravassard, P., Vermot, J., Voz, M. L., Mellitzer, G., and Gradwohl, G. (2010). Rfx6 is an Ngn3-dependent winged helix transcription factor required for pancreatic islet cell development. *Development* **137,** 203–212.
Stokes, G. G. (1851). On the effect of the internal friction of fluids on the motion of pendulums. *Trans. Camb. Philos. Soc.* **9,** 8–106.
Stone, H. A., Stroock, A. D., and Ajdari, A. (2004). Engineering flows in small devices: Microfluidics toward a lab-on-a-chip. *Annu. Rev. Fluid Mech.* **36,** 381–411.

Stubbs, J. L., Oishi, I., Belmonte, J. C. I., and Kintner, C. (2008). The forkhead protein Foxj1 specifies node-like cilia in Xenopus and zebrafish embryos. *Nat. Genet.* **40,** 1454–1460.

Sullivan-Brown, J., Schottenfeld, J., Okabe, N., Hostetter, C. L., Serluca, F. C., Thiberge, S. Y., and Burdine, R. D. (2008). Zebrafish mutations affecting cilia motility share similar cystic phenotypes and suggest a mechanism of cyst formation that differs from pkd2 morphants. *Dev. Biol.* **314,** 261–275.

Supatto, W., Débarre, D., Moulia, B., Brouzes, E., Martin, J. L., Farge, E., and Beaurepaire, E. (2005). In vivo modulation of morphogenetic movements in *Drosophila* embryos with femtosecond laser pulses. *Proc. Natl. Acad. Sci. USA* **102,** 1047–1052.

Supatto, W., Fraser, S. E., and Vermot, J. (2008). An all-optical approach for probing microscopic flows in living embryos. *Biophys. J.* **95,** L29–L31.

Tian, T., Zhao, L., Zhao, X. Y., Zhang, M., and Meng, A. M. (2009). A zebrafish gene trap line expresses GFP recapturing expression pattern of foxj1b. *J. Genet. Genomics* **36,** 581–589.

Vandaele, C., Coulon-Bublex, M., Couble, P., and Durand, B. (2001). *Drosophila* regulatory factor X is an embryonic type I sensory neuron marker also expressed in spermatids and in the brain of *Drosophila. Mech. Dev.* **103,** 159–162.

Vermot, J., Fraser, S., and Liebling, M. (2008). Fast fluorescence microscopy for imaging the dynamics of embryonic development. *HFSP J.* **2,** 143–155.

Vilfan, A., and Julicher, F. (2006). Hydrodynamic flow patterns and synchronization of beating cilia. *Phys. Rev. Lett.* **96,** 058102.

Vilfan, M., Potocnik, A., Kavcic, B., Osterman, N., Poberaj, I., Vilfan, A., and Babic, D. (2010). Self-assembled artificial cilia. *Proc. Natl. Acad. Sci. USA* **107,** 1844–1847.

White, R. D., and Grosh, K. (2005). Microengineered hydromechanical cochlear model. *Proc. Natl. Acad. Sci. USA* **102,** 1296–1301.

Wu, D., Freund, J. B., Fraser, S. E., Vermot, J. (2011). Mechanistic basis of otolith formation during teleost inner ear development. *Developmental Cell.* doi:10.1016/j.devcel.2010.12.006.

Yoganathan, A. P., He, Z., and Casey Jones, S. (2004). Fluid mechanics of heart valves. *Annu. Rev. Biomed. Eng.* **6,** 331–362.

Yu, X. W., Ng, C. P., Habacher, H., and Roy, S. (2008). Foxj1 transcription factors are master regulators of the motile ciliogenic program. *Nat. Genet.* **40,** 1445–1453.

Yu, X., Lau, D., Ng, C. P., and Roy, S. (2011). Cilia-driven fluid flow as an epigenetic cue for otolith biomineralization on sensory hair cells of the inner ear. *Development* **138,** 487–494.

Zahm, J. M., Pierrot, D., Hinnrasky, J., Fuchey, C., Chevillard, M., Gaillard, D., and Puchelle, E. (1990). Functional activity of ciliated outgrowths from cultured human nasal and tracheal epithelia. *Biorheology* **27,** 559–565.

Zamir, E. A., Czirok, A., Cui, C., Little, C. D., and Rongish, B. J. (2006). Mesodermal cell displacements during avian gastrulation are due to both individual cell-autonomous and convective tissue movements. *Proc. Natl. Acad. Sci. USA* **103,** 19806–19811.

CHAPTER THREE

Spontaneous Mechanical Oscillations: Implications for Developing Organisms

Karsten Kruse* *and* Daniel Riveline†

Contents

Abstract

Major transformations of cells during embryonic development are traditionally associated with the activation or inhibition of genes and with protein modifications. The contributions of mechanical properties intrinsic to the matter an organism is made of, however, are often overlooked. The emerging field "physics of living matter" is addressing active material properties of the cytoskeleton and tissues

* Theoretical Physics, Saarland University, Saarbrücken, Germany
† Laboratory of Cell Physics, Institut de Science et d'Ingénierie Supramoléculaires (ISIS, UMR 7006) and Institut de Génétique et de Biologie Moléculaire et Cellulaire (IGBMC, UMR 7104), Université de Strasbourg, France

Current Topics in Developmental Biology, Volume 95
ISSN 0070-2153, DOI: 10.1016/B978-0-12-385065-2.00003-7

like the spontaneous generation of stress, which may lead to shape changes and tissue flows, and their implications for embryonic development. Here, we discuss spontaneous mechanical oscillations to present some basic elements for understanding this physics, and we illustrate its application to developing embryos. We highlight the role of state diagrams to quantitatively probe the significance of the corresponding physical concepts for understanding development.

1. Introduction

Cellular shape transformations are essential during the development of an organism. These changes are ultimately driven by the cytoskeleton, a filamentous protein network that determines the mechanical properties of cells, and a substantial fraction of current research aims at understanding cytoskeletal organization during developmental processes. In this endeavor, the traditional approach of cell and developmental biology is to trace the ultimate cause of cytoskeletal changes back to signaling networks, which act like a program instructing the mechanical components of a cell or an embryo to execute the required transformations. Sure enough, for example, the suppression or modification of a gene can dramatically affect the development of an embryo. Further, the Rho pathway provides a versatile means to regulate cytoskeletal dynamics in response to external signals and the distribution of morphogens has a strong impact on the spatial organization of tissues during development and thus on the cytoskeleton, too. This approach, however, neglects the remarkable capabilities of the cytoskeleton and hence of tissues to spontaneously reorganize themselves. To appreciate the contribution of the intrinsic cytoskeletal activity to the reorganization of developing tissues, a thorough knowledge of the physical properties of active matter is crucial. We argue that only in this way a comprehensive and quantitative understanding of development will be possible.

Molecular motors play a key role in understanding the dynamics of living matter (Benazeraf *et al.*, 2010; Bertet *et al.*, 2004; Gally *et al.*, 2009; Howard and Hyman, 2009; Rauzi *et al.*, 2008; Wozniak and Chen, 2009). Fueled by the hydrolysis of ATP or GTP, motor proteins can generate stresses in the cytoskeleton which in turn lead to flows and thus to shape changes or movements of cells and tissues. While single molecule experiments resulted in a rather detailed understanding of how an individual motor enzyme acts, the rich collective action of many motors is still far from being fully explored and sometimes produces surprising and unintuitive phenomena. Concepts from statistical physics and nonlinear dynamics provide essential tools for understanding the collective behavior of molecular motors. These tools often require a high degree of formalization by writing equations, which are studied by analytical or numerical means.

An important goal of such an effort is to obtain a *state* (or *phase*) *diagram** that comprehensively describes qualitatively different states of a system as a function of molecular parameters, such as the density and activity of motor proteins. In addition, such an analysis aims at providing an explanation of quantitative features of a pattern, for example, its extension or the time needed for its formation.

In this short review, we illustrate the extraordinary properties of active matter by presenting mechanical oscillations observed in a variety of systems, ranging from the molecular level involving a few motor molecules up to the tissues of a developing embryo. We will argue that in spite of the vastly different length and time scales, the oscillations all have their origin in collective effects of motor enzymes that can spontaneously generate periodic motion and do not require regulation by a chemical oscillator. We first give some background on molecular motors and on the onset of spontaneous mechanical oscillations when many motors are mechanically coupled. With this property of motor ensembles in mind, we then discuss oscillatory phenomena in cells and developing embryos. Finally, we propose experiments to probe the applicability of these concepts and to test predicted new behaviors and we suggest a potential biological function of spontaneous mechanical oscillations.

2. Part I: Molecular Motors and Oscillations

2.1. Single molecular motors are rather well understood

There is a large variety of molecular motors generating translational or rotational motion upon the hydrolysis of a nucleotide such as ATP or GTP. Cytoskeletal molecular motors fall into the former class and interact with either actin filaments or microtubules (see Fig. 3.1). Myosin motors represent a major class of motors found in eukaryotic cells and are involved in muscle contraction, in cell motility, and in intracellular transport. It is, in fact, a superfamily of enzymes comprising roughly 20 families of related proteins interacting with actin filaments. Motors interacting with microtubules fall into two classes denoted as kinesins and dyneins. Among other processes, they are involved in directed intracellular transport of organelles and in the beating of cilia and flagella. Microtubules and actin filaments are polar objects with a fast growing plus- and a slowly growing minus-end and determine the direction of motion of the motor. For a given motor type, the direction of motion is fixed, seemingly prohibiting spontaneous oscillatory dynamics (Howard, 2001).

* see glossary

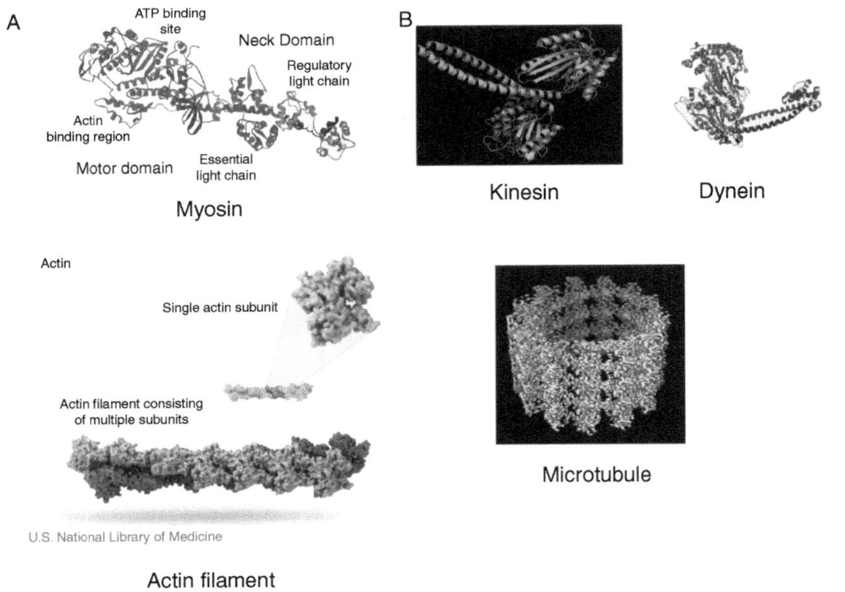

Figure 3.1 Structures of classical molecular motors exhibiting translational motions. (A) Myosin (top) and actin filament (bottom). (B) Kinesin and dynein (top) with microtubule (bottom). Even though the molecular structures of motor proteins are usually considered to be of prime importance, we argue that for collective behavior of motor ensembles only a few key features matter, while atomic details are largely irrelevant. Sources: http://dir.nhlbi.nih.gov/labs/lmc/cmm/myosinlab.asp, http://ghr.nlm.nih.gov/handbook/illustrations/actin, http://www.helsinki.fi/~pjojala/Kinesin.htm, and http://www.concord.org/publications/newsletter/2005-fall/friday.html.

Our understanding of molecular motors has profited enormously from *in vitro* assays (Sheetz *et al.*, 1984). Motility assays were designed to observe the directed motion of motors on glass coverslips, using optical microscopy. In conjunction with single molecule experiments employing optical or magnetic tweezers as well as atomic force microscopes, these methods led to a comprehensive characterization of the physical properties of a large number of motor types (Neuman and Nagy, 2008). The exerted forces have been found to be in the range of a few pico-Newtons, conformational changes associated with the hydrolysis of ATP are in the nanometer range, and the duty ratios, that is, the fraction of time a motor head is bound to a filament during a cycle of nucleotide hydrolysis, were found to differ largely between different motor types, ranging from a few percent up to more than 50%. In the latter case, a motor consisting of two heads is processive, meaning that it can move large distances without detachment. Full energy landscapes and mechanotransduction events have been measured and characterized using these methods; see, for example, Nishikawa *et al.* (2008). Consequently, rather detailed molecular pictures are used when explaining motor actions. As we will see below, however, detailed architectures play merely a subordinate role for explaining collective effects present in ensembles of many molecular motors such as oscillations. Instead, these phenomena can be quantitatively analyzed after appropriate coarse-graining, meaning that processes and structures on length and time scales below those relevant for the phenomenon under study are averaged out. For example, descriptions of the basic principles underlying directed motion reduce a motor to a point particle moving in a *periodic potential** (see Fig. 3.2 Terms marked with a * are explained in the glossary on p. 86.) that can switch between two states and neglects many internal *degrees of freedom** (Julicher and Prost, 1997a,b).

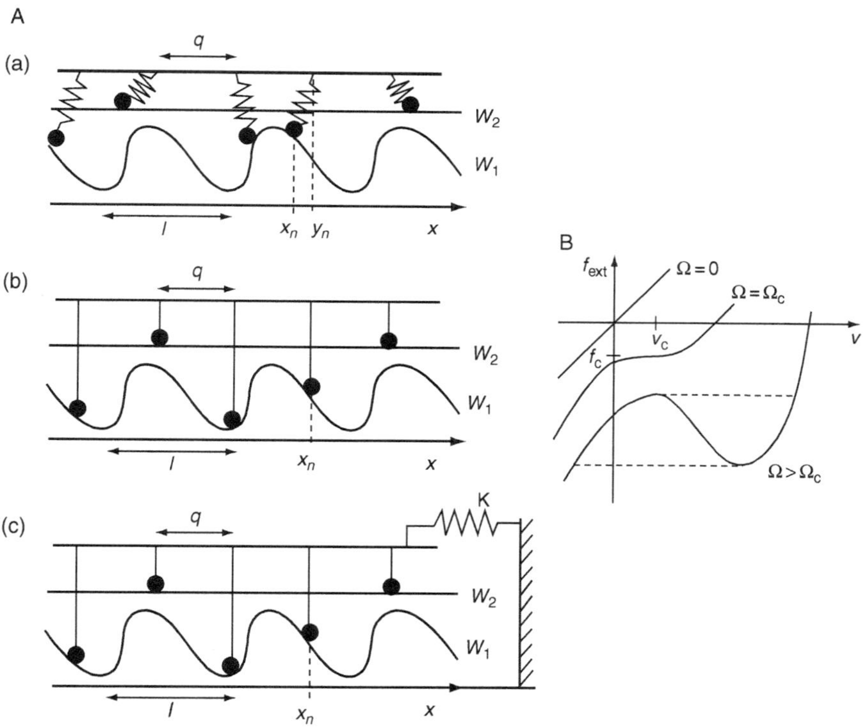

Figure 3.2 Generic descriptions of the dynamics of motor ensembles. (A) Schematic presentation of motor models presented in the text. Motor heads are represented as black dots that are connected to a rigid backbone, separated from each other by a distance q. Motor heads can be elastically (a) or rigidly coupled (b, c). The *interaction* of a motor head with the filament in different states is described by the potentials W_1 and W_2. These *potentials* are *periodic* with period l, reflecting the filament's polarity. Motor heads will always slide toward minima of a potential and switch stochastically between the two states. An isolated motor will move to the right. (a, b) The backbone is unrestricted. (c) The backbone is coupled to a spring. (B) Schematic representation of a *dynamic instability*. For an activity Ω than a critical value, the ensemble can move at two distinct velocities for the same external force f_{ext}. From (Julicher and Prost, 1997a,b).

2.2. Ensembles of motors behave differently from single motors

While single motor experiments have been extensively developed, much less attention has been paid to the analysis of effects present in collections of motors. This is surprising, since the behavior of motor ensembles cannot, in general, be inferred in a direct way from the behavior of single motor molecules. For example, although the direction of motion of a single motor is determined by the orientation of the actin filament or microtubule, collections of motors can exhibit bidirectional motion, which in connection with an elastic element can lead to spontaneous oscillations.

Experimentally, Riveline *et al.* revealed bidirectional motion by using an adapted actomyosin gliding assay (see Fig. 3.3A). Myosin heads were attached to a substrate and actin filaments were posed on top of this motor carpet, while their motion was restricted along some axis by microfabricated

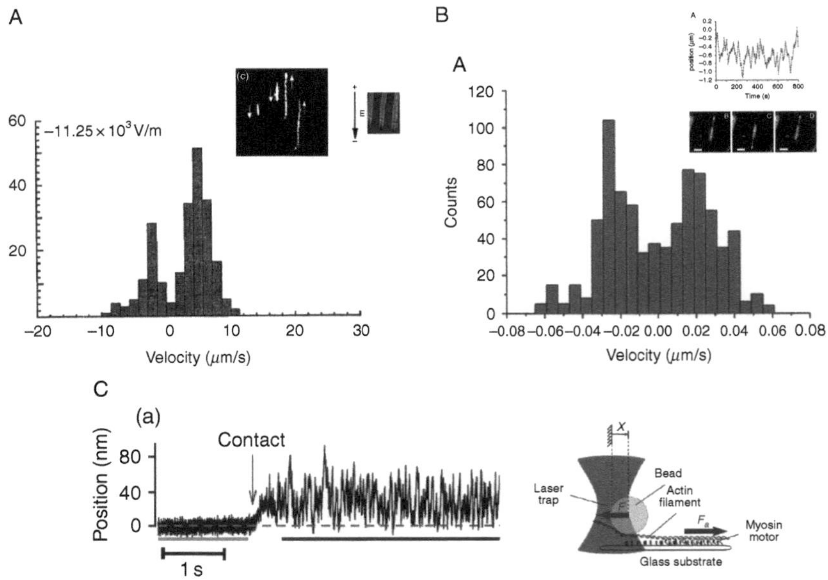

Figure 3.3 Dynamics of motor ensembles *in vitro*. (A) Bimodal velocity distribution for actin filaments confined in a micro-channel propelled by myosin motors attached on the substrate and subject to an external electrical field (Riveline *et al.*, 1998). Inset: traces of actin filaments (left), TEM of the microchannels. (B) Bimodal velocity distribution of an apolar actin bundle (Gilboa *et al.*, 2009). The velocities measured in (A) and (B) are different, because both experiments are distinct in their set-up, in particular, friction with the substrate and between motors as well as temperature differ. (C) For actin filaments moving on a carpet of myosin motors subject to a local external force exerted by optical tweezers (right), individual filaments oscillate (left) (Placais *et al.*, 2009).

channels (Riveline *et al.*, 1998). An electric field was used to impose a force parallel to the direction of motion of the filaments. The force–velocity relation of the filaments showed a region of coexistence in the *stalling force** region: Under the same conditions, filaments could move at two distinct velocities. Remarkably, the two velocities had opposite signs. Also in a more recent gliding assay, a bimodal velocity distribution was observed (Gilboa *et al.*, 2009). In this case, actin bundles of mixed polarity were used instead of single filaments (see Fig. 3.3B). Finally, Placais *et al.* used a substrate densely covered by myosin motors on which they posed an actin filament that was attached to a latex bead (Placais *et al.*, 2009). This bead was then trapped by optical tweezers. The tweezers provided a restoring force, which increased linearly as the actin filament was displaced (see Fig. 3.3C). The filament was found to oscillate with a frequency of a few tens of Hertz and with an amplitude of up to 10 nm. Coherence of the oscillations was lost after a few cycles, though. This was due to fluctuations in the system resulting from the finite stiffness of the actin filament and, more importantly, the finite number of motors interacting with the filament.

Two distinct mechanisms have been identified that can explain spontaneous mechanical oscillations induced by molecular motors (Grill *et al.*, 2005; Howard and Hyman, 2009; Julicher and Prost, 1997a,b; Vilfan and Frey, 2005). The first mechanism is based on the possibility of two coexisting motor speeds if many motors are connected to a common backbone (see Fig. 3.2A(a, b)) (Badoual *et al.*, 2002; Gillo *et al.*, 2009; Julicher and Prost, 1995). The velocity distribution is in this case expected to be bimodal, because fluctuations will induce switching between the two velocities. As in the experiments mentioned above, the two velocities can have opposite signs. This *dynamic instability** was predicted in (Julicher and Prost, 1995) using a coarse-grained two-state model for describing the motor dynamics. Consequently, the coexistence of two velocities of rigidly coupled motor enzymes is expected to be a generic phenomenon independent of many molecular details.

It is not difficult to see how coexistence of two velocities can lead to oscillations if the motor ensemble is connected to an elastic spring (Julicher and Prost, 1997a,b). To this end, consider the force–velocity relation, which is double-valued in a certain range of external forces. Assume that the motors move as to extend the spring (see Fig. 3.2A(c)). At some point, the interval of velocity coexistence ends and the direction of motion will be reversed. The spring will thus be shortened and the motors reenter the region of velocity coexistence, still compressing the spring. This is possible until the elastic force on the motors is again so large that the region of coexistence is left and the motors will change directions. This is a second example of a *dynamic instability*: there is a point at which the motor forces and the elastic force balance, but under a small perturbation, the system leaves this point and starts to oscillate spontaneously.

The onset of the instability depends on control parameters such as the activity of the motors (see Fig. 3.2B). The *state* (or *phase*) *diagram* summarizes for which values of the control parameters oscillations can be observed in the system and when it settles into a stationary state. *State diagrams* can be obtained for various combinations of control parameters. Comparing experimental and theoretical state diagrams is an essential tool for validating or falsifying possible mechanisms of collective behavior. We will come back to this issue below.

Let us now discuss the second mechanism for spontaneous motor oscillations, which relies on the force dependence of a motor's detachment rate (Grill *et al.*, 2005). Indeed, processive motors with a well-defined force–velocity relation show a detachment rate that depends on the applied force parallel to the polar filament they interact with (Schnitzer *et al.*, 2000). Assume that the detachment force increases with the applied force, and consider a motor-filament that is loading a spring while it moves along a filament. As the spring is loaded, the force on each motor increases. If one motor detaches from the filament, the remaining attached motors have to bear an increased force, which in turn makes their detachment more likely. Through this mechanical feedback, an avalanche of detachment events can be created, eventually leading to the detachment of all motors. The spring can then relax, the motors rebind, and the cycle starts anew. This mechanism does not rely on the coexistence of motor velocities of opposite signs. However, if two types of motors moving into opposite directions are involved, a similar tug-of-war mechanism can explain bidirectional motion of vesicles as observed in developing *Drosophila* (Muller *et al.*, 2008).

Recent work by Guérin *et al.* shows that the two oscillation mechanisms can be described in a unifying motor model where the strength of the motor linkage to the coupling backbone is an adjustable parameter (Guerin *et al.*, 2010). In what the authors call the weak pinning regime, oscillations can be generated by the first mechanism, while in what the authors call the strong pinning regime, oscillations can be generated by the second mechanism. Using parameters for the actomyosin system, the authors concluded that the second mechanism more appropriately describes spontaneous oscillations in that system.

2.3. Oscillations can be accompanied by cytoskeletal rearrangements

In the examples of mechanical oscillations discussed so far, the respective actin filaments or microtubules remained structurally intact. Still mechanical oscillations can also be coupled to the turnover of cytoskeletal filaments. One example is provided by the oscillations of nuclei in the fission yeast *Schizosaccharomyces pombe* prior to meiosis. There, microtubules are drawn along the cell wall by dyneins (Vogel *et al.*, 2009). As they are drawn toward the cell tip, they depolymerize. More complex examples are given by microtubule-depleted cells and by cell fragments (Bornens *et al.*, 1989; Costigliola *et al.*, 2010; Kapustina *et al.*, 2008; Paluch *et al.*, 2005; Pletjushkina *et al.*, 2001; Salbreux *et al.*, 2007; Weinreb *et al.*, 2006). Here, the actin cortex tears at one place and completely retracts due to the action of myosin. In the process of retraction, actin filaments disassemble and leave a growing membrane bulge behind that is essentially void of an actin cortex. As soon as most of the actin has disassembled, it reforms a cortex in the bulge and the process starts anew. The newly assembled cortex is most likely to tear at the point where myosin assembled during the previous retraction event, because there the motor-induced cortex tension is highest.

The assembly–disassembly oscillation is only one of many possible instabilities of the cytoskeletal network. *In vitro* reconstitutions of purified actin filaments or microtubules and associated motor proteins and observations of nucleus-free cell fragments have in addition revealed the possibility of such networks to form density bands, asters, vortices, or networks of motor rich foci from an initially homogenous state (Backouche *et al.*, 2006; Nedelec *et al.*, 1997; Schaller *et al.*, 2010; Surrey *et al.*, 2001).

2.4. Theoretical approaches to describe cytoskeletal behavior *in vitro*

Different models have been used to understand the mechanisms underlying these instabilities. Many of them aim at capturing details of processes on molecular scales. This is most apparent for particle-based stochastic simulations, where individual filaments and motor molecules are followed. The dynamics of these entities is governed by their physical properties as well as their biochemistry (Karsenti *et al.*, 2006). Similarly, "mesoscopic" descriptions, where the state of the system is given by averaged distributions of filaments and motors, employ simple dynamic rules that attempt to capture essential molecular properties; see, for example, Doubrovinski and Kruse (2007) and Liverpool and Marchetti (2003).

An alternative line of research focuses on the systems' behavior on large length and time scales. The ensuing *hydrodynamic theories** are based on conservation laws and symmetries only (Julicher *et al.*, 2007). That is, they neglect microscopic *degrees of freedom**, which in turn enter the description only through the values of a number of phenomenological parameters that control the large-scale dynamics. These parameters are analogous to the viscosity of simple fluids. For the *hydrodynamic theory* of motor-filament systems, the most salient parameters determine the coupling of the stresses in the system to the active processes driven by ATP-hydrolysis and the polar order. This approach has been used to study general physical properties of the cytoskeleton, like its behavior under shear. In addition, it has been used to study cell biological phenomena like the retrograde actin flow in lamellipodia (Kruse *et al.*, 2006) or spontaneous polarization of crawling cells (Callan-Jones *et al.*, 2008).

The advantage of hydrodynamic theories over microscopic approaches is their independence of microscopic details, many of which are unknown for the cytoskeleton. Since, the description depends merely on conservation laws and symmetries, it is applicable to a large class of systems, called *active polar gels**. This class comprises, notably, also tissues. Their behavior is thus also relevant for developing organisms. As a word of caution, let us mention that there are effects, which depend in an essential way on non-hydrodynamic variables. This is, for example, the case for spontaneous contraction–relaxation waves along myofibrils (Gunther and Kruse, 2007). In such a situation, the hydrodynamic descriptions must be extended to account for fast *degrees of freedom* or higher order coupling terms between the different quantities.

3. Part II: Oscillations Related to Myosin Motors in Cells and in Embryos

3.1. Cytoskeletal oscillations *in vivo*

Mechanical oscillations have been observed in a variety of *in vivo* systems, ranging from periodic back and forth movements of cytoskeletal elements in single cells to pulsed contractions in developing embryos. These phenomena have in common that they can result from a collection of many molecular motors acting against a resisting elastic force. Such oscillations were observed for actomyosin systems as well as for kinesins or dyneins interacting with microtubules. Consistent with the generic nature of the oscillation we discussed above, the nature of the motor responsible for oscillations *in vivo* is thus not unique. We will now detail these phenomena, each time stressing the generic features of the oscillatory behavior.

3.2. Oscillations *in vivo*: actomyosin

Insect flight muscles provide probably the earliest example of spontaneous mechanical oscillations that have been observed *in vivo*. In fact, the beating of wings occurs in some insects at a frequency that is larger than the maximal frequency at which neurons can fire action potentials (Josephson *et al.*, 2000). Muscles are composed of a highly organized array of actomyosin motors. The basic contractile unit is a sarcomere and it has indeed been found that sarcomeres can oscillate spontaneously under unphysiological conditions in the absence of calcium (Okamura and Ishiwata, 1988). Elastic elements are, in this case, provided by a number of structural proteins like titin or macromolecular assemblies like the Z-plane or the M-line.

In mammalian nonmuscle cells, myosin II and actin filaments can assemble into so-called stress fibers, which share with sarcomeres the linear arrangement of myosin and actin, but are less regular than the latter. Remarkably, stress fibers are encountered in a variety of situations: they form not only in single cells on a substrate and in neighboring cells in culture but also in developing embryos. These "mini-muscles" are notably required for cell motion. Further, it was shown that the cells probe their environment with these units (Delanoe-Ayari *et al.*, 2011): by applying force, cells can reinforce local adhesive contacts: The more extended the contact with a constant density of glue, the larger the local force applied by the cell (Balaban *et al.*, 2001; Grashoff *et al.*, 2010; Riveline *et al.*, 2001). This unusual phenomenon of mechanosensing was shown for focal contacts—contacts between cells and the extracellular matrix—and for cell–cell contacts (Brevier *et al.*, 2007; Brevier *et al.*, 2008; le Duc *et al.*, 2010; Liu *et al.*, 2010; Yonemura *et al.*, 2010)—contacts between neighboring cells.

The force applied by the cell on these adhesion areas was measured. It was reported that the force oscillates on timescales of a minute (Galbraith and Sheetz, 1997; see Fig. 3.4A), that is, a significantly larger time scale than the oscillations in isolated sarcomeres. This phenomenon is connected to stress fibers applying tension on focal contacts: the cells locally undergo cycles of pulling and detaching events that are consistent with the observed oscillations. Other oscillations related to myosin were also reported recently in single fibroblasts: first of all, the spatial localization of myosin II within the actomyosin network was observed to oscillate in time (Giannone *et al.*, 2007; Rossier *et al.*, 2010), second, thickness oscillations were observed (Kapustina *et al.*, 2008; Pletjushkina *et al.*, 2001).

To obtain undisputable evidence for spontaneous mechanical oscillations that are not imposed by some chemical oscillation is indeed hard *in vivo*. The reason is that, in a living cell or organism, it is impossible to control all influences on a mechanical subsystem. Sure enough, for some systems, models have been found that do not involve a chemical feedback, but do semiquantitatively reproduce the observations. Examples are given by spindle oscillations during asymmetric cell division (Grill *et al.*, 2005; Kozlowski *et al.*, 2007), chromosome oscillations during mitosis (Campas and Sens, 2006), and spontaneous oscillations of muscle sarcomeres (Gunther and Kruse, 2007). Further, two- and three-dimensional beat patterns of sperm flagella can be obtained by the theory of *internally driven filaments** (Camalet and Julicher, 2000; Camalet *et al.*, 1999; Hilfinger and Julicher, 2008; Hilfinger *et al.*, 2009; Riedel-Kruse *et al.*, 2007). However, quantitative agreement between a model solution and an isolated measurement or even an average over many measurements is not enough. We argue that for a real comparison between experiment and theory, system parameters have to be varied in a controlled way in both approaches. Put in a different way, experimental and theoretical state diagrams have to be compared.

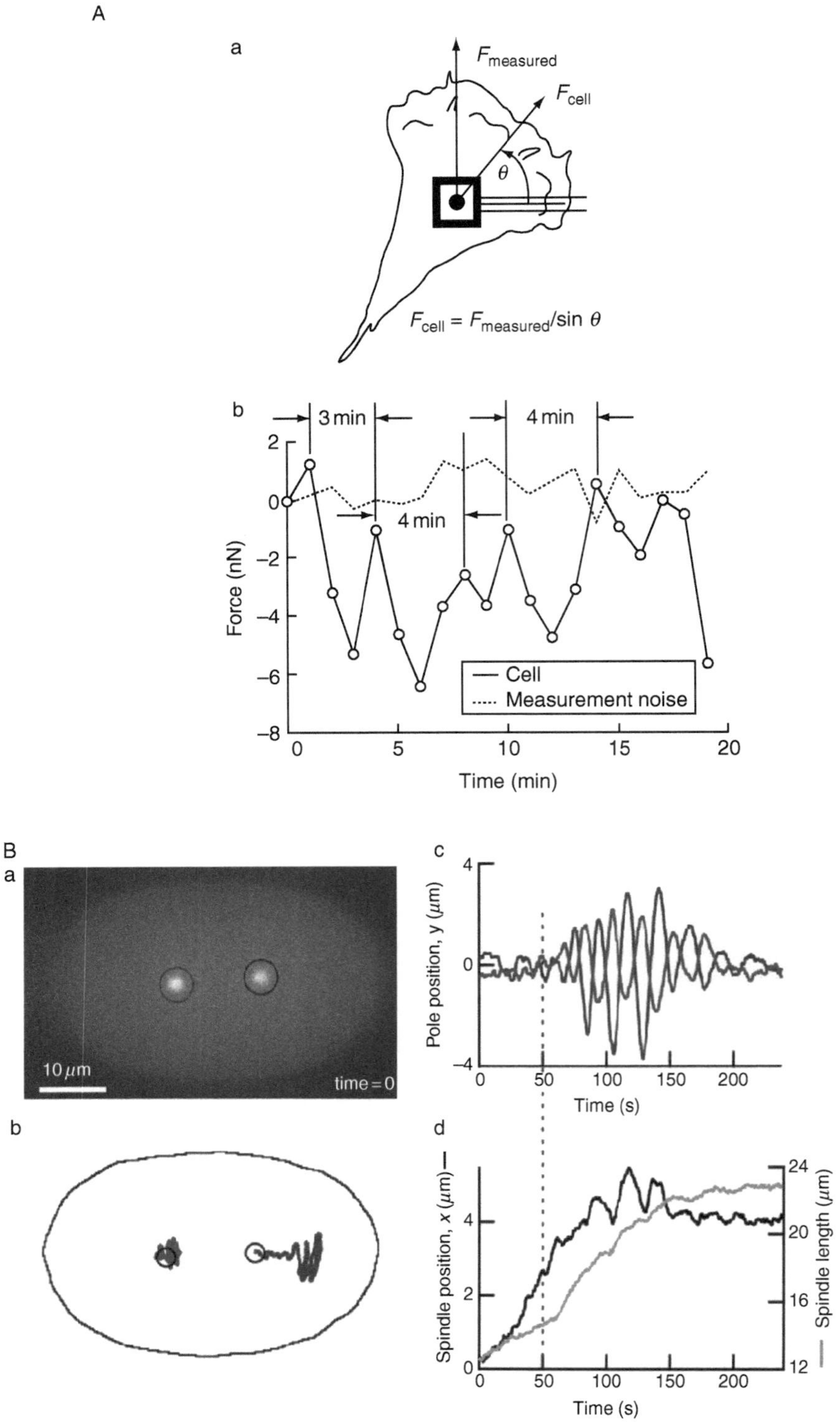

A
a
$F_{measured}$
F_{cell}
θ
$F_{cell} = F_{measured}/\sin\theta$
b
3 min
4 min
4 min
Force (nN)
2
0
−2
−4
−6
−8
0
5
10
15
20
Time (min)
Cell
Measurement noise
B
a
10 μm
time = 0
b
c
Pole position, y (μm)
4
0
−4
0
50
100
150
200
Time (s)
d
Spindle position, x (μm)
4
2
0
Spindle length (μm)
24
20
16
12
0
50
100
150
200
Time (s)

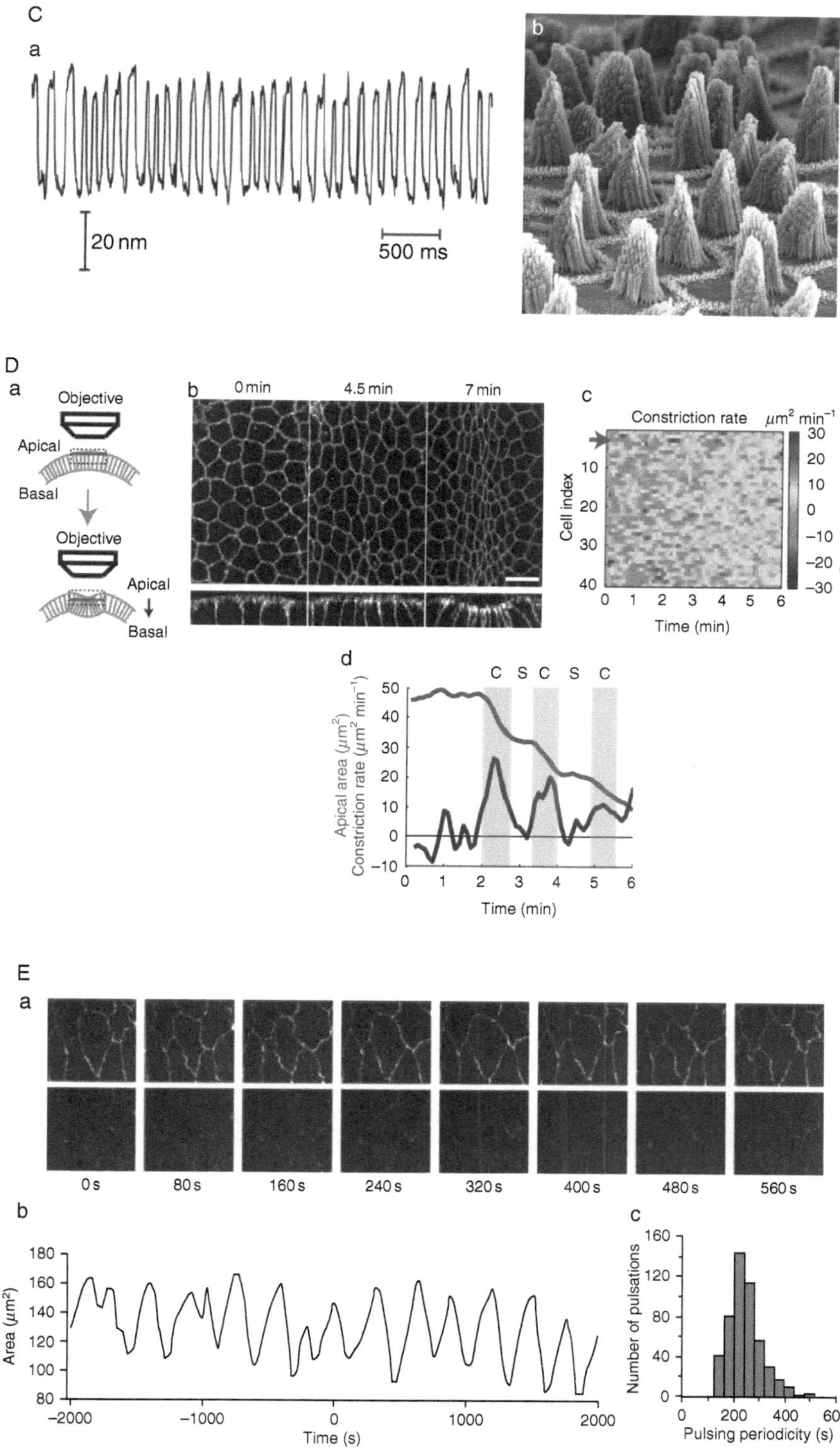

Figure 3.4 (Continued)

3.3. Spontaneous cytoskeletal waves

Intimately related to temporal cytoskeletal oscillations are cytoskeletal waves. An increasing number of works show that such waves are ubiquitous in living cells. Several types of waves associated with the actin cytoskeleton can be distinguished. Spontaneous polymerization waves have been observed under various conditions in a variety of cell types (Asano *et al.*, 2009; Bretschneider *et al.*, 2004, 2009; Vicker, 2002; Weiner *et al.*, 2007). A common feature of these waves is that they all seem to involve complexes, for example, the Scar/WAVE complex, that nucleate new actin filaments as well as a negative feedback through which existing actin filaments inactivate the nucleating complexes. Different theoretical works have explored various specific mechanisms and showed that the interplay of actin filaments and nucleating components as indicated indeed suffices to generate spontaneous waves (Carlsson, 2010; Doubrovinski and Kruse, 2008; Weiner *et al.*, 2007; Whitelam *et al.*, 2009). However, for the time being, the connection of these theoretical studies to experiments remains rather weak. Let us note that while molecular motors are not necessary to generate spontaneous polymerization waves, there are mechanisms that depend in an essential way on directed transport. In particular, it has been shown theoretically that waves can be generated in the absence of a negative feedback from existing filaments on the nucleating proteins provided that the nucleators are transported along existing filaments by motors (Doubrovinski and Kruse, 2007).

Motors play an essential role in another type of cytoskeletal waves, namely contraction waves. Such waves have been observed in spreading fibroblast (Dobereiner *et al.*, 2006; Giannone *et al.*, 2007). Another example of contraction waves is provided by spiral waves in the cytoplasm of the true slime mold *Physarum polycephalum* or the thickness oscillations in fibroblast mentioned above (Kapustina *et al.*, 2008; Pletjushkina *et al.*, 2001), which

Figure 3.4 Spontaneous oscillations *in vivo*. (A) Individual fibroblasts, schematically represented in (a) exert an oscillatory force on their environment (b) (Galbraith and Sheetz, 1997); (B) Spindle oscillations during the first asymmetric cell division in a *C. elegans* embryo: (a) fluorescence image indicating the positions of the two spindle poles, (b) traces of the positions of the spindle poles in the cell, (c, d) elongation of the spindle poles from the cell's symmetry axis are periodic (Pecreaux *et al.*, 2006); (C) Hair bundles of cells in the inner ear (Hudspeth, 2008): ensembles of cilia are shown (b), their spontaneous oscillation (a) is reported with a glass fiber attached to the hair bundle's top (Martin *et al.*, 2001). (D) Apical constriction in developing *Drosophila*: (a) schematic illustration of apical constriction and of the set up, (b) subsequent snapshots of the tissue with membranes stained fluorescently, (c) the apical constriction rate of some cells varies periodically with time, (d) oscillation in the constriction rate associated with a decrease in the apical area (Martin *et al.*, 2009). (E) Dorsal closure in *Drosophila*: (a) Subsequent snapshots of dorsal tissue with membrane stained cells, (b) area of a cell as a function of time shows periodic variations, (c) histogram of recorded pulsation periods (Solon *et al.*, 2009). (See Color Insert.)

really are standing waves. Also, relaxation waves observed in myofibrils (Sasaki *et al.*, 2005) fall into this class. A detailed understanding of contraction waves is still missing, although density waves have been found in models of bundles of filaments that are cross-linked by molecular motors (Guthardt Torres *et al.*, 2010) and in models of myofibrils (Gunther and Kruse, 2007). Similar principles might lead to contraction waves observed during embryonic development to which we turn now.

3.4. Mechanical oscillations during embryonic development

The probably best-studied oscillation occurring in developing embryos is observed during somitogenesis (Ozbudak and Pourquie, 2008): Gene expression profiles change periodically in a certain region from which cells leave constantly into the direction of the forming spine. The temporal period is thus transformed into a spatially periodic signal. In spite of associated tissue movements, this oscillator is believed to result from enzymatic reactions without any mechanical component. Nevertheless, spontaneous mechanical oscillations do play a role also during developmental processes (Fig. 3.4).

A striking example is provided by mitotic spindle oscillations (see Fig. 3.4B) during the first asymmetric cell division in the developing nematode *Caenorhabditis elegans* (Albertson, 1984). There, force generators—presumably dynein motors—are located along the cell's plasma membrane and pull on the microtubules emanating from a centrosome. Some microtubules are not attached to motors and bend when the centrosome is moved toward the membrane providing an elastic restoring force. Theoretical analysis predicted that oscillations should appear spontaneously when the number of motors on the membrane exceeds a critical value (Grill *et al.*, 2005). Experimentally, the activity of cortical force generators has been reduced using RNAi and indeed a critical number of motors was found to be necessary before oscillations started (Pecreaux *et al.*, 2006). This is an example of an experimental exploration of a system's state diagram. As mentioned before, overall agreement between experimental and theoretical state diagrams are indispensable for verifying a theoretically proposed mechanism.

Of particular relevance for development are processes, where several mechanical oscillators are mechanically coupled. We have already mentioned systems of coupled oscillators outside of development. In isolated myofibrils from skeletal or cardiac muscle, each sarcomere can spontaneously oscillate (Okamura and Ishiwata, 1988). Their mechanical coupling leads to coherent contraction–relaxation waves along the myofibrils (Sasaki *et al.*, 2005). As for the individual oscillators, no further chemical or other regulation is needed as these phenomena occur in front of a chemically homogenous and constant background. The axonemal structure of eukaryotic cilia and flagella is in some sense very similar to the myofibril: ensembles of molecular motors can induce elastic deformations that feed back on the action of motors. In contrast to myofibrils, though, axonemes show lateral instead of longitudinal deformations. Further, an axoneme does not consist of a discrete chain of mechanical oscillators. Rather dynein motors are distributed along the whole structure and the elastic restoring force results from bending it. Two and three-dimensional beating patterns can readily be explained by spontaneous motor oscillations. On a larger scale, cilia and flagella can synchronize their beating through hydrodynamic coupling by the surrounding fluid (Riedel *et al.*, 2005). Metachronal waves on the surfaces of ciliates likely result from this mechanism. Similarly, hair bundles are coupled hydrodynamically and mechanically in the inner ear (Barral *et al.*, 2010; Hudspeth, 2008; see Fig. 3.4C). Hydrodynamic coupling is probably important during mammalian development, where the synchronized beating of cilia leads to a fluid flow that is essential for determining the left–right asymmetry of the body (Drescher *et al.*, 2009; Nonaka *et al.*, 1998; Vilfan and Frey, 2005).

Also, periodic contractions of actomyosin networks occur in developing embryos. This is illustrated by an oscillatory phenomenon present during gastrulation in *Drosophila*. Apical constriction of ventral cells allows the formation of a ventral furrow. It was shown that this major tissue transformation is not occurring in a continuous manner: the contraction of the actomyosin network is pulsed during apical constriction (Martin *et al.*, 2009). This oscillation was reported using fluorescent markers for myosins, cadherins, and cell membrane markers. It was shown that the actomyosin network is driving the pulsed constriction. Cells contract asynchronously; still, the tissue as a whole alternates between a phase of contraction where some cells decrease in volume and phases of pause where the cell–cell junctions reorganize; they are followed again by contraction of some ventral cells. These cycles of stretching and constriction events (see Fig. 3.4D) are reminiscent of the spontaneous relaxation oscillation in sarcomeres mentioned above (Okamura and Ishiwata, 1988). The temporal period of the oscillations is on the order of a minute and thus slower than the sarcomere oscillations. Rather, the time scale is similar to that of the oscillations in single fibroblasts quoted above (Galbraith and Sheetz, 1997).

This observation suggests an underlying mechanism similar to the *in vitro* oscillations quoted above: the actomyosin network applies a force on the elastic cell–cell junctions. When the deformation of the tissue reaches a maximum, the contraction stops, and the cell–cell junctions reorganize and thereby relax the stress. The actomyosin network can then again apply a load on the boundaries. The spatial organization of the actomyosin network in the tissue is obviously more complicated than in the one-dimensional *in vitro* motility assay showing oscillations. It is nevertheless exhibiting the same features described qualitatively with the spring opposing the action of coupled motors.

One more example of actomyosin oscillations was demonstrated in *Drosophila* (Fig. 3.4E) during dorsal closure, another important developmental event (Solon *et al.*, 2009). Led by the *amnioserosa* cells, an epithelial gap is closed during this major shape transformation that is generally viewed as a good model for studying wound healing. Using real time imaging of myosin, of actin associated proteins, and of cell membranes, it was shown that the *amnioserosa* cells undergo oscillations, exhibiting the same features as the oscillations during apical constrictions. Similarly, their period is on a minute timescale and they show an actomyosin dependence. This pulsed tissue movement is causing the displacement of the neighboring tissue, the so-called *leading edge* of the opening. While tension builds up in this tissue, actin cables are stabilizing the new organization, relaxing the stress exerted by the *amnioserosa* cells. Subsequently, these cells increase the tension again. The repetition of this cycle allows the actual closure. Numerical simulations were performed to test the potential reproduction of such transformation *in silico*: *amnioserosa* cells were undergoing pulsed shape transformations, and the overall tissue shape was monitored. Based on these mechanical properties alone, the simulations reproduced the main spatial and temporal features of dorsal closure. Again, a comparison of experimental and theoretical state diagrams would be necessary to identify the molecular mechanism underlying the periodic motion.

Altogether, these oscillations related single cell behavior to that of developing embryos and display several conserved features: spatial amplitudes on the micrometer scale, temporal periods on the minute timescale, and a dependence on the actin cytoskeleton. The similarity with the *in vitro* experiments showing mechanical instabilities is therefore appealing and suggests a common physical mechanism for the diverse ensemble of phenomena: the collection of myosin motors build up stress in each cell and hence in the tissue. Various intra- and intercellular elastic components resist to the resulting force until they yield and new stress can be generated. As a result, in each of the situations, the cells would go through phases of contractions and relaxations, in the same way as single filaments were shown to oscillate *in vitro* when interacting with a collection of motors.

The similarity of amplitudes and periods of oscillations *in vitro* and *in vivo* allows us to conclude for the central role played by the collective effects of molecular motors. However, even if the similarities between experiments *in vitro* and *in vivo* are appealing, more work needs to be done to correlate rigorous oscillations in both situations. As pointed out above, systematic analysis of the state diagrams will be required for this: frequencies and amplitudes of oscillations should be scanned in motility assays and in embryos, by varying motor activity, for example. After establishing the state diagram, the comparison will be done quantitatively and give profound insight into the collective nature of the oscillation. Obviously, the determination of a state diagram might pose a considerable challenge in a developing organism.

3.5. Possible functions of mechanical oscillations during development

It is worth trying to suggest a potential function of mechanical oscillations during development. First of all, with the stress fibers, cells have a way of periodically probing the resistance of their environments. If the matrix or the neighboring cells oppose a resistance, then cells can reinforce the contacts. If not, contacts remain as they were. As such, the reinforcement may play a key role in cell fate determination. For example, it has been shown that cell shapes and divisions were intimately connected to focal contacts distributions (Thery *et al.*, 2006). By undergoing oscillations, the stress fibers would allow the cells to probe the mechanical environment with a time resolution of minutes. This checkpoint would be a signal constituted by mechanical properties of living matter. In addition, the relaxation phase could have a specific role. During this phase, tension is released, giving potentially the opportunity to other motors to bind and to the stress fiber to apply different forces. In addition, the low duty ratio of motors such as myosin may require this relaxation phase for further binding of motors and force application. This argument assumes that the amplitudes of forces are varying from cycles to cycles during the oscillation.

Remarkably, it was shown that also the differentiation of stem cells was related to the resistance of the environment (Engler *et al.*, 2006). Here the probing mechanism would also have an information component, by giving the cells the appropriate mechanical bias to differentiate cells into the proper cellular type. These arguments can be extrapolated to tissues: cells would collectively probe their environments to see if they can undergo a developmental transition. There is a clear advantage in periodic probing instead of a permanent one: in addition to elastic properties, the cell can in this way also probe viscous properties of the environment. This information is not given by a continuous force application. For an oscillatory exploration of the mechanical properties of the environment, the cytoskeleton must be mechanically linked to the environment, for example, by anchoring proteins, and the actin filaments must be organized such that motors can induce contraction. In filament bundles like stress fibers, this likely requires filaments of mixed polarization. Also passive cross-links, for example, by α-actinin or fimbrin, need to have a finite lifetime. Further, we speculate that a too high filament turnover would severely limit the possibility of a cell to probe its environment in an oscillatory fashion.

As we have seen above, like in the case of apical constrictions in *Drosophila* (Martin *et al.*, 2009), changes in the morphology of tissues can result from pulsed contractions. This suggests other possible functions of such mechanical oscillations. It might be difficult to rearrange the tissue and at the same to maintain the necessary contractile stresses. Alternations between the two phases—rearrangement and contraction—would result in oscillations. Furthermore, an organism might seek to evolve in such a way as to lock the tissue at intermediate stages, so as to avoid having it fall back to the original state if something goes wrong on the way. We propose that reinforced adhesion between cells during periods of maximal stress could be a mechanism for such a locking phenomenon: stress fibers would not relax back to their original positions thanks to the reinforced adhesion between neighboring cells. Instead, further forces would be generated in the modified tissue by newly formed stress. In turn, the new contraction would promote further evolution of the tissues to continue the modification. This locking mechanism would be well represented by the classical *ratchet and pawl of R. Feynman** with no necessary changes in the force from cycles to cycles. Note that in situations where tissues are not changing their shapes during the contractions, such a locking mechanism could be absent.

Most of the frequencies of oscillations reported here were of the order of minutes. Could there be a biological significance to this timescale? It is not possible to give a definite answer of course, but it is tempting to suggest the following. On timescales of a second or below, fluctuations of chemical reactions are occurring: a probing would be challenging, as it would examine potential changes related to noise. In contrast, the typical cellular reorganizations occur on timescales of a minute or longer: a local probing at the minute timescale would be the appropriate timing to allow the cell to check the connectivity and resistance of its environment.

4. Conclusions

We have presented oscillatory mechanisms documented *in vitro* and *in vivo* for a variety of systems. We emphasized oscillations generated by the actomyosin system, since they have been well demonstrated in motility assays, in single cells, and in developing embryos. As we indicated above, though, spontaneous oscillations are a generic property of molecular motors. Thus, we expect that further oscillations also associated with other motors will be unraveled in the future. Obviously, the observation of mechanical cellular oscillations is not sufficient to conclude for a collective behavior of motors. As we have repeatedly stated in the context of actomyosin oscillations, a systematic experimental acquisition of a system's state diagram will be required to appropriately assess the nature of an oscillatory phenomenon. Notably, mechanical control parameters need to be varied, for example, the density of active motors, their quantified activities, or the system's elastic properties. In tissues, the restraining force of neighboring cells could be reduced by acting on the adhesive contacts. Simple read-outs to quantify the effects of systematic parameter changes are provided by the oscillation frequency and amplitude. While this way of probing and explaining the transformation of cell and tissue shapes is new and demanding, it has the power of explaining developmental transitions in situations, where purely genetic approaches fail to give satisfying answers.

GLOSSARY

Active polar gel
A gel is an ensemble of polymers that are either covalently or noncovalently linked. If the polymers are linear, that is, filaments, and structurally polar, that is, the two ends of a filament are distinguishable, the ensemble can macroscopically exhibit in certain regions a preferred orientation. The gel is then said to be polar. Such a gel is active, if it is internally driven by chemical reactions, for example, the hydrolysis of ATP.

Degrees of freedom
Independent characteristic features of a system that can change with time. Examples of microscopic degrees of freedom are the positions and velocities of all atoms making up a protein. The density and the polarization of a gel are examples of macroscopic degrees of freedom.

Dynamic instability
Upon a (small) perturbation, a system can relax back to its original state or it can evolve into a new state. In the latter case, the original state is said to be dynamically unstable and the system presents a dynamic instability.

Hydrodynamic theory
A purely phenomenological description of material properties that is valid only on large length and time scales—processes occurring on microscopic scales are averaged out. Filament density and polarization are two hydrodynamic variables for the cytoskeleton. Water was the first material described in such a way, hence the name.

Feynman's ratchet
A device that consists of a circular ratchet and a pawl such that the ratchet can only turn in one direction. This device was introduced by R. P. Feynman to discuss the possibility of rectifying random molecular motion such as to extract work from a heat bath.

Interaction potential
The interaction of a motor protein with its associated filament is characterized by the interaction energy: the stronger the interaction, the lower the interaction energy. The variation of the interaction energy along the filament is given by the interaction potential. It depends on the internal state of the motor protein.

Internally driven filament
Filamentous protrusion that is internally deformed by the action of molecular motors, for example, filopodia and axonemes.

Periodic potential
Cytoskeletal filaments are periodic—the same structure repeats every 8 nm for a microtubule and every 37 nm for an actin filament—and the interaction potential reflects this periodicity as the interaction energy is determined by local properties of the filament.

Stalling force
The magnitude of the force that one needs to apply on a motor molecule to make it stop.

State or phase
Given constant conditions, a system will eventually evolve into a well-defined state, like the solid, liquid, and gas phases for water. For different values of the system parameters, qualitatively different states (also called phases) can emerge. A state (or phase) diagram presents these states as a function of the system parameters.

ACKNOWLEDGMENTS

We thank our former and current collaborators on these topics. Because of space limitations and goals for this highlight, we quote only a small fraction of papers related to the topics. We apologize to the nonquoted authors, who have also contributed significantly to this emerging field.

REFERENCES

Albertson, D. G. (1984). Formation of the first cleavage spindle in nematode embryos. *Dev. Biol.* **101,** 61–72.

Asano, Y., Jimenez-Dalmaroni, A., Liverpool, T. B., Marchetti, M. C., Giomi, L., Kiger, A., Duke, T., and Baum, B. (2009). Pak3 inhibits local actin filament formation to regulate global cell polarity. *HFSP J.* **3,** 194–203.

Backouche, F., Haviv, L., Groswasser, D., and Bernheim-Groswasser, A. (2006). Active gels: Dynamics of patterning and self-organization. *Phys. Biol.* **3,** 264–273.

Badoual, M., Julicher, F., and Prost, J. (2002). Bidirectional cooperative motion of molecular motors. *Proc. Natl. Acad. Sci. USA* **99,** 6696–6701.

Balaban, N. Q., Schwarz, U. S., Riveline, D., Goichberg, P., Tzur, G., Sabanay, I., Mahalu, D., Safran, S., Bershadsky, A., Addadi, L., and Geiger, B. (2001). Force and focal adhesion assembly: A close relationship studied using elastic micropatterned substrates. *Nat. Cell Biol.* **3,** 466–472.

Barral, J., Dierkes, K., Lindner, B., Julicher, F., and Martin, P. (2010). Coupling a sensory hair-cell bundle to cyber clones enhances nonlinear amplification. *Proc. Natl. Acad. Sci. USA* **107,** 8079–8084.

Benazeraf, B., Francois, P., Baker, R. E., Denans, N., Little, C. D., and Pourquie, O. (2010). A random cell motility gradient downstream of FGF controls elongation of an amniote embryo. *Nature* **466,** 248–252.

Bertet, C., Sulak, L., and Lecuit, T. (2004). Myosin-dependent junction remodelling controls planar cell intercalation and axis elongation. *Nature* **429,** 667–671.

Bornens, M., Paintrand, M., and Celati, C. (1989). The cortical microfilament system of lymphoblasts displays a periodic oscillatory activity in the absence of microtubules: Implications for cell polarity. *J. Cell Biol.* **109,** 1071–1083.

Bretschneider, T., Diez, S., Anderson, K., Heuser, J., Clarke, M., Muller-Taubenberger, A., Kohler, J., and Gerisch, G. (2004). Dynamic actin patterns and Arp2/3 assembly at the substrate-attached surface of motile cells. *Curr. Biol.* **14,** 1–10.

Bretschneider, T., Anderson, K., Ecke, M., Muller-Taubenberger, A., Schroth-Diez, B., Ishikawa-Ankerhold, H. C., and Gerisch, G. (2009). The three-dimensional dynamics of actin waves, a model of cytoskeletal self-organization. *Biophys. J.* **96,** 2888–2900.

Brevier, J., Vallade, M., and Riveline, D. (2007). Force-extension relationship of cell-cell contacts. *Phys. Rev. Lett.* **98,** 268101.

Brevier, J., Montero, D., Svitkina, T., and Riveline, D. (2008). The asymmetric self-assembly mechanism of adherens junctions: A cellular push-pull unit. *Phys. Biol.* **5,** 016005.

Callan-Jones, A. C., Joanny, J. F., and Prost, J. (2008). Viscous-fingering-like instability of cell fragments. *Phys. Rev. Lett.* **100,** article no: 258106.

Campas, O., and Sens, P. (2006). Chromosome oscillations in mitosis. *Phys. Rev. Lett.* **97,** article no: 128102.

Camalet, S., and Julicher, F. (2000). Generic aspects of axonemal beating. *New J. Phys.* **2,** 1–23.

Camalet, S., Julicher, F., and Prost, J. (1999). Self-organized beating and swimming of internally driven filaments. *Phys. Rev. Lett.* **82,** 1590–1593.

Carlsson, A. E. (2010). Dendritic actin filament nucleation causes traveling waves and patches. *Phys. Rev. Lett.* **104,** 228102.

Costigliola, N., Kapustina, M. T., Weinreb, G. E., Monteith, A., Rajfur, Z., Elston, T. C., and Jacobson, K. (2010). RhoA regulates calcium-independent periodic contractions of the cell cortex. *Biophys. J.* **99,** 1053–1063.

Delanoë-Ayari, H., Brevier, J., and Riveline, D. (2011). Scaling concepts in cell physics: paradigms for cell adhesion. *Soft Matter* **7,** 824–829.

Dobereiner, H. G., Dubin-Thaler, B. J., Hofman, J. M., Xenias, H. S., Sims, T. N., Giannone, G., Dustin, M. L., Wiggins, C. H., and Sheetz, M. P. (2006). Lateral membrane waves constitute a universal dynamic pattern of motile cells. *Phys. Rev. Lett.* **97,** 038102.

Doubrovinski, K., and Kruse, K. (2007). Self-organization of treadmilling filaments. *Phys. Rev. Lett.* **99,** 228104.

Doubrovinski, K., and Kruse, K. (2008). Self-organization in systems of treadmilling filaments. *Eur. Phys. J. E Soft Matter* **31,** 95–104.

Drescher, K., Leptos, K. C., Tuval, I., Ishikawa, T., Pedley, T. J., and Goldstein, R. E. (2009). Dancing volvox: Hydrodynamic bound states of swimming algae. *Phys. Rev. Lett.* **102,** 168101.

Engler, A. J., Sen, S., Sweeney, H. L., and Discher, D. E. (2006). Matrix elasticity directs stem cell lineage specification. *Cell* **126,** 677–689.

Galbraith, C. G., and Sheetz, M. P. (1997). A micromachined device provides a new bend on fibroblast traction forces. *Proc. Natl. Acad. Sci. USA* **94,** 9114–9118.

Gally, C., Wissler, F., Zahreddine, H., Quintin, S., Landmann, F., and Labouesse, M. (2009). Myosin II regulation during C. elegans embryonic elongation: LET-502/ROCK, MRCK-1 and PAK-1, three kinases with different roles. *Development* **136,** 3109–3119.

Giannone, G., Dubin-Thaler, B. J., Rossier, O., Cai, Y., Chaga, O., Jiang, G., Beaver, W., Dobereiner, H. G., Freund, Y., Borisy, G., and Sheetz, M. P. (2007). Lamellipodial actin mechanically links myosin activity with adhesion-site formation. *Cell* **128,** 561–575.

Gilboa, B., Gillo, D., Farago, O., and Bernheim-Groswasser, A. (2009). Bidirectional cooperative motion of myosin-II motors on actin tracks with randomly alternating polarities. *Soft Matter* **5,** 2223–2231.

Gillo, D., Gur, B., Bernheim-Groswasser, A., and Farago, O. (2009). Cooperative molecular motors moving back and forth. *Phys. Rev. E Stat. Nonlin. Soft Matter Phys.* **80,** 021929.

Grashoff, C., Hoffman, B. D., Brenner, M. D., Zhou, R., Parsons, M., Yang, M. T., McLean, M. A., Sligar, S. G., Chen, C. S., Ha, T., and Schwartz, M. A. (2010). Measuring mechanical tension across vinculin reveals regulation of focal adhesion dynamics. *Nature* **466,** 263–266.

Grill, S. W., Kruse, K., and Julicher, F. (2005). Theory of mitotic spindle oscillations. *Phys. Rev. Lett.* **94,** 108104.

Guerin, T., Prost, J., Martin, P., and Joanny, J. F. (2010). Coordination and collective properties of molecular motors: Theory. *Curr. Opin. Cell Biol.* **22,** 14–20.

Gunther, S., and Kruse, K. (2007). Spontaneous waves in muscle fibres. *New J. Phys.* **9,** 417.

Guthardt Torres, P., Doubrovinski, K., and Kruse, K. (2010). Filament turnover stabilizes contractile cytoskeletal structures. *Europhys. Lett.* **91,** 68003.

Hilfinger, A., and Julicher, F, (2008). The chirality of ciliary beats. *Phys. Biol.* **5,** 016003.

Hilfinger, A., Chattopadhyay, A. K., Julicher, F. (2009). Nonlinear dynamics of cilia and flagella. *Phys. Rev. E* **79,** 051918.

Howard, J. (2001). Mechanics of Motor Proteins and the Cytoskeleton. Sinauer Associates, Sunderland, Massachusetts.

Howard, J., and Hyman, A. A. (2009). Growth, fluctuation and switching at microtubule plus ends. *Nat. Rev. Mol. Cell Biol.* **10,** 569–574.

Hudspeth, A. J. (2008). Making an effort to listen: Mechanical amplification in the ear. *Neuron* **59,** 530–545.

Josephson, R. K., Malamud, J. G., and Stokes, D. R. (2000). Asynchronous muscle: A primer. *J. Exp. Biol.* **203,** 2713–2722.

Julicher, F., and Prost, J. (1995). Cooperative molecular motors. *Phys. Rev. Lett.* **75,** 2618–2621.

Julicher, A., and Prost, J. (1997a). Modeling molecular motors. *Rev. Mod. Phys.* **69,** 1269.

Julicher, F., and Prost, J. (1997b). Spontaneous oscillations of collective molecular motors. *Phys. Rev. Lett.* **78,** 4510–4513.

Julicher, F., Kruse, K., Prost, J., and Joanny, J. F. (2007). Active behavior of the cytoskeleton. *Phys. Rep.-Rev. Sect. Phys. Lett.* **449,** 3–28.

Kapustina, M., Weinreb, G. E., Costigliola, N., Rajfur, Z., Jacobson, K., and Elston, T. C. (2008). Mechanical and biochemical modeling of cortical oscillations in spreading cells. *Biophys. J.* **94,** 4605–4620.

Karsenti, E., Nedelec, F., and Surrey, T. (2006). Modelling microtubule patterns. *Nat. Cell Biol.* **8,** 1204–1211.

Kozlowski, C., Srayko, M., and Nedelec, F. (2007). Cortical microtubule contacts position the spindle in *C. elegans* embryos. *Cell* **129,** 499–510.

Kruse, K., Joanny, J. F., Julicher, F., and Prost, J. (2006). Contractility and retrograde flow in lamellipodium motion. *Phys. Biol.* **3,** 130–137.

le Duc, Q., Shi, Q., Blonk, I., Sonnenberg, A., Wang, N., Leckband, D., and de Rooij, J. (2010). Vinculin potentiates E-cadherin mechanosensing and is recruited to actin-anchored sites within adherens junctions in a myosin II-dependent manner. *J. Cell Biol.* **189,** 1107–1115.

Liu, Z., Tan, J. L., Cohen, D. M., Yang, M. T., Sniadecki, N. J., Ruiz, S. A., Nelson, C. M., and Chen, C. S. (2010). Mechanical tugging force regulates the size of cell-cell junctions. *Proc. Natl. Acad. Sci. USA* **107,** 9944–9949.

Liverpool, T. B., and Marchetti, M. C. (2003). Instabilities of isotropic solutions of active polar filaments. *Phys. Rev. Lett.* **90,** 138102.

Martin, P., Hudspeth, A. J., and Julicher, F. (2001). Comparison of a hair bundle's spontaneous oscillations with its response to mechanical stimulation reveals the underlying active process. *Proc. Natl. Acad. Sci. USA* **98,** 14380–14385.

Martin, A. C., Kaschube, M., and Wieschaus, E. F. (2009). Pulsed contractions of an actin-myosin network drive apical constriction. *Nature* **457,** 495–499.

Muller, M. J., Klumpp, S., and Lipowsky, R. (2008). Tug-of-war as a cooperative mechanism for bidirectional cargo transport by molecular motors. *Proc. Natl. Acad. Sci. USA* **105,** 4609–4614.

Nedelec, F. J., Surrey, T., Maggs, A. C., and Leibler, S. (1997). Self-organization of microtubules and motors. *Nature* **389,** 305–308.

Neuman, K. C., and Nagy, A. (2008). Single-molecule force spectroscopy: Optical tweezers, magnetic tweezers and atomic force microscopy. *Nat. Methods* **5,** 491–505.

Nishikawa, M., Takagi, H., Shibata, T., Iwane, A. H., and Yanagida, T. (2008). Fluctuation analysis of mechanochemical coupling depending on the type of biomolecular motors. *Phys. Rev. Lett.* **101,** 128103.

Nonaka, S., Tanaka, Y., Okada, Y., Takeda, S., Harada, A., Kanai, Y., Kido, M., and Hirokawa, N. (1998). Randomization of left-right asymmetry due to loss of nodal cilia generating leftward flow of extraembryonic fluid in mice lacking KIF3B motor protein. *Cell* **95,** 829–837.

Okamura, N., and Ishiwata, S. (1988). Spontaneous oscillatory contraction of sarcomeres in skeletal myofibrils. *J. Muscle Res. Cell Motil.* **9,** 111–119.

Ozbudak, E. M., and Pourquie, O. (2008). The vertebrate segmentation clock: The tip of the iceberg. *Curr. Opin. Genet. Dev.* **18,** 317–323.

Paluch, E., Piel, M., Prost, J., Bornens, M., and Sykes, C. (2005). Cortical actomyosin breakage triggers shape oscillations in cells and cell fragments. *Biophys. J.* **89,** 724–733.

Pecreaux, J., Roper, J. C., Kruse, K., Julicher, F., Hyman, A. A., Grill, S. W., and Howard, J. (2006). Spindle oscillations during asymmetric cell division require a threshold number of active cortical force generators. *Curr. Biol.* **16,** 2111–2122.

Placais, P. Y., Balland, M., Guerin, T., Joanny, J. F., and Martin, P. (2009). Spontaneous oscillations of a minimal actomyosin system under elastic loading. *Phys. Rev. Lett.* **103,** 158102.

Pletjushkina, O. J., Rajfur, Z., Pomorski, P., Oliver, T. N., Vasiliev, J. M., and Jacobson, K. A. (2001). Induction of cortical oscillations in spreading cells by depolymerization of microtubules. *Cell Motil. Cytoskeleton* **48,** 235–244.

Rauzi, M., Verant, P., Lecuit, T., and Lenne, P. F. (2008). Nature and anisotropy of cortical forces orienting *Drosophila* tissue morphogenesis. *Nat. Cell Biol.* **10,** 1401–1410.

Riedel, I. H., Kruse, K., and Howard, J. (2005). A self-organized vortex array of hydrodynamically entrained sperm cells. *Science* **309,** 300–303.

Riedel-Kruse, I. H., Hilfinger, A., Howard, J., *et al.* (2007). How molecular motors shape the flagellar beat. *Hfsp J.* **1,** 192–208.

Riveline, D., Ott, A., Julicher, F., Winkelmann, D. A., Cardoso, O., Lacapere, J. J., Magnusdottir, S., Viovy, J. L., Gorre-Talini, L., and Prost, J. (1998). Acting on actin: The electric motility assay. *Eur. Biophys. J.* **27,** 403–408.

Riveline, D., Zamir, E., Balaban, N. Q., Schwarz, U. S., Ishizaki, T., Narumiya, S., Kam, Z., Geiger, B., and Bershadsky, A. D. (2001). Focal contacts as mechanosensors: Externally applied local mechanical force induces growth of focal contacts by an mDia1-dependent and ROCK-independent mechanism. *J. Cell Biol.* **153,** 1175–1186.

Rossier, O. M., Gauthier, N., Biais, N., Vonnegut, W., Fardin, M. A., Avigan, P., Heller, E. R., Mathur, A., Ghassemi, S., Koeckert, M. S., Hone, J. C., and Sheetz, M. P. (2010). Force generated by actomyosin contraction builds bridges between adhesive contacts. *EMBO J.* **29,** 1055–1068.

Salbreux, G., Joanny, J. F., Prost, J., and Pullarkat, P. (2007). Shape oscillations of non-adhering fibroblast cells. *Phys. Biol.* **4,** 268–284.
Sasaki, D., Fujita, H., Fukuda, N., Kurihara, S., and Ishiwata, S. (2005). Auto-oscillations of skinned myocardium correlating with heartbeat. *J. Muscle Res. Cell Motil.* **26,** 93–101.
Schaller, V., Weber, C., Semmrich, C., Frey, E., and Bausch, A. R. (2010). Polar patterns of driven filaments. *Nature* **467,** 73–77.
Schnitzer, M. J., Visscher, K., and Block, S. M. (2000). Force production by single kinesin motors. *Nat. Cell Biol.* **2,** 718–723.
Sheetz, M. P., Chasan, R., and Spudich, J. A. (1984). ATP-dependent movement of myosin in vitro: Characterization of a quantitative assay. *J. Cell Biol.* **99,** 1867–1871.
Solon, J., Kaya-Copur, A., Colombelli, J., and Brunner, D. (2009). Pulsed forces timed by a ratchet-like mechanism drive directed tissue movement during dorsal closure. *Cell* **137,** 1331–1342.
Surrey, T., Nedelec, F., Leibler, S., and Karsenti, E. (2001). Physical properties determining self-organization of motors and microtubules. *Science* **292,** 1167–1171.
Thery, M., Racine, V., Piel, M., Pepin, A., Dimitrov, A., Chen, Y., Sibarita, J. B., and Bornens, M. (2006). Anisotropy of cell adhesive microenvironment governs cell internal organization and orientation of polarity. *Proc. Natl. Acad. Sci. USA* **103,** 19771–19776.
Vicker, M. G. (2002). Eukaryotic cell locomotion depends on the propagation of self-organized reaction-diffusion waves and oscillations of actin filament assembly. *Exp. Cell Res.* **275,** 54–66.
Vilfan, A., and Frey, E. (2005). Oscillations in molecular motor assemblies. *J. Phys. Condens. Matter* **17,** S3901–S3911.
Vogel, S. K., Pavin, N., Maghelli, N., Julicher, F., and Tolic-Norrelykke, I. M. (2009). Self-organization of dynein motors generates meiotic nuclear oscillations. *PLoS Biol.* **7,** 918–928.
Weiner, O. D., Marganski, W. A., Wu, L. F., Altschuler, S. J., and Kirschner, M. W. (2007). An actin-based wave generator organizes cell motility. *PLoS Biol.* **5,** e221.
Weinreb, G. E., Elston, T. C., and Jacobson, K. (2006). Causal mapping as a tool to mechanistically interpret phenomena in cell motility: Application to cortical oscillations in spreading cells. *Cell Motil. Cytoskeleton* **63,** 523–532.
Whitelam, S., Bretschneider, T., and Burroughs, N. J. (2009). Transformation from spots to waves in a model of actin pattern formation. *Phys. Rev. Lett.* **102,** 198103.
Wozniak, M. A., and Chen, C. S. (2009). Mechanotransduction in development: A growing role for contractility. *Nat. Rev. Mol. Cell Biol.* **10,** 34–43.
Yonemura, S., Wada, Y., Watanabe, T., Nagafuchi, A., and Shibata, M. (2010). alpha-Catenin as a tension transducer that induces adherens junction development. *Nat. Cell Biol.* **12,** 533–542.

CHAPTER FOUR

Cortical Forces in Cell Shape Changes and Tissue Morphogenesis

Matteo Rauzi[1] *and* Pierre-François Lenne

Contents

IBDML, UMR6216 CNRS-Université de la Méditerranée, Campus de Luminy, Case 907, 13288 Marseille Cedex 09, France
[1] Present address: EMBL, Meyerhofstrasse 1, 69117, Heidelberg, Germany

Current Topics in Developmental Biology, Volume 95
ISSN 0070-2153, DOI: 10.1016/B978-0-12-385065-2.00004-9

Abstract

Cortical forces drive a variety of cell shape changes and cell movements during tissue morphogenesis. While the molecular components underlying these forces have been largely identified, how they assemble and spatially and temporally organize at cell surfaces to promote cell shape changes in developing tissues are open questions. We present here different key aspects of cortical forces: their physical nature, some rules governing their emergence, and how their deployment at cell surfaces drives important morphogenetic movements in epithelia. We review a wide range of literature combining genetic/molecular, biophysical and modeling approaches, which explore essential features of cortical force generation and transmission in tissues.

1. Introduction

Elucidating the forces that form and reshape muticellular structures is integral to our understanding of development. During tissue morphogenesis, cells change shapes and give rise, collectively, to a myriad of forms. The ability of cells to change their shape relies mainly on forces that are produced at cell surfaces, and are transmitted trough cell interfaces. These forces, called cortical forces, are generated in the cell cortex which is a 50-nm to 2-μm thick layer of cytoskeleton under the plasma membrane, rich in actin filaments, Myosin II, and actin-binding proteins. Cortical forces build up from a range of molecular mechanisms including Myosin II and actin filaments assembly, which are spatially and temporally controlled in the cell. Actomyosin networks produce force by active contraction. Understanding how cortical forces emerge from the assembly and contraction of actomyosin networks coupled to adhesion structures is a central issue in cell and developmental biology. In this context, a growing number of studies focus on the mechanisms and role of cortical forces in early tissue morphogenesis, in which assembly and dynamics of actomyosin networks play a central role.

This chapter focuses on different aspects of cortical forces, which are crucial to understand their origins and the developmental mechanisms they drive. The first section presents the molecular building blocks of cortical forces, namely the molecular Myosin II and actin filaments, which self-assemble into dynamic networks at cell surfaces. We emphasize also in this section the role of adhesion structures, which are anchoring points of cortical forces. The second section is devoted to the role of cortical forces in controlling cell shapes and cell shape changes during epithelial morphogenesis. We review studies, which quantitatively describe how cortical forces determine cell shapes and tissue topology, and to some extent, the pattern of cell shape changes. We then explore morphogenetic movements in which the dynamic nature of cortical forces has been recently revealed. The spatiotemporal dynamics of actomyosin networks share striking similarities in these systems, for example, pulses and flows, but yield different cell movements and cell shape changes, such as cell apical constriction and cell intercalation. Probing cortical forces *in vivo* require specific tools, which allow to address mechanical properties of cells in a tissue. In the last section, we will introduce the physical and technical grounds of the most widespread technique in this area, laser dissection, and discuss the pros and cons of the different experimental strategies recently used.

2. Molecular Origins of Cortical Forces

2.1. Myosin II assembly

Non-muscle Myosin II (Myosin II called hereafter) is a hexamer composed of two heavy chains, two essential light chains, and two regulatory light chains (RLCs). Each of the two heavy chains includes a globular head domain that binds F-actin and ATP in the presence of actin filaments, and undergoes a mechanochemical cycle of binding, hydrolysis, and release of ATP. These steps are tightly coupled to filament binding, conformational change, and force production. Each heavy chain continues into a tail domain in which heptad repeat sequences promote dimerization by interacting to form a rod-like α-helical coiled coil. Additional interactions of antiparallel coiled–coiled domains mediate the self-assembly of Myosin II into bipolar minifilaments, containing a few dozen of individual motor heads. While single two-headed Myosin II is nonprocessive machines, Myosin II minifilaments are highly processive. The mechanical properties of these contractile force generation units depend on ATPase rate of individual heads, their duty cycle, the minifilament assembly state, and the local density and orientation of actin filaments.

Myosin II of smooth muscle as well as nonmuscle cells is primarily regulated by phosphorylation of its RLCs. When RLC is unphosphorylated, Myosin II adopts a folded conformation in which (i) binding of the two heads to one another mutually prevents ATPase activity and actin binding and (ii) head–tail interactions inhibit minifilament assembly. Upon RLC phosphorylation, Myosin II unfolds into a conformation, which relieves (i) and (ii). First demonstrated *in vitro*, this conformational change is likely to occur also *in vivo*.

A growing body of evidence indicates that phosphorylation of the heavy chain provides an additional level of regulation (for a review see Vicente-Manzanares *et al.*, 2009). Indeed, numerous phosphorylation sites have been identified in tail of heavy chains (in vertebrate muscle myosin II), which favor assembly and/or disassembly.

How does Myosin II localize in the cell? Localization of Myosin II is strongly dependent on binding sites, including F-actin. Mobile Myosin II will tend to accumulate at regions of higher F-actin density. Depolymerization of F-actin or removal of the Myosin II binding domain to actin was shown to retrieve Myosin II to the cortex of *Dictyostellium* (Zang and Spudich, 1998) and *Drosophila* epithelial cells (Bertet *et al.*, 2009; Homem and Peifer, 2009). However, during cytokinesis, F-actin is not required for Myosin II recruitment (Foe and von Dassow, 2008; Zang and Spudich, 1998), and accessory proteins at the cortex could also determine Myosin II localization. Whatever the binding sites, assembly state of Myosin II could affect its localization as affinity of minifilaments to actin network depends on the number of Myosin heads. In addition to these local effects, cortical flows driven by Myosin II itself can contribute its localization (see below).

2.2. Actin assembly at the cell cortex

The actin cytoskeleton at the cell periphery consists of highly organized networks of F-actin, coupled with the plasma membrane. Actin filament growth relies first on the pool of actin monomers, which add to the barbed ends of existing filaments, allowing their fast growth. Because cytoplasm contains a high concentration of actin monomers, regulation is essential and different modes of regulation involve proteins, which distribute at the cortex. Among the large number of proteins that regulate actin assembly and especially initiate new actin filaments, actin-related protein Arp2/3 complex and formins are the best known (for a review see Pollard, 2007). Arp2/3 complex produces branched filaments to push forward the leading edge of motile cells and for endocytosis. Arp2/3 complex initiates the new filament to the preexisting actin network. The new filament elongates until a capping protein associates at the barbed end and inhibits growth. Formins nucleate unbranched filaments and remain associated with their barbed ends as they elongate, preventing the attachment of capping proteins. Formins promote the formation of actin bundles found in filipodia and cytokinetic contractile rings.

The Arp2/3 complex is intrinsically inactive. Thus, regulatory proteins called nucleation-promoting factor (NPFs) are required for nucleation of a new actin branch from a mother filament and an actin monomer. Wiskott–Aldrich syndrome protein (WASp) and Scar/WAVE proteins are the best-characterized members of the NPFs family. Intramolecular interactions between C-VCA region and N-terminal region of WASp autoinhibit WASp. Rho-family GTPases, Cdc 42 and Rac, cooperate with phosphatidylinositol 4,5-bisphosphate (PIP2) to retrieve WASp autoinhibition and activate WASp, and therefore Arp2/3. WAVE proteins form heteropentameric complexes (Wave complexes). There has been controversy on Wave constitutive activity but there is now a growing body of circumstantial evidence that WAVE complex is intrinsically inactive, like WASp, and that prenylated Rac-GTP, acidic phospholipids, and a specific state of phosphorylation of WAVE are simultaneously required for its activation.

Importantly, both WAVE and WASp require coincident and cooperative signals at the plasma membrane to promote Arp2/3 activity and actin nucleation. This could allow cells to tightly control the spatial and temporal deployment of cortical branched actin networks. In motile cells, Arp2/3-dependent actin nucleation at the vicinity of protruding membrane will favor the growth of filaments in the direction of movement. In adjacent cells contacting each other, Arp2/3-dependent actin nucleation at cell surfaces will create branched networks, which can serve tissue development mechanics.

Formins are multidomain proteins with two major functional regions: the N-terminal region, which is important for *in vivo* localization and the C-terminal region, which promotes actin assembly (for a review see Chesarone *et al.*, 2010). The C-terminal region consists of the formin homology (FH1) and FH2 domains. Although the precise mechanisms of actin nucleation by formins have remained elusive, FH2 domain is thought to catalyze nucleation of actin filaments. It remains attached to the newly formed filament and moves processively at the barbed end during filament growth, shielding it from capping proteins (Pruyne *et al.*, 2002). Meanwhile, FH1 domain, which recruits profilin-actin complexes, cooperates to deliver actin monomers to the growing barbed ends. Formins thus stimulate the assembly of long unbranched filaments, which grow rapidly (2 μm/s *in vivo*; Higashida *et al.*, 2004).

In the cytosol, formin dimers are autoinhibited by intramolecular interactions between their N-terminal diaphanous inhibitory domain (DID) and N-terminal diaphanous autoregulatory domain (DAD). Autoinhibition is retrieved by Rho proteins (Rose *et al.*, 2005); the cell cortex promotes the assembly of unbranched filaments, cables, cytokinetic rings, and stress fibers. As formins have the ability to remain attached to the filaments, cells can organize the points of force production by specific localization of formin regulation. During cytokinesis in yeast, several dozen specific landmarks at cell cortex (nodes formed by Myosin II molecules and formins) were shown to promote the assembly of the contractile ring (Vavylonis *et al.*, 2008; Wu *et al.*, 2006). Numerical simulations and experiments support a mechanism whereby Myosin II in a node capture and exert force on actin filaments, which are assembled from adjacent nodes. One can speculate that other processes, such as the formation of actin bundles in motile cells and epithelia, can involve similar modes of actin assembly. In *Drosophila* morphogenesis, formin helps coordinate adhesion and contractility of the actomyosin networks at adherens junction (Homem and Peifer, 2009). Activation of formins in cells, where adherens junctions are planar polarized, promotes actomyosin organization at cell borders where adherens junctions are enriched. Understanding the regulation of formins will be crucial.

Despite a higher intrinsic nucleation activity than Arp2/3 complex, it was shown (Buttery *et al.*, 2007; Martin and Chang, 2006) that some formins cooperate with cofactors to promote localized actin assembly.

This draws a first parallel between Arp2/3 complex and formin, which can be extended further as both share common elements in their regulatory systems. Switching between branched and unbranched assembly was shown to occur at the leading edge of motile cells, where Arp2/3-mediated branched filaments can be converted into arrays of longer filaments producing bundles that support contractile force generation (Hotulainen and Lappalainen, 2006).

Whether this mechanism is used in other processes is unknown but it suggests that the cells can use switchable and tunable modes of F-actin assembly, which are available for force production.

2.3. F-actin cross-linkers and the mechanical properties of actomyosin networks

Additional levels of F-actin assembly rely on F-actin cross-linkers, which are essential for the emergence of contractility in actomyosin networks. They must be present to provide sites for mechanical anchorage. While Arp2/3 complex and formins nucleate and assemble actin in branched and unbranched filament, cross-linkers can organize filaments into different higher-order structures: loose/tight networks, orthogonal networks, parallel/antiparallel bundles. *In vitro* studies have demonstrated the wide range of mechanical properties that cross-linkers confer to actomyosin networks. Whatever the origins of forces (internally generated or externally applied), actomyosin networks have complex mechanical properties, which are dependent on timescale. To some extent, actomyosin networks behave like viscoelastic materials. On fast time scales relative to actin turnover and cross-links, they resist to deformation like springs and restore their shape after the force is released. At slow strain rates, they can flow like fluids, given that cross-links in the networks have enough time to bind/unbind and to allow actin networks reshaping. Branched actin networks *in vitro* are intrinsically very stiff with elastic moduli of 1000–10,000 Pa (Chaudhuri *et al.*, 2007). *In vivo* measurements indicate comparable properties.

Stiff networks are important to produce and resist force at the cell periphery of crawling cells (Mullins *et al.*, 1998; Pollard and Borisy, 2003; Svitkina and Borisy, 1999). In contrast, unbranched actin networks are intrinsically soft, unless organized in higher-order structures by cross-linkers. Cross-linkers affect the organization of actin networks depending on their kinetics and geometry (for a review see Fletcher and Mullins, 2010). Due to its strong coupling to actin-binding sites, the cross-linker fascin preferentially organizes formin-mediated actin filaments into rigid bundles to generate protrusive forces in filipodia. In contrast, actin cross-linkers such as α-catenin can stabilize either orthogonal networks or parallel bundles, depending on the kinetics of interactions (Wachsstock *et al.*, 1994).

In addition, crossed-linked F-actin networks exhibit strong nonlinear stiffening with strain (Gardel *et al.*, 2004; Storm *et al.*, 2005), as a result of entropic elasticity (associated with a decrease of available configurations). For example, F-actin network cross-linked by filamin is very soft in their linear regime (1 Pa), yet they stiffen by three orders of magnitude in response to externally applied stress (Gardel *et al.*, 2006; Kasza *et al.*, 2009; Wagner *et al.*, 2006). This behavior resembles that of living cells in response to shear stress (Trepat *et al.*, 2004).

Actomyosin networks are not passive biopolymer gels and internally generated forces by Myosin II that affect their mechanical properties. It is shown *in vitro* that, in the absence of cross-linkers, Myosin II is able to fluidize actomyosin networks by active filament sliding thereby reducing their apparent viscosity (Humphrey *et al.*, 2002; Le Goff *et al.*, 2002). In presence of cross-linkers such as filamins which are large and compliant, Myosin II motors can stiffen the networks by more than two orders of magnitudes by pulling on actin filaments (Koenderink *et al.*, 2009). This supports the observation that Myosin II prestress contributes to cell stiffness *in vivo* (Matzke *et al.*, 2001; Wang *et al.*, 2002).

How can the mechanical properties of actomyosin networks impact on the generation and transmission of forces? Actomyosin networks need to be sufficiently stiff to generate and transmit force but also to sustain the forces it produces. However, to accommodate cell shape changes, they must be not too stiff. The precise control of network stiffness afforded by actin assembly, Myosin II activity, and cross-linkers suggests that the cell may spatially and temporally control its stiffness at different levels.

In timescales of few tens of seconds to minutes, which are relevant to morphogenetic movements, actomysin networks behave like fluids: they flow and are able to transport materials, including their own constituents, at long distances. For example, cortical flows driven by Myosin II motors drive the asymmetry of the first mitotic division in *Caenorhabditis elegans* zygotes (Goldstein and Hird, 1996; Hird and White, 1993; Munro *et al.*, 2004). In motile cells, retrograde F-actin flow and myosin II activity within the leading cell edge deliver F-actin to the lamella where they can seed graded polarized filaments. A growing body of evidence indicates that these flows are driven by gradients in actomyosin contractility. In absence of flows, actomyosin contractility determines the cortical tension. However, if actomyosin contractility drives flows, these flows contribute also to cortical tension (Mayer *et al.*, 2010; Rauzi *et al.*, 2010).

2.4. Cortical forces and adhesion structures in epithelia

Building up forces in actomyosin cortical networks require anchoring points in the networks but also at cell surfaces to produce cell shape changes. In epithelial cells, cell–cell adhesion structures are essential to transmit internally generated forces to other cells through cell surfaces or to extracellular matrix, and to reshape cell contours.

Between the different molecular components that mediate cell adhesion, we present below cadherin-based structures which promote cell–cell adhesion (for integrin-based structures see recent reviews Albiges-Rizo *et al.*, 2009; Geiger *et al.*, 2009; Dubash *et al.*, 2009). Observations in both cultured epithelial mammalian cells (Angres *et al.*, 1996 ; Kametani and Takeichi, 2007) and in early epithelia of nonvertebrates (Cavey *et al.*, 2008b; Harris and Peifer, 2004; Müller and Wieschaus, 1996; Tepass and Hartenstein, 1994) indicated that E-cadherin forms dense protein clusters, which are thought to represent clusters of homophilic dimers in transassociation (Kametani and Takeichi, 2007). At cell junctions, E-cadherin binds to the cytoplasmic protein β-catenin. α-catenin binds β-catenin, and mediates interactions with the actomyosin cytoskeleton (Abe and Takeichi, 2008; Cavey *et al.*, 2008b). The links between E-cadherin/β-catenin and actin trough α-catenin are dynamic (Drees *et al.*, 2005; Yamada *et al.*, 2005), yet several studies have demonstrated that α-catenin is responsible for the mechanical coupling between E-cadherin clusters and actin at adherens junctions. A recent report shows that α-catenin recruits vinculin, another main actin-binding protein of adherens junctions, through force-dependent changes in α-catenin conformation (Yonemura *et al.*, 2010). Unfolding of α-catenin would expose cryptic sites for vinculin binding, as it has been shown for talin in integrin-mediated adhesion (del Rio *et al.*, 2009). This could allow biochemical amplification upon application of force.

Forces modify adhesion. Conversely, adhesion is able to change forces by regulating actomyosin assembly. During the formation of cell–cell junctions, actin bundles stabilize the cadherin clusters at the end of filipodia contacting cells (Vasioukhin *et al.*, 2000). In turn, cadherin clusters can control actin assembly through Arp2/3 complex (Helwani *et al.*, 2004; Kovacs *et al.*, 2002; Verma *et al.*, 2004) and formins. Formin-1 binds directly α-catenin, whereas Arp2/3 complex binds β-catenin in competition with α-catenin (Drees *et al.*, 2005).

3. Cortical Forces Controlling Cell Shapes During Tissue Morphogenesis

In the previous section, we explored how actomyosin networks generate forces at the cell cortex and how their mechanical coupling to adhesion clusters allows efficient transmission of forces at cell interfaces. Cortical forces depend on local properties of the actomyosin networks and of their interactions with the plasma membrane but also on more global properties at the cell level such as elasticity (e.g., actin ring behaving as spring at cell periphery) and viscous flows. During the past decade, few studies delineate the local and global contributions of cortical forces to the cell shapes in tissues. Most examples are found in epithelia (Farhadifar *et al.*, 2007; Kafer *et al.*, 2007; Rauzi *et al.*, 2008), where cells are geometrically constrained and form a compact layer by cell–cell adhesion. While past works have focused on the role of adhesion molecules in determining the shape of cells by local interactions, more recent studies have shown that actomyosin networks also control cell shape and contribute both local and global properties of cortical forces.

We present here few studies on cell shapes and cell shape changes in epithelia. They focus on junctional (local) and cellular (global) features, which are set by both cortical and adhesion forces. These studies (except Sherrard *et al.*, 2010) assume a steady-state distribution of forces, thereby ignoring dynamics processes which we will discuss in more detail in the Section 3.1.

3.1. The *Drosophila* ommatidium

The *Drosophila* retina is an epithelium with a spatial periodic structure rising from a repeated fundamental module named *ommatidium* composed by a group of 20 cells stereotypically configured. The *ommatidium* consists of a central unit composed by four cone cells (setting over a cluster of photoreceptor cells), surrounded by two larger primary pigment cells. This central unit is then embedded in a hexagonal matrix, constituted by secondary and tertiary pigment cells and bristle.

What cortical properties are necessary to have cells arranged in such a precise pattern? A previous study compared patterns formed by one to six cone cells (at the apical site) in a *Rough eye* (*Roi*) mutant to patterns formed by aggregates of one to six soap bubbles (Hayashi and Carthew, 2004): the resemblance is striking. Plateau in 1873 had developed a set of rules describing the pattern of soap bubbles aggregates (Plateau, 1873). The physical justification to these patterns was based on the principle of free energy minimization. If we consider equal to γ the energy necessary to increase the surface of a soap film of a unit area, γS (with S the surface area of the film) represents the total surface energy of the film. Since γ is constant, minimizing the surface energy corresponds in minimizing S. This correlation thus suggested that also cells do aggregate in a way that minimizes the overall surface area. Hayashi *et al.* also revealed the important role of adhesive proteins (N-cadherin expressed between junctions of cone cells and E-cadherin expressed along all junctions) in cone cell pattern formation. Nevertheless, cell and soap bubbles differ greatly from one another in their internal composition and in the way they form contacts: while bubbles are in contact through a single soap film which is enveloping the whole bubble aggregate, two cells join their lipid leaflet via transmembrane adhesion proteins that spread the surface of contact. Thus, adhesive cells in proximity have the general tendency to increase the surface of contact and not to minimize the overall surface. However, the cortical actomyosin network responds mechanically to cell surface extension, and tends to reduce the surface of contact between cells (Lecuit and Lenne, 2007). To identify the physical principles underlying cone cells pattern, Käfer *et al.* compared the patterns observed *in vivo* with *in silico* predictions (Kafer *et al.*, 2007). The authors show that local adhesive properties of cells cannot account alone for cone cell shapes (this differs from other studies focusing on larger cell aggregates comprising 10^2–10^4 cells, in which properties of adhesion-based tension were sufficient to explain tissue rounding and cell sorting Brodland and Chen, 2000; Glazier and Graner, 1993; Kafer *et al.*, 2006; Maree and Hogeweg, 2001). Cells at the cortex are enriched by a cytoskeleton network having elastic properties. Adhesion and cortical elasticity have opposite contributions to surface energy: while cell adhesion tends to decrease the surface energy increasing the surface of contact between cells, the cortical elasticity increases the surface energy decreasing the surface of contact. Käfer *et al.* proposed that the energy of the cellular networks writes as the sum of adhesion energy, perimeter elastic energy, and area elastic energy:

$$E = \sum_{\text{interfaces}} \underbrace{-J_{ij} l_{ij}}_{\text{adhesion}} + \sum_{\text{cells}} \left(\underbrace{\frac{1}{2} k_i^{\mathrm{P}} \left(P_i - P_i^0\right)^2}_{\text{perimeter elasticity}} + \underbrace{\frac{1}{2} k_i^{\mathrm{A}} \left(A_i - A_i^0\right)^2}_{\text{area elasticity}} \right), \quad (1)$$

where $-J_{ij} l_{ij}$ denotes the adhesion energy between cells i and j with a contact length l_{ij}, k_i^{P} the elastic perimeter modulus of cell i, P_i its actual perimeter, P_i^0 its preferred perimeter, k_i^{A} its elastic perimeter modulus, A_i its actual area and A_i^0 its preferred area. By minimizing this energy, Käfer *et al.* could obtain *in silico* cone cells pattern similar to *in vivo* patterns. The authors indicate that the cortical elastic forces acting at the cell perimeter produce cell shape changes, which in turn can modify the elastic forces. This suggests the existence of a mechanical feedback between cell shape and cortical forces in tissues.

3.2. The proliferating *Drosophila* wing

In a growing tissue, cells pack developing characteristic shapes and sizes. The actomyosin cytoskeleton and the adhesive properties of cells mechanically control cell packing dynamics and cell morphological properties (Farhadifar *et al.*, 2007; Kafer *et al.*, 2007; Landsberg *et al.*, 2009; Major and Irvine, 2006). A biological model often used to study cell packing geometries is the developing *Drosophila* imaginal wing disk, a growing tissue that constitutes the perspective wing of the *Drosophila* fly. This tissue is well suited for this type of study since it rapidly proliferates—increasing from 4 to 50000 cells—is rather easy to isolate, cultivate, image, and make clone analysis. How cell shape, size, and packing dynamics are controlled during tissue growth?

A study by Gibson *et al.* (2006) has analyzed the statistics of neighbor cell number in proliferating tissues. The authors found that the number of n sided cells is remarkably well predicted by a simple mathematical model using a topological rule of cell allocation after cell division. From a physical point of view, Farhadifar *et al.* have investigated how cell mechanics impact on cell shape, surface area, and neighbor number distribution in the third instar larval wing disk of *Drosophila* (Farhadifar *et al.*, 2007). The epithelial tissue is modeled in the plane of cell junctions—cell apical site—as a 2D network in which cells are represented as 2D polygons. A stable configuration of the 2D network corresponds to a force balance at vertices (geometrical point from which more than two junctions radiate), and the only perturbation driving the system out of equilibrium is cell division. Stationary and stable network configurations (i.e., satisfying mechanical balance) were determined as local minima of an energy function:

$$E = \sum_{\text{interfaces}} \underbrace{\sigma_{ij} l_{ij}}_{\text{line energy}} + \sum_{\text{cells}} \left(\underbrace{\frac{1}{2} k_i^{\text{A}} \left(A_i - A_i^0\right)^2}_{\text{area elasticity}} + \underbrace{\frac{1}{2} \Gamma_i^{\text{P}} P_i^2}_{\text{perimeter contractiliy}} \right), \quad (2)$$

where σ_{ij} denotes the line tensions at junctions between individual cells, and Γ_i^{P} is a coefficient which reflects contractility of cell perimeter. Note that this energy function is very similar to that given in Eq. (1).

The first term corresponds to the local adhesive and contractile properties of junctions determined by adhesive proteins (e.g., E-cadherin) and the actin cytoskeleton, respectively. The second is referred to a term of cell area elasticity and to a term of cell perimeter contractility reflecting the overall actomyosing ring spanning the apical region like a purse string

Farhadifar *et al.* explored different combinations of cell and junctional forces by comparing *in silico* simulations and *in vivo* analysis of cell packing features. To further probe the mechanical properties of cells, the authors used UV laser ablation to "cut" cell bonds and analyzed their relaxation. This final experiment allowed the authors to come down to a few combinations of cortical forces matching *in vivo* analysis. Also this study (along with the work of Käfer J. *et al.*) concludes that cortical elasticity plays an important role in cell mechanics and thus cell packing properties. The phase diagram, obtained from *in silico* simulations (Fig. 4.1A), also predicts interesting features that could possibly govern cell behaviors: for instance, increased adhesion along cell boundaries is predicted to possibly facilitate cell rearrangements, which would make the tissues behave like liquids. In an attempt to understand the mechanical history of tissues *in vivo* at the light of simulations, Farhadifar *et al.* have estimated the number of cells neighbor exchanges (called T1 process) and cell removals (T2) (Fig. 4.1B) *in vivo* and found that they were in agreement with predictions. Nevertheless, it will be important to characterize cell properties throughout the developmental history of a tissue since the molecular machinery that reorganizes the epithelial contacts and the cytoskeleton is constantly at work.

Whether the developing imaginal wing disk of *Drosophila* is a proliferating tissue with homogenous mechanical properties is a matter of debate, given that it is divided into four compartments: anterior/posterior—A/P—and dorsal/ventral—D/V—(Fig. 4.2A). These compartments are defined by two supracellular actomyosin cables, perpendicular one to the other, running across the tissue along cell junctions establishing a smooth interface between adjacent cell populations (Major and Irvine, 2005, 2006). Compartment boundaries are thought to play a major role to separate cells with distinct fates, helping to maintain them within their correct location (Fig. 4.2A′). To challenge the role of these actomyosin cables in the imaginal disk, Major *et al.* analyzed mutants having defective formation of F-actin cable delimiting D/V compartments. This was possible by using Notch mutants. High levels of Notch are detected at the boundary between D/V compartments, and Notch signaling is necessary to establish the corresponding F-actin cable. In the absence of the prominent F-actin cable delimiting the D/V compartments, the boundary becomes rough but no cell mixing was reported (Fig. 4.2A″). Major *et al.* probed also the function of Myosin II by analyzing a *zip* (a gene encoding Myosin II heavy chain) allelic combination that permitted recovery of third instar wing imaginal disk (note that in this mutant Myosin II functionality is affected over the whole disk and not only at the boundary between compartments). In this mutant, the boundary between D/V compartments was also reported to be rough like in the Notch mutant. In addition, cells belonging to

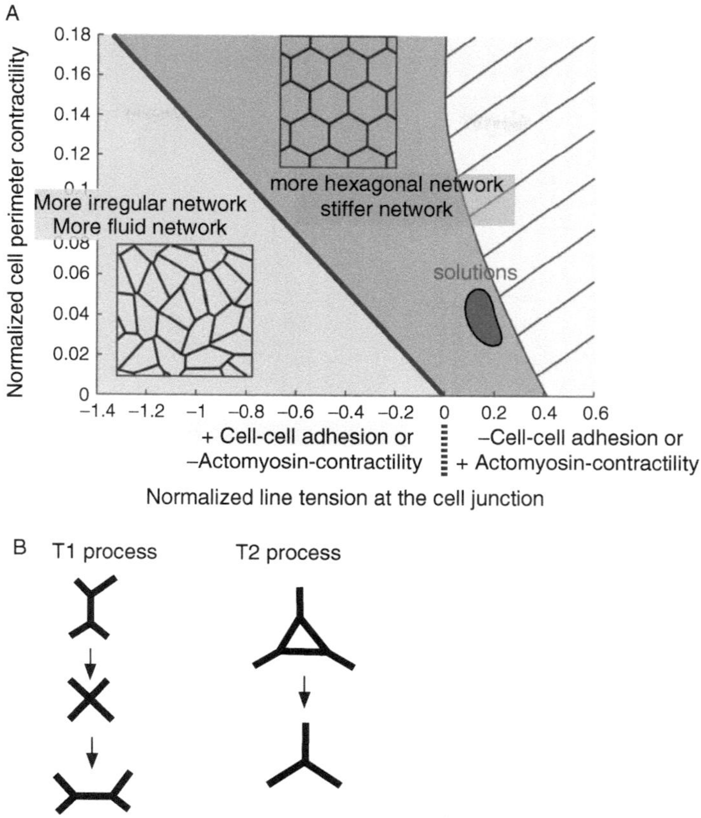

Figure 4.1 (A) Phase diagram of cell packing geometry as a function of cell perimeter contractility and line tension at cell junction (vertex model). *In silico* simulations predict more irregular and fluid cell networks for lower line tension and more regular and stiffer cell networks for higher line tension. Solutions (red region) are the cases which best fit *in vivo* data. (B) Cartoon showing two possible cell rearrangements driven by cell intercalation (T1 transition) or by cells apoptosis (T2 transition). Modified with permission from Farhadifar *et al.* (2007). (For interpretation of the references to color in this figure legend, the reader is referred to the Web version of this chapter.)

different compartments were able to intermix (Fig. 4.2A‴). Finally, these data bring evidence that the actomyosin supracellular cable, defining boundaries between compartments, plays a major role in providing a smooth frontier. In following studies (on the wing disk and on other model systems), other scientists measured higher tension along F-actin supracellular cables delimiting compartments (Landsberg *et al.*, 2009; Monier *et al.*, 2010; Solon *et al.*, 2009) by using laser ablation and chromophore-assisted light inactivation—CALI—(both techniques are presented in a following section). This suggests that high cortical forces driven by

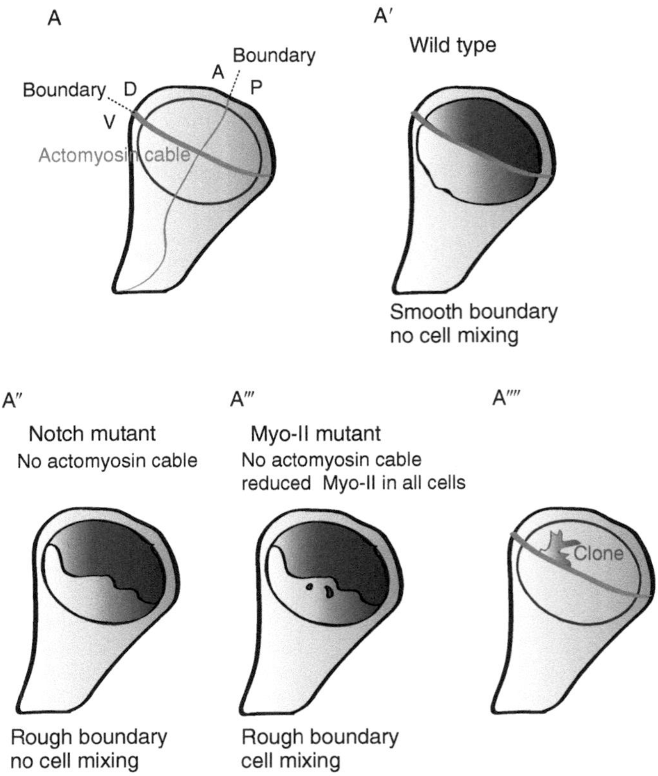

Figure 4.2 (A) Cartoon depicting the imaginal wing disk of *Drosophila* partitioned in four compartments. The compartment boundaries are delimited by thick actomyosin cables (green). (A′) In the wild type case, cells dorsally (red) and ventrally (yellow) positioned do not intermix and are separated by a smooth boundary (green). (A″) In a Notch mutant, the actomyosin cable, delimiting the boundary between dorsal and ventral compartments, is absent. This produces a rough boundary but no cell intermixing. (A‴) In a *zip* mutant (gene encoding the Myosin II heavy chain), Myosin II activity is compromised. This produces a rough boundary between dorsal and ventral compartments plus cell mixing. (A‴) Clone cells, in general, tend not to intermix neither with cells of other compartments nor with cells of the same compartment. (For interpretation of the references to color in this figure legend, the reader is referred to the Web version of this chapter.)

Myosin II contractility along compartment boundaries could be responsible for boundary straightness and smoothness. If the actin cable is responsible for boundary smoothness and not for cell separation, what role could play the smoothness of the boundary in the wing development? The imaginal disk of *Drosophila* is a tissue that grows and develops in a direction perpendicular to the plane of the tissue itself like a rubber glove that takes form by blowing inside of it. A smooth separation between dorsal and ventral tissues may function as scaffolding that could play a major role in directing wing growth thus providing the appropriate architectural structure for the formation of a straight wingblade. An initial rough boundary could instead lead, in the perpendicular direction, to a wiggly wingblade. This could be an interesting hypothesis to test. What prevents cells from one compartment from intermixing with cells of another compartment is still not clear. The formation of defective F-actin supracellular cables (e.g., in Notch mutant) is not sufficient to allow cell intermixing. Clone analysis also shows that there is almost no intermixing even within cells belonging to the same compartment: a marked clone of cells will almost always remain together in a coherent group (Fig. 4.2A‴). Cells do intermix instead when affecting Myosin II (e.g., in a *zip* mutant) over the whole imaginal wing disk. This could change the mechanical properties of cells allowing a greater number of T1 transitions to occur. Cultured cell studies show that elevated Myosin II can impair cell adhesiveness (Ivanov *et al.*, 2004; Sahai and Marshall, 2002) and *in silico* data indicate that higher adhesiveness can fluidify tissues (Farhadifar *et al.*, 2007). This emphasizes the complex coupling between actomyosin mechanics and adhesion which both contribute to control tissue mechanics through cortical forces.

3.3. Germband elongation of the *Drosophila* embryo

A tissue can elongate in one direction if, for instance, cells forming the tissue elongate in concert in the same direction. Tissue elongation can occur also if cells can spatially rearrange. An example of cell rearrangements driving tissue elongation is cell intercalation. During intercalation, cells exchange neighbors making the tissue converge in one direction and extend in the perpendicular direction. Cell intercalation happens, for example, during gastrulation and neurulation and can happen in both epithelial and nonepithelial cells (Ettensohn, 1985; Irvine and Wieschaus, 1994; Jacobson and Gordon, 1976; Keller, 1978; Warga and Kimmel, 1990). A remarkable example of intercalating tissue is the elongating germband of the *Drosophila* embryo (Fig. 4.3A). Irvine *et al.* suggested that differential adhesion between groups of cells could drive cell rearrangement (Irvine and Wieschaus, 1994), as hypothesized before by Steinberg to explain tissue sorting (Steinberg, 1963). A study by Bertet *et al.* highlighted for the first time the cellular process allowing cells to rearrange by exchanging neighbors: cells remodel their junction in a polarized fashion so that junctions parallel to the dorsal/ventral axis (vertical junctions) shrink bringing four cells in contact and then expand in a direction parallel to the anterior/posterior axis so that more dorsal and ventral cells form new contacts (Bertet *et al.*, 2004) (Fig. 4.3A′). During this process, Myosin II enriches along vertical junctions (Bertet *et al.*, 2004; Zallen and Wieschaus, 2004) (Fig. 4.3A″). Bertet *et al.* probed the activity of Myosin II during cell intercalation showing that Myosin II is

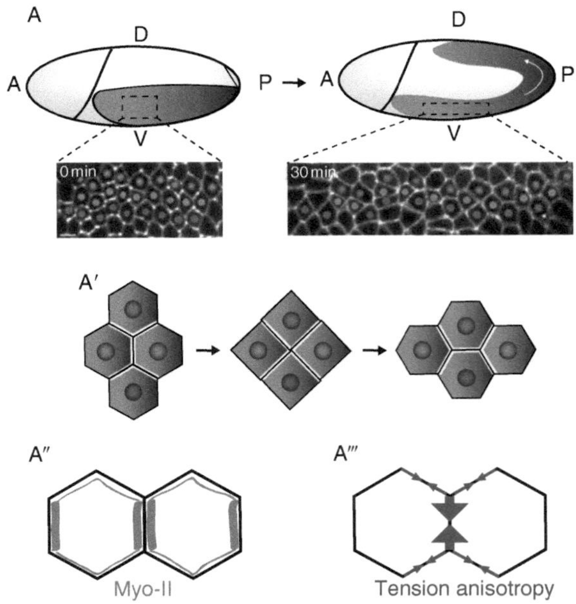

Figure 4.3 (A) Cartoon depicting a *Drosophila* embryo during gastrulation. The germ-band (GB) converges in one direction extending in the perpendicular direction. GB convergence-extension is driven by a cell rearrangement named cell intercalation. Cell intercalation is polarized along the anterior/posterior axis. (A′) Cartoon depicting the three main steps of cell intercalation: (1) a junction shrinks (2) bringing four cells in close proximity; (3) finally a new junction expands forming a new contact with more dorsal and ventral cells. (A″) Cartoon depicting Myosin II polarity along junctions parallel to the dorsal/ventral axis. (A‴) Cartoon showing a model of tension anisotropy (Myosin II based) driving cell intercalation. Modified with permission from Bertet *et al.* (2004).

necessary for junction remodeling. This study suggested that the contractile activity of Myosin II might create a local tension that orients the disassembly of junctions. This hypothesis was then tested by a quantitative comparison between *in vivo* data and *in silico* predictions and laser subcellular dissection (Rauzi *et al.*, 2008). Rauzi *et al.* revealed an anisotropy of cortical forces along cell junctions controlled by Myosin II: this was measured to be a factor 2 along vertical junctions (junctions with greater density of Myosin II) compared to other junctions. Cortical forces were inferred by laser dissection experiments: disruption of the actomyosin network underlying a given junction modified the balance of forces and produced junction relaxation, whose speed is indicative of cortical tension (see Section 5). The authors designed a model based on the local (junctional) and global (cellular) natures of cortical forces. The cellular network configurations during tissue shape changes were described as the succession of local minima of an energy function, which incorporates the anisotropic mechanical features of junctions (i.e., dependent on orientation):

$$E = \underbrace{\sum_{\text{interfaces}} \sigma(\theta_{ij}) l_{ij}}_{\text{line energy}} + \sum_{\text{cells}} \left(\underbrace{\frac{1}{2} k_i^P \left(\sum_j \beta(\theta_{ij}) l_{ij} - P_i^0 \right)^2}_{\text{perimeter elasticity}} + \underbrace{\frac{1}{2} k_i^A (A_i - A_i^0)^2}_{\text{area elasticity}} \right), \quad (3)$$

where $\sigma(\theta_{ij})$ is the line tension at junctions with an orientation θ_{ij}, and $\beta(\theta_{ij})$ defines the cortical elasticity state of junctions with a given orientation. The overall framework reveals that the anisotropy of subcellular forces at the cortex is sufficient to drive cell intercalation and orient tissue elongation (Fig. 4.3A‴). The model also reveals the importance of cortical elastic forces in the junctional shrinking process.

Fluctuations of cell vertices were implemented *in silico*, which reflect *in vivo* movements of cell vertices. The simulations suggest that vertex fluctuations prevent cells from being trapped in local energy minima, helping junction remodeling. Note that cell shape fluctuations have been reported in other epithelia, and cortical forces are likely to be important in these movements. It will be interesting in the future to investigate the nature of these fluctuations, probing their effective role in cell intercalation.

In some cases, consecutive junctions can shrink bringing more than four cells in contact. These higher-order structures are named rosettes and can resolve creating new multiple contacts along the elongating axis (Blankenship *et al.*, 2006). The origin of the rosettes is a matter of debate. While we explained these structures simply as a result of individual T1 transitions (Rauzi *et al.*, 2008), Zallen and colleagues consider rosettes to be emergent mechanical entities *per se* (Blankenship *et al.*, 2006; Fernandez-Gonzalez *et al.*, 2009). In the latter report, rosettes are hypothesized to stem from multicellular contractile structures in which intracellular Myosin II filaments are functionally associated across cells. Based on this hypothesis, the authors studied the molecular and mechanical properties of supracellular actomyosin cables that are formed half-way through intercalation. Determining whether these cables effectively generate rosettes will require further investigation. At the end of cell intercalation, actomyosin cables, parallel to the dorsal/ventral axis, are present along the parasegmental boundaries (boundaries of lineage separation; Monier *et al.*, 2010). Thinner actomyosin cables, seen during the second half of cell intercalation, could presumably be precursors of the final "parasegmental cables."

Cell intercalation is composed of three successive steps: (1) a cell junction shrinks until (2) it is reduced to a punctual junction where four cells come in contact; finally (3) a new junction expands along the direction perpendicular to the initial shrinking junction. The process going from (2) to (3) is still not well understood. In our simulations, this process is a consequence of the minimization of the energy of the system (Rauzi *et al.*, 2008). The *in silico* model does not take into account the viscous properties of the tissue, which are crucial to understand the system dynamics. Active process could be necessary to expand junctions to complete intercalation and maintain the necessary intercalation rate for proper tissue elongation. Future works should focus on this last stage of intercalation, which is necessary for efficient tissue elongation.

3.4. The ascidian endoderm invagination

Sheets of epithelial cells can adopt diverse shapes. A recurrent example, that constitutes one of the fundamental building blocks in morphogenesis, is when tissues bend and fold to form a pit or groove. This process is commonly known as tissue invagination. Previous studies have shown that a series of coordinated cell shape changes plays a major role in tissue invagination (Hardin and Keller, 1988; Kam *et al.*, 1991; Leptin and Grunewald, 1990; Sweeton *et al.*, 1991): one example is constriction of the apical surface of cells that produces cell wedging and tissue flattening and bending. Although tissue invagination has been characterized to a certain extent, its mechanical basis remains still poorly understood.

What are the molecular origins and the distribution of forces that make cells change their shape and how these forces are integrated within the invaginating tissue? Diverse actors were hypothesized to be responsible for cell invagination (reviewed in Keller *et al.*, 2003): an example is cell shape change of the invaginating tissue driven by differential adhesiveness, acto-myosin contractility, and microtubules activity. Tissue-extrinsic forces were also hypothesized to contribute to invagination (e.g., forces driven by tissue epiboly). Computer simulation of tissue invagination shows that, given specific boundary conditions, multiple scenarios are possible (Clausi and Brodland, 1993; Conte *et al.*, 2008, 2009; Odell *et al.*, 1981) but strong experimental evidence supporting one of these scenarios is still missing. Apical constriction of mesodermal cells has been intensively studied and characterized in various works (Martin *et al.*, 2009). This process was shown to play a necessary role in tissue flattening and bending. Theoretical modeling has shown that apical constriction can be sufficient for the furrow formation step (Odell *et al.*, 1981). However, more recent studies on *Drosophila* and also on *Xenopus* suggest that apical constriction may not be an essential process for the final step of effective mesoderm internalization (reviewed in Leptin, 2005). This indicates that other, as yet unknown, forces must be responsible for the formation of the furrow.

A recent work has shed new light on this process probing the mechanical properties of cells responsible for the endoderm invagination in the ascidian gastrulating embryo (Sherrard *et al.*, 2010) (Fig. 4.7). During the invagination of the 10 cells monolayer plate, ascidians (Urochordata) are built up by around 100 cells. Ascidians are also rather transparent making 3D imaging, in fixed and *in vivo* samples, very efficient. This biological system is thus very suitable for developmental studies and for *in silico* simulations since the overall embryo can be analyzed at once (all three endoderm, mesoderm, and ectoderm tissues) and each cell can be studied as a specific entity located in a defined volume within the whole embryo. By using 4D morphometric and protein distribution analyses over wild type and mutated embryos, computer simulation and laser ablation, this study reveals an actomyosin-based mechanism driving endoderm invagination that can be subdivided in two steps. During the first step, 1P-myosin is enriched at the apical surface of endoderm cells and produces a cortical force driving apical constriction. Apical recruitment of 1P-myosin is shown to be Rho/Rho-kinase-dependent. During the second step, Rho/Rho-kinase-independent lateral enrichment of 1P-myosin drives apicobasal shortening, while Rho/Rho-kinase-dependent circumapical enrichment of 1P and 2P-myosin prevents the apical surface to expand back. During the second step, actual tissue invagination is produced. This second-step process is named by the authors "collared rounding." Furthermore, computer simulation prediction along with laser ablation experiment shows that mesectoderm tends to resist endoderm invagination. In the end, the overall framework shows how timed differential contractility along the entire cortex of cells plays a major role in tissue invagination (Fig. 4.4). Cell apicobasal shortening is not a universally shared shape change of invaginating cells: for instance, during *Xenopus* embryo gastrulation and neurulation in chick and mouse, no apicobasal shortening is reported. Interestingly, the authors suggest that this could be a specific mode of functioning for tissues that undergo fast invagination (e.g., 45 min for ascidians embryos and 30 min for *Drosophila* embryos).

In numerous studies aiming to better understand tissue morphogenesis, the three-dimensional structure of cells is often simplified in two dimensions (e.g., morphogenetic studies presented previously in the text analyze one single plane intersecting cell adherens junctions). This study shows how extremely important can be to analyze the three-dimensional properties of cells in time to achieve a better understanding of the morphogenesis of tissues.

4. Dynamic Spatiotemporal Distribution of Cortical Forces and Cell Shape Changes

Cortical forces, necessary to sculpt tissue shape, can be studied at different space and time scales. For instance, average cellular properties and dynamics can be studied at low time resolution (e.g., 30 s period or more): in these conditions steady-state distribution of forces are generally sufficient to explain cell and tissue shape changes. However, supramolecular

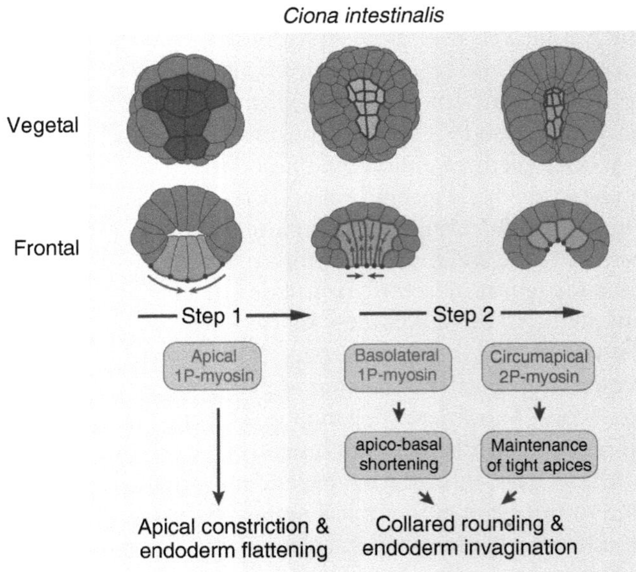

Figure 4.4 Cartoon depicting a two-step model reasoning the endoderm invagination of the *Ciona Intestinalis* embryo. In the first step, endoderm cells apically constrict flattening the tissue and mesectoderm cells shrink apicobasally (violet arrows). In a second step, endoderm shrinks apicobasally pulling the tissue inward and thus producing a groove. While 1P-myosin (red) plays a major role in cell deformation (apical constriction and apicobasal shortening), 2P-myosin (blue) acts to maintain apical constriction. During endoderm invagination, the mesectoderm feeds back resisting to the endoderm inward movement (black arrows). Modified with permission from Sherrard *et al.* (2010). (For interpretation of the references to color in this figure legend, the reader is referred to the Web version of this chapter.)

structures underlying cortical forces have faster dynamical features and produce measurable cell shape changes in time windows inferior to 1 s. High space and time resolution imaging can thus help to bridge the gap between cell and molecular level. In this context, recent studies have shed light on the dynamics of cortical forces.

As presented below, most of these studies report a pulsatile actomyosin distribution at the cell cortex in remodeling tissues responsible for cell shape changes, which proceed by successive steps lasting less than few minutes.

4.1. Actomyosin pulsed contractions in *Drosophila* mesoderm invagination

One of the first steps in *Drosophila* gastrulation is the internalization of the prospective mesoderm and formation of the ventral furrow. The ventral furrow is formed by a stripe of cells along the ventral side of the embryo that is 18 cells wide and 60 cells long. These cells invaginate, forming a "tube" in the interior part of the embryo. Cells in the tube then disassemble and reassemble in a single cell layer beneath the ectoderm (Leptin and Grunewald, 1990).

During the first phase of mesoderm invagination, cells randomly constrict apically, producing tissue flattening and bending (Leptin, 1995) (Fig. 4.5A). This process was thought to be driven by junctional contraction in a purse-string like fashion (reviewed in Lecuit and Lenne, 2007). Martin *et al.* revisited this process by analyzing the correlation of actomyosin distribution with cell shape changes (Martin *et al.*, 2009). The authors could show that cell constriction proceeds in two alternating steps: (1) the apical surface is partially reduced and (2) the reduced apical surface is stabilized. The authors revealed the existence of an actomyosin network spanning the medial apical region of cells and presenting a pulsating stochastic behavior. Pulses of actomyosin are driven by the coalescence of Myosin II clusters toward the center of the medial apical region. These are shown to be responsible for cell apical constriction. The medial apical network pulls inwards on discrete junctional sites enriched of E-cadherin

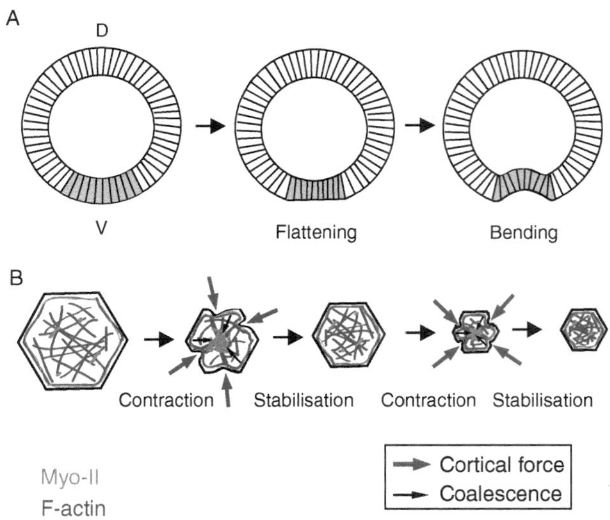

Figure 4.5 (A) Cartoon showing the first steps characterizing mesoderm invagination: (1) tissue flattening and (2) tissue bending. D and V indicate dorsal and ventral sites, respectively. (B) Representation of a mesoderm cell during apical constriction: the cell undergoes successive phases of apical surface contraction and stabilization driven by coalescence (black arrows) of Myosin II (green) patches over the medial apical actin network (red) producing cortical forces (red arrows). (For interpretation of the references to color in this figure legend, the reader is referred to the Web version of this chapter.)

reducing the apical surface of the cell and imposing a rough junctional contour. In the second step, Myosin II coalescence dissolves but the cell maintains its reduced apical surface and the junctional contour is smoothened (Fig. 4.5B). If adherens junctions are disrupted, Myosin II coalescences form, but no shape change occurs (Dawes-Hoang *et al.*, 2005). An important architectural feature that allows this ratchet-like mechanism to work is thus the coupling of the medial meshwork to adherens junctions. How can the overall cell apical contour move in average toward the medial region during constriction even though the medial apical actomyosin network seems to pull on few discrete sites along junctions? A recent study shows that the mesoderm fails to invaginate if the RNA encoding for the α-cat or ß-cat unit is inhibited (Martin *et al.*, 2010). These units are supposed to link the junctional actomyosin network to adherent E-cadherin sites. More specifically, Myosin II medioapical coalescence in α-cat or ß-cat mutants forms membrane tethers while the cell enlarges apically (Martin *et al.*, 2010). This suggests that the junctional actomyosin coupled to E-cad plays a major role in maintaining apical integrity, thus transferring forces from few discrete sites to the overall junctional contour. For an efficient apical constriction, the apical contracted surface is then stabilized. How does the stabilizing process take place? The junctional actomyosin ring could be remodeled, acting as a purse-string-like contractor smoothening the junctional contour and providing cell surface stabilization. It is important to elucidate the different roles of junctional versus medial apical actomyosin networks and how the two are coupled to work as a ratchet.

4.2. Actomyosin pulsed contractions in *Drosophila* dorsal closure

During embryonic development, different tissues, sharing boundaries, change their shape: this causes tissues to interact generating extratissue forces that can pull, push, stretch, and compress neighboring tissues. A clear example of tissue interaction is the dorsal closure of the *Drosophila* embryo. This process consists in the dorsal retraction of the amnioserosa tissue accompanied by the epiboly of the surrounding and contacting epidermis. Different actors contribute mechanically to dorsal closure: (1) amnioserosa cell contraction (Gorfinkiel *et al.*, 2009; Hutson *et al.*, 2003b; Kiehart *et al.*, 2000; Schock and Perrimon, 2002; Solon *et al.*, 2009), (2) supracellular actomyosin cable (Franke *et al.*, 2005; Jacinto *et al.*, 2002; Kiehart *et al.*, 2000; Solon *et al.*, 2009; Young *et al.*, 1991), and (3) protrusion formation (Jacinto *et al.*, 2000; Millard and Martin, 2008) at the epidermis boundary, (4) canti[1] zippering (Hutson *et al.*, 2003b; Jacinto *et al.*, 2000), and (5)

[1] During dorsal closure the amnioserosa is eye shaped and the canti are the two extremities of the "eye."

amnioserosa cell apoptosis (Toyama *et al.*, 2008a). Given the great number of mechanical players, dorsal closure is thus a complex process to disentangle. Closure lasts up to 2 h: in this time delay, each player acts in different time windows and, in each time window, possibly adopting different roles. A challenging task is thus to decipher the contribution of each player at a given time. Few studies have identified and characterized the different force contributions to dorsal closure (Hutson *et al.*, 2003b; Solon *et al.*, 2009). The work of Solon *et al.* is a good example of how the system can be dissected and studied. The authors focus their analysis on the first two steps of dorsal closure ((1) and (2)) and try to determine how these two steps can be coordinated. Solon *et al.* show that amnioserosa cells, before and during the onset of dorsal closure, contract and relax their apical surface producing an intrinsic oscillatory movement with a period of 3–4 min. An actomyosin meshwork spanning the medial apical region of amnioserosa cell was reported in a study by Franke *et al.* (Franke *et al.*, 2005). Ma *et al.* used laser ablation to probe the contractile properties of the meshwork revealing a tensile sheet spanning the medial apical region of amnioserosa cells (Ma *et al.*, 2009). Recent studies show also that pulses of actomyosin form and vanish in these cells correlating with cell contraction and expansion (Blanchard *et al.*, 2010; David *et al.*, 2010) (Fig. 4.6). Interestingly, actomyosin densification not only forms in the medial apical region, but also tends to flow preferentially in a direction parallel to the dorsal/ventral axis (David *et al.*, 2010) (Fig. 4.6). The cell junctional outline is smooth during the phase of cell expansion while it is rough and folded in the phase of contraction. At the beginning of dorsal closure also a supracellular actomyosin cable forms in epidermis cells located at the leading edge (Fig. 4.6). Increasing levels of actin are reported in this first phase in concomitance with the amnioserosa/epidermis boundary (leading edge) straightening (Solon *et al.*, 2009). Laser ablation experiments show that this actomyosin cable is acting as a tensile structure and that the force is mostly oriented along the direction of the cable. When the actomyosin cable forms, amnioserosa cells, positioned closer to the leading edge, first attenuate and then stop their oscillation persisting in a contracted state but with a junction outline smooth and straight. How are these cortical actomyosin forces coordinated to drive dorsal closure? The study from Solon *et al.* suggests a ratchet mechanism at the tissue level where amnioserosa pulsations are the active force while the epidermis supracellular cable orients and sustains the force toward the dorsal midline like a gear that directs the energy produced by the oscillating pistons of a motor. The coordination of the two actors is thus necessary for efficient dorsal closure. Blanchard *et al.*, studying the same process, report an increasing amount of Myosin II in amnioserosa cells in both the medial apical region and in the junctional ring (Blanchard *et al.*, 2010). In mutants overexpressing Myosin II in amnioserosa cells, precocious dorsal closure has been reported. In addition, mutants that exhibit defects in actin cable

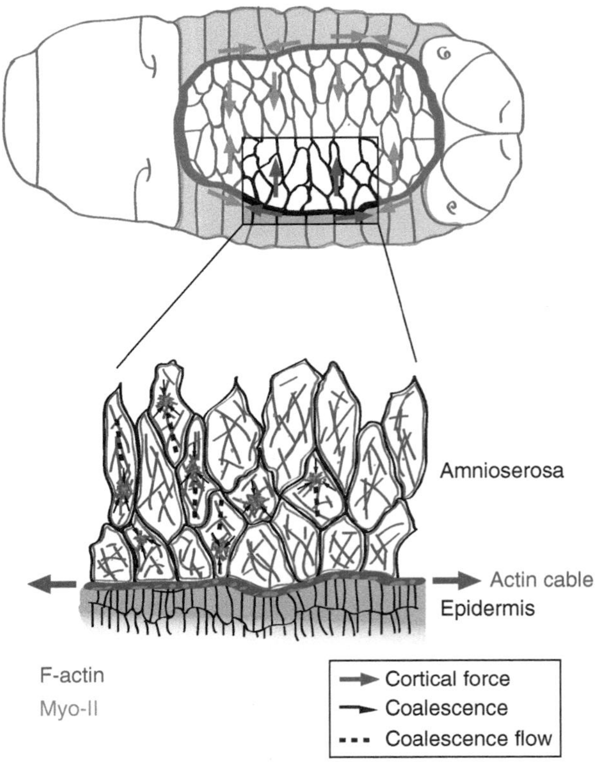

Figure 4.6 Cartoon representing a *Drosophila* embryo during dorsal closure. In beige the epidermis and in white the amnioserosa tissue. Red arrows show the force contribution of both the epidermis and the amnioserosa tissue. The zoom shows details of amnioserosa cells. These cells have a junctional and a medial apical actomyosin network (F-actin in red and Myosin II in green). The apical surface of these cells undergoes contraction and expansion. Apical surface oscillation is concomitant with coalescence of actomyosin (black arrows) that flows preferentially in a direction parallel to the dorsal/ventral axis (dashed line). An actomyosin supracellular cable (red and green horizontal stripe) is formed in epidermis cells at the boundary between epidermis and amnioserosa tissue generating a cortical force directed along its length (red arrows). (For interpretation of the references to color in this figure legend, the reader is referred to the Web version of this chapter.)

formation still show similar Myosin II recruitments and timings of oscillation damping suggesting that the damping is an intrinsic property of amnioserosa cells. A different scenario is thus suggested: Myosin II pool in the medial apical region of amnioserosa cell produces apical surface oscillations while the junctional pool provides the cortical tension necessary to straighten junctions and to stabilize cell apical surface. In the end, a ratchet mechanism intrinsic to amnioserosa cells is proposed. Extracellular and intracellular ratchets, which are, respectively, due to the supracellular actin cable and junctional Myosin II pool, are thus possible and their cooperation could make dorsal closure a robust process with redundant contributions both at tissue and cell scale.

Are amnioserosa cell oscillations coordinated within the tissue? Solon *et al.*, by using cross-correlation analysis, show that cells tend to oscillate in-phase or antiphase with neighboring cells with a preference for the latter (Solon *et al.*, 2009). In addition, cells can quickly invert their oscillation phase. Laser ablation performed at the boundary between two amnioserosa cells produces oscillation arrest for the two cells ablated and for their close neighbors. This evidence indicates a high degree of mechanical coupling between neighboring cells. Blanchard *et al.* report an even larger scale correlation of oscillations with in-phase oscillation in row of cells in one direction and antiphase correlation in the perpendicular direction (Blanchard *et al.*, 2010). David *et al.* show that the actomyosin densification and contraction in one cell is followed by the formation of the same in neighboring cells (in accordance with Solon *et al.* correlation analysis) (David *et al.*, 2010). Still it is not clear how actomyosin densification in the medial apical region of amnioserosa cell tends to flow in a direction parallel to the dorsal/ventral axis. Davis *et al.* speculate that a feedback between the epidermis and amnioserosa cells orienting the flow. What role plays the directed flow is still an intriguing open question that deserves much attention for future studies.

4.3. Actomyosin pulsed contraction in *Xenopus* convergence-extension

The convergence-extension of the dorsal axial and paraxial mesoderm in vertebrate embryos (e.g., in the *Xenopus* embryo) is driven by the polarized intercalation of cells like during germband extension in the *Drosophila* embryo. Nevertheless, the mechanics that govern this process seems quite different in the two cases mainly because tissues of one and the other do not have the same properties. While in the *Drosophila*, cells form a homogeneous sheet of columnar epithelial cells, in the *Xenopus*, for instance, the tissue is formed by mesenchymal protruding cells lacking apical junctions. In the *Xenopus* embryo, convergence-extension plays an important role in the blastopore closure and in the elongation of the body axis. Dorsal explants of *Xenopus* embryo in tissue culture can undergo convergence-extension producing a pushing force of 1.2 μN before buckling (Keller and Danilchik, 1988; Moore, 1994). This clearly shows that forces driving convergence-extension are at least partially generated within the intercalating tissue. Filo–lamellipodia protrusions play a major role in cell intercalation. These are formed in a polarized fashion at the two extremities displaying cycles of extension and retraction (Keller *et al.*, 1989, 2000; Shih and Keller, 1992). Protrusion possibly exerts forces on the substratum generating the traction that elongates mesenchymal cells stretching and wedging them between one another. A recent study from Skoglund *et al.* reports on a cortical actin network Myosin IIB dependent and shows that this network plays a major role in cell intercalation (Skoglund *et al.*, 2008). The network consists of actin cables meeting at nodes within cells and at cell–cell focal contacts. Skoglund *et al.* propose that these tensile elements counteract the stretching and thinning of cells driven by the protrusion activity limiting cell elongation. The actomyosin network thus transmits tension through the tissue developing arcs of tension between nodes at the adhesion sites. In this way, the overall tissue during cell intercalation becomes stiffer probably in order to build up the necessary strength to pull along the axis of convergence (Moore *et al.*, 1995; Zhou *et al.*, 2009). Skoglund *et al.* also show that the actin network is mostly directed along the axis of convergent forces, and that actomyosin cables show an oscillatory behavior by continuously elongating and shortening. The authors interpret this as possible active contraction that could bring together cells wedging between one another along the convergent axis. Further experiments, using subcellular laser ablation for instance, should be done to test this hypothesis. This work shows that pulsatile contractions are reported not only in invertebrate but also in vertebrate suggesting a possible universal mode of functioning underlying tissue morphogenesis.

4.4. Actomyosin flows and pulsed contraction during cell intercalation in the *Drosophila* embryo

In the early *Drosophila* embryo, a ring of actin, localized at the cell apical cortex, is conventionally thought to play a major role in cell shape changes in various processes of tissue morphogenesis (e.g., mesoderm invagination initiated by cell apical constriction and germband extension driven by cell intercalation, reviewed in Lecuit and Lenne, 2007). This idea is based on a purse-string model. As presented above, recent studies have shown that this is not always the case. Martin *et al.* have shown that an actomyosin meshwork spanning the medial region of cells is mainly responsible for apical constriction in the invaginating mesoderm during *Drososphila* embryo gastrulation (discussed previously).

In a recent paper (Rauzi *et al.*, 2010), Rauzi *et al.* show that in the ectoderm, elongating by cell intercalation, cells have a medial apical actomyosin meshwork coalescing in a pulsatile manner. This meshwork not only shares similarities to the one revealed in the mesoderm by Martin *et al.* but also presents strikingly different behaviors. In mesoderm cells, the meshwork, after few minutes of pulsed constrictions medially directed, persistently contracts towards the center of the apical cell surface; in contrast in the ectoderm the meshwork flows along the cell medial apical surface toward the cell junctions with actomyosin coalescences (pulses) forming periodically throughout germband extension. Rauzi *et al.* push the analysis further revealing a role of the actomyosin flow in junction shrinkage (first step of cell intercalation, Fig. 4.7A). The authors show that when the flow approaches a junction, the junction partially shrinks (Fig. 4.7B). The flow then continues and the actomyosin pulse fuses with the junctional actomyosin meshwork (Fig. 4.7B). Actomyosin fusion is shown to be necessary for junction stabilization. This process is iterated resulting in a ratchet mechanism of periodic shrinkage/stabilization in which the junction length is

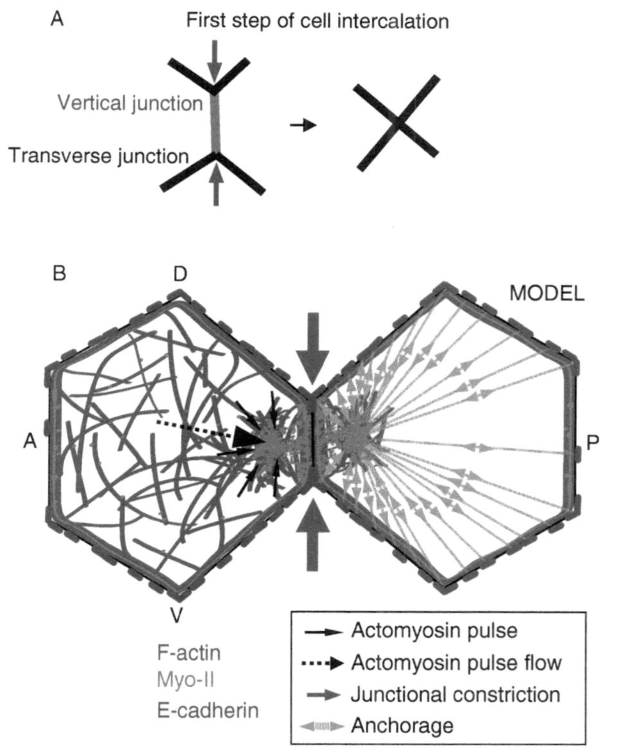

Figure 4.7 (A) Cartoon representing the first step of cell intercalation: a junction shrinks (the vertical junction in this example) bringing four cells in contact. (B) *Left.* Cartoon showing the coalescence of the medial apical F-actin (red) and Myosin II (green) network, which flows toward a junction parallel (dashed arrow) to the dorsal/ventral axis (black line). Clusters of E-cadherin (blue) are depicted at the cell junctions. This produces a cortical constricting force (red arrows) that shrinks the junction length. The actomyosin pulse then fuses to the junctional actomyosin cortex. This ensures junction length stabilization after shrinkage. *Right.* E-cad is less abundant on vertical than transverse junctions. This produces an imbalance of anchorage points (orange arrow), which would orient the flow of the actomyosin pulse. (For interpretation of the references to color in this figure legend, the reader is referred to the Web version of this chapter.)

reduced in steps until complete shrinkage. By analyzing the directionality of the flow, Rauzi *et al.* show that the medial apical actomyosin flows in a polarized fashion toward junctions parallel to the dorso-ventral axis in agreement with the polarized junction actomyosin recruitment and shrinking (Bertet *et al.*, 2004; Rauzi *et al.*, 2008; Zallen and Wieschaus, 2004). With these findings, the authors shed light on the origin of the junctional actomyosin planar cell polarity (PCP) in intercalating cells during germband extension. What could direct the actomyosin flow? One previous study reported that the distribution of E-cadherin along cell junctions during the process of cell intercalation is anisotropic (Zallen and Wieschaus, 2004). E-cadherin is shown to be distributed in a planar polarized fashion with more E-cad on transverse junction compared to vertical ones. Thus, we can conclude that actomyosin pulses tend to flow preferentially toward junctions with lower level of E-cad. Rauzi *et al.* extend this performing a correlation analysis of E-cad intensity at transverse over vertical junctions and Myosin II flow. The authors show that a pulse of Myosin II flows toward a vertical junction when E-cad maximum anisotropy is reached. In addition, E-cad and α-cat mutants lacking E-cad polarity did not show flow polarity, thus no PCP Myosin II recruitment and no cell intercalation. How can E-cadherin distribution control actomyosin flow? E-cad could play a role in forming barriers that would inhibit actomyosin flow in specific regions of the cell apical cortex. Alternatively, it could form sites of anchorage for the actomyosin meshwork that could generate an imbalance of cortical forces and thus a flow. To test these two possibilities, the authors perturbed the medial apical actomyosin meshwork by using laser ablation. After ablating part of a forming and flowing actomyosin pulse, the remaining pulse bit always flowed radially away from the point of ablation even toward transverse junctions. This experiment better supports a model based on unbalanced cortical forces driven by anchorage anisotropy (Fig. 4.7, orange lines of anchorage).

E-cad role in epithelial remodeling has mostly been considered only from the standpoint of adhesion, which determines the stability of cell contacts. Rauzi *et al.* indicate that E-cad complexes also play a pivotal role in controlling the spatial–temporal pattern of actomyosin contractile activity during epithelial morphogenesis. How E-cad PCP is controlled is still not well understood. Endocytosis could control E-cad distribution along the cell cortex. This in turn could be regulated by a polarized distribution of endocytic proteins (e.g., clathrin). Another possibility could be a mechanism based on membrane tension. Different studies have shown that a decrease in membrane tension stimulates endocytosis, while increases in tension stimulate secretion (Raucher and Sheetz, 1999a,b; Sheetz and Dai, 1996). Future studies will be necessary to test these hypotheses.

Pulse formations and flow dynamics seem to be a major mode of functioning in these cells where pulse frequency could dictate the speed at which the processes can advance while flow the spatial orientation of cell deformation. Speckles of Myosin II that coalesce and persist in the central medial apical region drive isotropic apical constriction in the mesoderm of the developing *Drosophila* embryo (Martin *et al.*, 2009), while flow of the actomyosin meshwork in the *C. elegans* monocellular system reshapes the cortex in a polarized fashion (Munro *et al.*, 2004). How the actin meshwork is organized and how is linked to the cortex could explain flow directionality. Thus, future investigation will be required to determine how actomyosin properties can, at different scales, drive flow, cell shape changes and tissue remodeling.

5. Methods

Physicists and engineers have developed several tools to probe the mechanics of cells, for example, laser ablation, optical and magnetic tweezers, atomic force microscopy, micropipette aspiration, etc. Most of these techniques can be applied to single cells but not directly to tissues or cells within living embryos. An exception is laser ablation that has been used in different works studying cell and tissue mechanics to better understand cortical forces and tissue morphogenesis (Desprat *et al.*, 2008; Farhadifar *et al.*, 2007; Fernandez-Gonzalez *et al.*, 2009; Grill *et al.*, 2003; Hutson *et al.*, 2003a; Landsberg *et al.*, 2009; Martin *et al.*, 2010; Rauzi *et al.*, 2008; Sherrard *et al.*, 2010; Solon *et al.*, 2009; Toyama *et al.*, 2008b). The aim of this section is to give a general overview of laser–tissue interactions and to focus on plasma-induced ablation, technique particularly suited for subcellular studies in living embryos. With this paragraph, we hope to give the reader a chance to understand the principles, and pros and cons of the different laser ablation techniques that have been used extensively and that are becoming more and more popular in developmental biology studies focused on biomechanics. For a deeper and more detailed presentation on laser–tissue interactions, we recommend the book of Niemz, which has been an important source of information for us to write this section (Niemz, 2004).

5.1. General Principles of laser–tissue interaction

When light encounters matter, three main processes are generated: light can be reflected from or absorbed within or transmitted throughout the sample. These interactions depend on the properties of light and of the material interfering with light propagation. Biological samples are particularly complex materials since they are heterogeneous in composition, density, and structure. If we consider developing embryos for instance, all properties are also varying rapidly over time (even in less than a minute). Characterizing

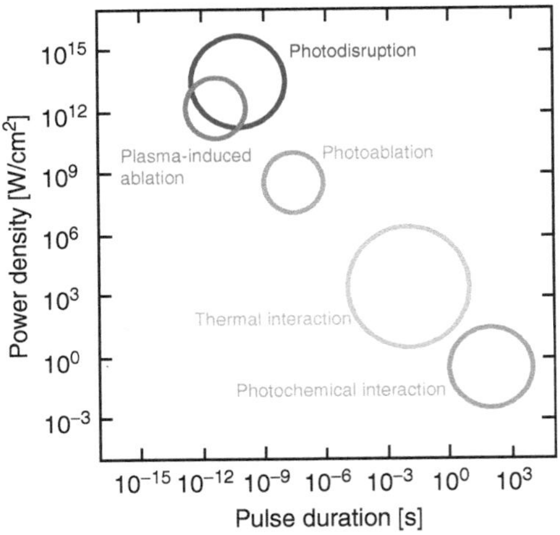

Figure 4.8 Five categories of laser–tissue interaction depending on pulse duration and power density. Adapted from Niemz (2004).

the interaction of light with these specimens can thus become a hard, even though still important, task.

Four main light parameters can be defined: wavelength, power density, temporal compaction of photons (low compaction, i.e., continuous wave (CW) vs. high compaction, i.e., pulsed lasers), and pulse repetition rate. Depending on the wavelength of the laser used, for instance, different absorption coefficients and penetration depths inside the bulk of the tissue can be achieved. Then, depending on the efficient power intensity and on the temporal compaction of photons (light pulse width), we can classify five categories of laser–tissue interactions: photochemical interaction, thermal interaction, photoablation, plasma-induced ablation, photodisruption (Fig. 4.8). We will see also how laser pulse repetition rate is an important parameter that can be tuned to achieve specific laser–tissue interactions. Laser light has a high degree of spatial coherence and can be therefore focused down to the diffraction limit, thereby achieving a very high irradiance.

Photochemical interaction takes place at very low light power densities and long exposure times going from second to CW. An example of such a process is photosynthesis. Other example is the light induced reaction of an excited chromophore. Thermal interaction groups many subcategories that have in common the local increase of temperature. This process can be achieved both by using CW or pulsed lasers and it accounts for hyperthermia, coagulation, vaporization, carbonization, and melting of tissues ordered in function of temperature increase.

Photoablation can be achieved at power densities around 10^7–10^8 W/cm^2 and laser pulse duration in the nanosecond range. An important feature of such process is the precision in tissue etching with no thermal damage on the adjacent tissue. In order to better explain what photoablation is about, we consider, for instance, as target a polymeric structure. Polymers are the result of the binding between monomers that are held together by a strong attractive force. Laser irradiation breaks the bond between monomers that pass from an attractive to a repulsive state. This process is associated with a change of volume occupied by each monomer resulting in material etching. In general, only UV lasers provide enough energy for bond dissociation. Photoablation is initiated at a certain threshold power density, and etching depth in the target increases by increasing the energy density.

Plasma (free-electron cloud) is formed and optical breakdown[2] occurs when applying light power densities exceeding 10^{11} W/cm^2 in solids and in fluids (10^{14} W/cm^2 in air). Such high power densities applied to tissues result in an extremely high electric field that has the same order of magnitude of the average atomic or intramolecular Coulomb electric field. Plasma-induced ablation results in clean cuts with no evidence of any thermal or mechanical damage. Pulsed laser in the nanosecond range (Q-switch technology) and pico/femtosecond lasers (mode lock technology) can both generate microplasma but through different processes. Nanosecond pulses generate free electrons through a thermionic emission, that is, thermal ionization. For pico/femto pulses, free electrons are released through a multiphoton absorption given by the high electric field induced by the intense laser pulse. The term multiphoton ionization stems from the fact that the energy necessary for ionization is provided by a simultaneous (coherent) absorption of several photons. This is achievable only for pico/femtosecond exciting light pulses since photons are compressed in a short period of time. Plasma energies and the temperature produced are higher for nanosecond than for pico/femto pulses since the threshold energy needed for plasma formation is higher for the former than for the latter. Thus, higher energy will produce side effects apart from ionization (Hutson and Ma, 2007). How does plasma formation take place exactly? A free electron released by the process of thermionic emission or multiphoton ionization absorbs more photons and accelerates colliding with another atom ionizing it. Now, two free electrons with low kinetic energy absorb more photons and accelerate once again. The process is repeated in loop giving rise to the so cold electron avalanche growth. The basic process of photon absorption and electron acceleration is named inverse Bremsstrahlung.[3] Thus, a free–free electron process takes place, in which a free electron is present before and after photon absorption. One important feature of plasma-induced ablation is that it can be performed on samples that are not pigmented,

[2] Optical breakdown refers to the absorption process of UV, visible, or IR light by the steamed plasma.

[3] An example of Bremsstrahlung process is fluorescence in which an electron is accelerated within an atom and finally releases its energy in the form of a photon.

thus not particularly absorbent (e.g., transparent sample like embryonic tissues). This is possible because optical breakdown takes place, a process by which energy is absorbed by the plasma itself. In this way, the plasma behaves like a trap for succeeding laser photons thus constituting a shield for the underlying tissue. Such process is also known as plasma shielding.

In general, tissues tends to absorb more UV light than visible or IR light. But if we now take into consideration the absorption coefficient of the tissue α and the absorption coefficient of the plasma α_{pl} the former tends to be inferior to the latter by few orders of magnitude; so we can write: $\alpha_{\mathrm{pl}} \gg \alpha$. Thus, if a plasma is steamed, an enhanced absorption and efficient ablation occur also for visible and IR radiations. α_{pl} is defined as follows:

$$\alpha_{\mathrm{pl}} = \frac{\nu_{\mathrm{ei}}}{nc}\frac{\omega_{\mathrm{pl}}^2}{\omega^2},$$

where ν_{ei} is the mean collision rate of free electrons and ions, n the index of refraction, c the speed of light, ω_{pl} the plasma frequency which is proportional to the density of free electrons, and ω the frequency of radiation. This equation shows that in case of plasma formation the absorption is even more enhanced for radiation in the IR region of the spectrum since:

$$\lambda_{\mathrm{IR}} > \lambda_{\mathrm{UV}} \rightarrow \omega_{\mathrm{IR}} < \omega_{\mathrm{UV}} \rightarrow \alpha_{\mathrm{pl}}^{(\mathrm{IR})} > \alpha_{\mathrm{pl}}^{(\mathrm{UV})}.$$

As free-electron density increases massively, during plasma growth, photon scattering is enhanced giving rise to a quench of the electron avalanche process. The threshold electron density over which such process gets critical and further energy is no more converted into plasma is obtained when the plasma frequency becomes equal to the frequency of the incident electric wave, that is, $\omega_{\mathrm{pl}} = \omega$.

For IR radiation, the electron density threshold is thus lower meaning that a smaller plasma will be generated compared to UV for instance.

Optical breakdown is the process by which a generated plasma starts absorbing incoming light and eventually starts growing generating mechanical shock waves. In fluids, shock waves can result in cavitation bubbles formation. When higher pulse energies are used, these mechanical effects are predominant. Tissue perturbation becomes mechanical and not simply plasma induced. Mechanical forces tend to propagate to adjacent zones while plasma-induced ablation is confined to the region of optical breakdown. In the case of nanosecond pulses, the mechanical damage can span over millimeters from the central core of the plasma. In general, purely plasma-induced ablation is never observed in the nanosecond pulse range since the threshold energy for plasma formation is too high: in this specific case even by using energies at the very threshold, plasma-induced ablation is associated with strong mechanical perturbation (photodisruption). In contrast pico- and femtosecond pulses allow exposing the target to very high peak power intensities but for a very short time period: this leads to a very low transfer of energy (of the order of nJ for femtosecond pulses). In this later case, breakdown is still achieved producing low energy plasma and no disruptive effects. In photo disruption, three processes take place one after the other: first plasma formation, followed by a shock wave and finally a cavitation bubble. In the regime of photodisruption, the energy of the plasma generated is two or more orders of magnitude higher than the case of plasma-induced ablation. As a consequence, three main effects take place: plasma shielding, photon scattering, and multiple plasma generation (a cascade of plasmas going from the focal spot toward the direction of the laser source). Optical breakdown induces a high density of free electrons with high kinetic energy. This is associated with the high rise of plasma temperature. Accelerated free electrons tend to diffuse in the surrounding medium. When the inert ions follow, after a time delay, mass is displaced giving rise to shock wave. Cavitation bubbles are then generated inside soft matter. Because of the high plasma temperature, water is transformed into vapor. The bubble expands producing work against the surrounding medium: kinetic energy is stored in the expanded cavitation bubble under the form of potential energy. The bubble finally can implode as a result of the outer pressure. The bubble content (typically water vapor and carbon oxides) is strongly compressed, thus pressure and temperature inside the bubble rise again resulting in cavitation bubble rebounds. The same process can continue in loops until all energy is dissipated and all gases are dissolved in the surrounding medium.

Different laser pulse repetition rates can generate different types of ablations. At a kilohertz regime, pulses are separated one from the other at time periods of the order of milliseconds. The time gap is usually considered large enough to exclude any type of cumulative processes (Schaffer *et al.*, 2001). In the case of megahertz range, pulses are usually separated by hundreds or even tens of nanoseconds. The proximity between pulses can lead to cumulative processes that can in turn give rise to, for example, the increase in temperature of the sample (Schaffer *et al.*, 2003). Materials such as fused silica have a typical heat diffusion time of the order of 1 μs on a distance of 1 μm. In such specific case, kilohertz pulses will act independently to generate disruption (Fig. 4.9A), while megahertz pulses will produce a cumulative damage due to a rapid energy deposition (Fig. 4.9B). For biological samples, this is still not clear. Previous works concluded that damage at megahertz frequencies in biological samples is thermally induced (Vogel and Venugopalan, 2003), but more recent theoretical works suggest that temperature rise would be of just a few degrees Celsius (Vogel *et al.*, 2005). This would mean that megahertz ablation would be mainly a consequence of free-electron-induced bond breaking

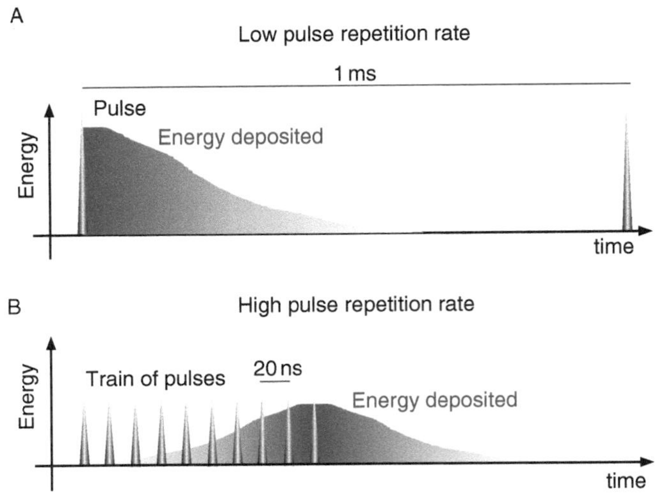

Figure 4.9 Energy deposition (red) for (A) low (KHz) and (B) high (MHz) pulse repetition rates lasers. (For interpretation of the references to color in this figure legend, the reader is referred to the Web version of this chapter.)

(Vogel and Venugopalan, 2003). Differently, in the kilohertz range, damage is created by microsecond explosions that are thermoelastically generated (Glezer and Mazur, 1997; Vogel *et al.*, 2005). In summary, megahertz ablations deposit lower subthreshold pulse energies compared to kilohertz and have higher spatial resolution due to higher precision in the ablation mechanism (free-electron-induced bond breaking compared to microsecond explosions). At megahertz frequencies, heat could be cause of more extended damage and this should be experimentally investigated. Surgeries in the kilohertz range deliver higher peak power intensities depositing lower energies in total. Damages in the kilohertz frequencies are unlikely to extend much beyond the focal volume because pulses are greatly separated one from the other preventing energy from accumulating. Some works suggest that megahertz repetition rates would be better suited for performing ablation in nonliving biomaterials or fixed samples, while kilohertz ablations for living samples (Chung and Mazur, 2009). Such suggestion is still not experimentally confirmed and is still a matter of discussion.

5.2. Comparison between ultra-violet (UV) and near-infrared (NIR) laser ablation

The wavelength used to perform laser ablation is a critical parameter since photons, coming from a UV and NIR light source for instance, do not carry the same energy. UV photons carry at least five times more energy than NIR photons. At first glance, one would thus suppose that UV laser sources would be more suitable for laser surgery. This is not true in general and in the following, some general principles of light absorption are presented to explain why.

The absorption coefficient α of a material is proportional to the beam intensity I (Fig. 4.10A) with k the number of photons necessary to produce free electrons (i.e., to cover the energy gap between bound and ionized state or, in simple words, to ablate Fig. 4.10B). If only one photon is necessary to overcome the energy gap, absorption is linear. If k photons (with $k > 1$) are required to overcome the gap, k photons must be simultaneously absorbed at the same location. Such event has a nonlinear probability, which is proportional to I^5. Let us take as example water ionization. With a UV laser, one single UV photon is sufficient to ionize water: the absorption coefficient is proportional to I. With an NIR 800 nm wavelength femtosecond pulse laser, ionization is achieved with $k = 5$. The absorption of

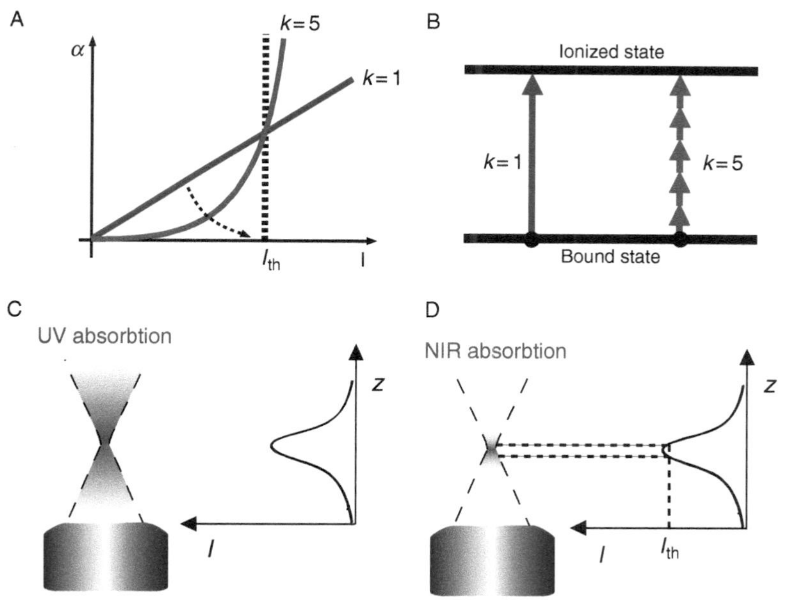

Figure 4.10 Light can be absorbed by a material to produce free electrons. (A) Graph showing absorption in function of the laser intensity I with k numbers of photons absorbed to bridge the energy gap between ionized and bound state. A threshold regime is achieved when the number of absorbed photon necessary to bridge the energy gap increases. (B) Diagram showing the number of absorbed photons k at different wavelengths (UV in blue and NIR in red) necessary to produce molecular bond breaking or matter ionization. Confinement of absorption and damage in linear (C) and nonlinear (C) regimes. Blue and red shadowing shows the volume of damage for UV and NIR laser sources, respectively. (For interpretation of the references to color in this figure legend, the reader is referred to the Web version of this chapter.)

water α will be thus proportional to I^5. As the order of absorption increases, the system becomes threshold-dependent (Fig. 4.10A): no or little absorption is produced for laser intensities below a threshold I_{th}. Absorption becomes efficient only for $I > I_{th}$. When injecting a parallel laser beam into the back aperture of an objective, the beam is focused at the objective focal plane. In the case of linear absorption, there is no intensity threshold: absorption and damage encompasses all the volume exposed to the laser beam starting from regions of the target closer to the light source (Fig. 4.10C). In the case of nonlinear absorption, absorption and damage is produced only at the very focal point of the objective, where the cross-section of the beam is minimum and the laser intensity is thus maximum (Fig. 4.10D). In order to reduce damage to the focal volume, the linear absorption of the sample at the laser wavelength must be zero since linear absorption typically dominates the nonlinear one. UV light is generally linearly absorbed in transparent samples that is why it is usually used for photoablation and surface etching. By using the same UV power needed for photoablation, NIR lasers (having lower energy photons) with pulse duration in the nanosecond range, for instance, could only produce thermal effects. NIR nanosecond pulses can produce photoablation only at higher power producing irreversible damages over great volumes, because NIR photons are not sufficiently compacted. Picosecond and more efficiently femtosecond pulsed NIR lasers can produce very fine ablations in the bulk by strongly compressing photons in time: in this way, the probability of photons being simultaneously absorbed is strongly increased. In these conditions, small amounts of energy (of the order of some nJ) are sufficient to perform clean ablations with high resolution in the bulk with no collateral damages. Ablation in this regime is plasma-induced.

The basic condition to obtain plasma-induced ablations with NIR short pulse lasers is to reach very high intensities. A simple way to satisfy this condition is also to carefully choose the laser focusing parameters. Enhancing the focusing conditions (thus by using objectives with high numerical apertures NA > 1) is a simple and efficient way to reach the plasma intensity threshold: this allows exposing the sample to even smaller amounts of energy and in general performing highly spatially resolved ablations. The objective should have also a good transmission at the wavelength used and chromatic aberrations corrected.

In this first part of Section 5, we have presented a very general and broad overview on laser–tissue interaction. This is not sufficient to define in detail what a laser beam will produce interacting with a very complex system as for instance the cell cytoskeleton within a living embryo. It is thus necessary to characterize this for each specific case. In the following, several examples of laser ablation characterization are presented mainly in reference to a study by Rauzi *et al.* on cell mechanics during germband extension of the developing *Drosophila* embryo.

5.3. Characterization of NIR femtosecond pulsed laser ablation on subcellular actomyosin networks in developing embryo

Laser ablation can be a very powerful tool, yet, understanding what ablation does to a tissue or to a cell is not trivial. Interpretation can be quite straightforward when ablating thick actin bundles in single flat cells, using ablations at high resolution (e.g., plasma-induced ablation with IR femtosecond pulsed laser Kumar *et al.*, 2006). Interpretation can be more complex when performing ablation on living embryos and even more so when targeting large protein complexes (e.g., the protein complexes building up adherens junctions) (Cavey *et al.*, 2008b; Farhadifar *et al.*, 2007; Fernandez-Gonzalez *et al.*, 2009; Landsberg *et al.*, 2009; Rauzi *et al.*, 2008). To better characterize laser ablation damage is important to perform further experiments. In the following, some published and unpublished examples are presented on the developing *Drosophila* embryo. An NIR femtosecond pulsed laser with megahertz repetition rate is used in all following examples. The mean power used is around 250 mW after the objective and the exposure time is between 3 and 1.5 ms.

The plasma formed can have a broad light spectrum: at the end of electron avalanche growth electrons recombine to atoms at different electronic states thus releasing the acquired energy under the form of light within a large range of wavelengths. To prove the real nature of ablation, a spectrometer coupled to the microscope can be used to detect possible light coming from the target at the time of ablation. Here, we show an example in which a *Drosophila* embryo, having the actomyosin cytoskeleton tagged with GFP, was taken as sample. By using low IR laser power, the spectrum detected has the outline of an emitting GFP (Fig. 4.11A): a two-photon absorption process thus takes place. The intensity detected scales linearly with the square of the incident beam power (Fig. 4.11B) obeying to the two-photon absorption equation:

$$I_{\mathrm{em}} \propto \sqrt{P_{\mathrm{ex}}},$$

where I_{em} is the emission intensity and P_{ex} the laser exciting power. When reaching laser exciting power around 250 mW at the level of the target, the emitted intensity is strongly increased (Fig. 4.11B) and the spectrum detected does not match any longer the outline of the GFP spectrum (Fig. 4.11A). This threshold process can be interpreted as the signature of a forming plasma.

A challenging experiment is measuring the size of the ablation spot. This depends not only on the properties of the laser and the objective used but also on the target. The target, in the case of a study by Rauzi *et al.* for example, is a dense actomyosin network localized on the apical lateral cortex of cells at the level of cell junctions during stage 7 of the developing *Drosophila* embryo. Such cortical network is rather thin along the *XY* plane

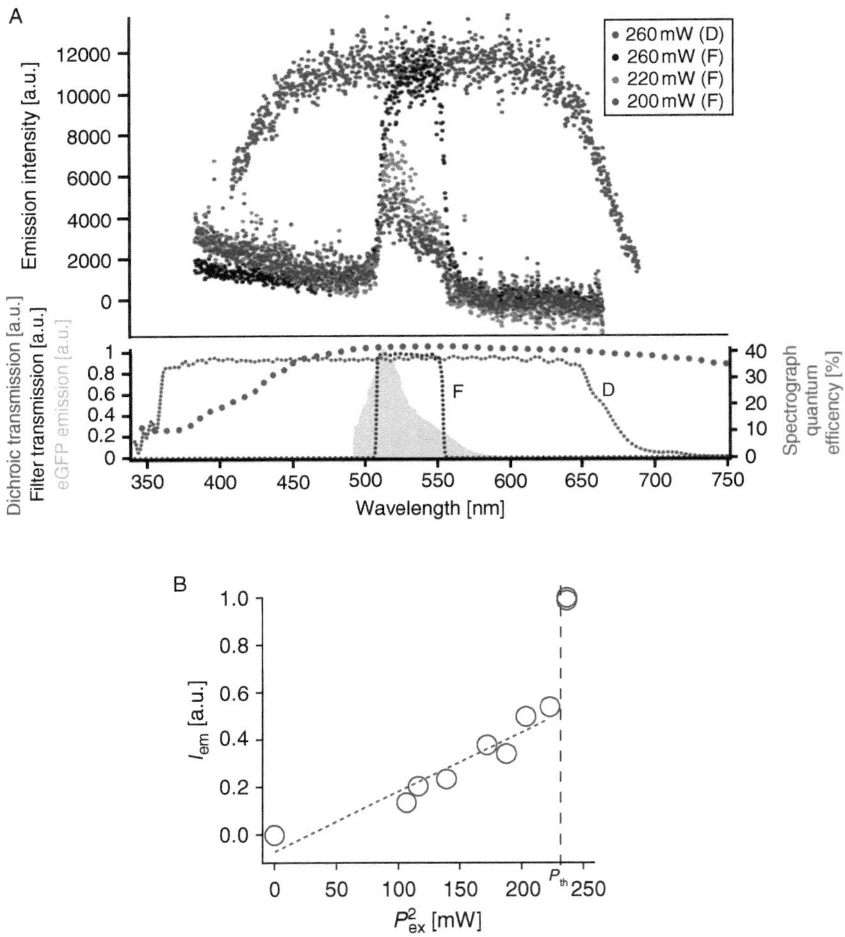

Figure 4.11 (A) Top graph: spectrograph showing light emitted by the ablated cell actomyosin cortex during cell intercalation in the developing *Drosophila* embryo (stage 7). Purple, green, black, and red are scatter plot of the emission curve for exciting power of 200, 220, 260, and 260 mW, respectively. The black and red scatter plots differ in the filters used to acquire photons coming from the sample: the black scatter plot is obtained using a band-pass filter (F) while the red scatter plot is obtained by using a dichroic (D). Bottom graph: spectra of the dichroic and band-pass filter used to detect photon coming from the ablated sample. Also the spectrum of eGFP and the quantum efficiency of the spectrograph are shown. (B) Graph showing the emitted intensity light I_{em} from the target detected by the spectrograph during ablation in function of the square exciting laser power used P_{ex}^2. I_{em} scales linearly with P_{ex}^2 until a power threshold I_{th} of ~240 mW is reached. (See Color Insert.)

(<250 nm) making it a nonideal structure to measure the dimensions of a punctual ablation that has theoretically a similar or even greater size. 40 min after stage 7, cytokinesis takes place. During cytokinesis, a ring of actomyosin (known as cytokinetic ring) divides daughter cells. This ring is 3 μm wide in average making it a similar and also ideal target to measure the wound dimensions. By performing five to six punctual ablations side by side, the ring is sufficiently cut to open up tearing apart the remaining actomyosin filaments (Fig. 4.12A). This experiment gives a first rough estimation of the cut: less than 500 nm. By performing a single ablation on the ring, the puncture can be measured: Fig. 4.12B and B‴ shows the width of the puncture being 800 nm along one direction and 400 nm in a perpendicular direction. The fact that the wound is not isotropic is explained by the fact that the ring is contractile leading to an immediate stretch of the perforation in directions dependent on the ring orientation. The energy deposited by the laser decreases spatially from the central focus of the objective and since the actomyosin is marked with a GFP protein, we can thus assume that a portion around the ablated region is just photobleached. We can thus confirm the wound width being <400 nm in diameter. This experiment still does not reveal the depth of the ablation. Theoretical analysis predict that such type of ablation with an IR laser is <1 μm in depth (the optical resolution of the objective is lower on the *Z* axis than in the *XY*).

Following protein redistribution after ablation can be very informative: laser ablation effects can be more precisely characterized and protein properties and protein–protein interactions revealed. An example is presented in Fig. 4.13. After a punctual ablation in the center of a cell–cell junction, actin retracts from the wounded site toward the vertices (points where three junctions meet). The junction is finally depleted from actin (Fig. 4.13A). The same happens for E-cad (Fig. 4.13B). Surprisingly, ablation seems not to affect other proteins not associated to the adherens junction complex, for example, the membrane protein VSVG (Fig. 4.13C). Altogether, these experiments show how the adherens junctions disassemble, revealing a tethering mechanism linking E-cad proteins to actin filaments at the cortex (Cavey *et al.*, 2008a).

When analyzing the contribution of contractile fibers and adhesion proteins to cell shape, the cell envelope (the membrane) must be preserved. If the cell is under pressure, a strong perforation of the membrane can lead to abundant cytoplasmic leakages. The cytoplasmic leakage produces cell deformation and a misplacement of all sorts of proteins inside the cell. In these conditions, it is difficult if not impossible to analyze the contribution of a single-specific subcellular component to cell shape since the whole system is affected. In living embryos, a strong perforation usually causes cells to "fall" from the tissue within few minutes (Fig. 4.14A). Rauzi *et al.* put much effort to define the response of the plasma membrane to ablation. Membrane is not only important to preserve cell integrity but also membrane itself could bear tensile loads. If this is the case, for each ablation, it is

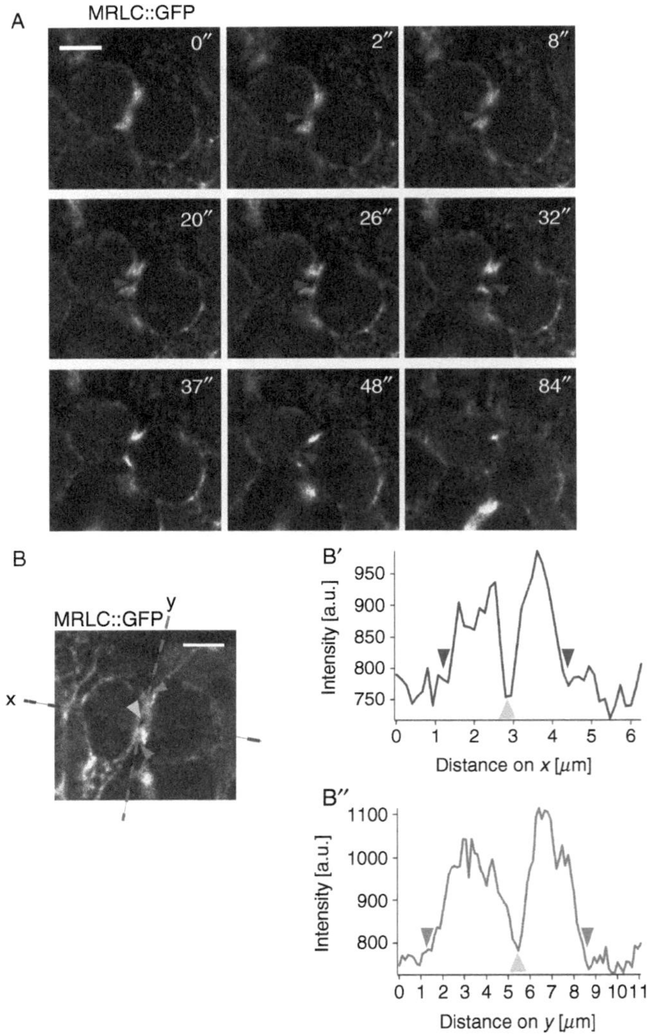

Figure 4.12 Sequence of confocal images showing Myosin II distribution in a dividing cell of the elongating epithelium of the *Drosophila* embryo. Six ablations (red arrowheads) are performed one next to the other to "cut" completely the actomyosin cytokinetic ring. Scale bar: 5 μm. (B) A confocal image showing Myosin II distribution during cell division: a punctual ablation is performed in the center of the mitotic actomyosin network (yellow arrowhead). The punctual ablation produces a stretched hole 400 nm wide (B′) (relative to the red axis *x* shown in B) and 800 nm long (B") (relative to the green axis *y* shown in B). (For interpretation of the references to color in this figure legend, the reader is referred to the Web version of this chapter.)

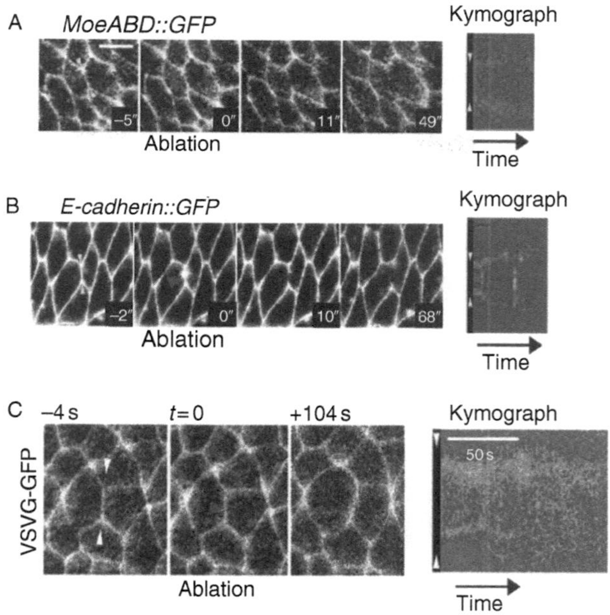

Figure 4.13 A punctual ablation in the center of a cell junction depletes actin (A) and E-cadherin (B) without affecting other proteins not associated to the adherens junction complex (C). Kymographs show the distribution of each protein along the ablated junction in function of time. A and B are adapted from Rauzi *et al.* (2008) and C from Cavey *et al.* (2008a).

also essential to define the extent of membrane perforation. The authors devised a protocol to detect cytoplasmic leakages as a consequence of membrane perforation. They injected a caged-fluorescent probe in the embryo during cellularization (stage at which cells are still not completely surrounded by the membrane). This probe diffuses inside the embryo in cells. At the end of cellularization, each cell is finally loaded with a reservoir of such a probe. After ablating a subcellular target, possible leakages are immediately detected by uncaging the probe (Fig. 4.14B and C). The probe is a very tiny molecule with a large diffusion coefficient (200 $\mu m^2/s$); even very small holes can be immediately detected. Further controls were done by doing the uncaging before ablation. When no leakages were detected, the membrane was considered intact.

5.4. Ratio measurements of cortical forces

Ratios between cortical forces at different subcellular locations for instance can be measured using laser ablation. If we consider a system in a quasi-equilibrium state in time scales of few seconds or more, a local ablation in a

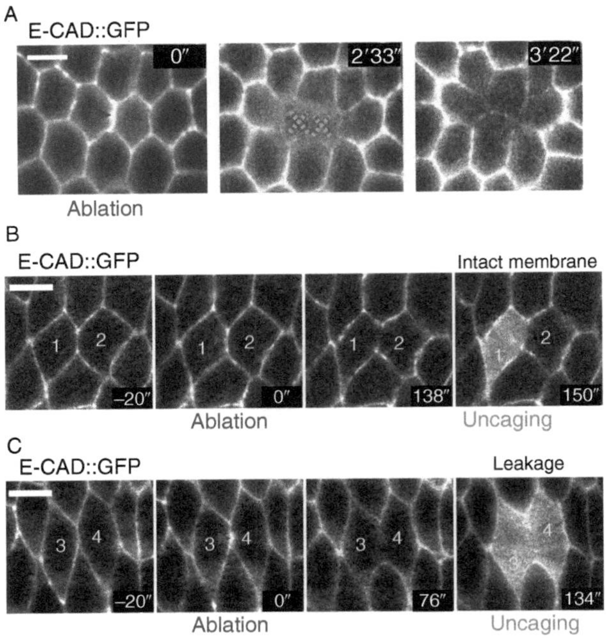

Figure 4.14 (A) After exposing a cell–cell junctions to the NIR laser source for 30 ms (red arrowhead)—corresponding to 10 times more the usual exposure time—cells are damaged and delaminate from the embryonic epithelial tissue. Scale bar: 5 μm. (B and C) All cells are loaded with a caged probe used to detect possible holes in the cell membrane after performing laser ablation. (B) After ablation (red arrowhead), no leakages are detected between cell 1 and 2 since the caged probe, uncaged (green arrowhead) in cell 1, does not diffuse out from the cell. (C) Using higher laser power or longer laser exposure time is possible to form holes in the cell membrane. After laser ablation, the caged molecule is uncaged (green arrowhead) in cell 3 and it leaks in the neighboring cell 4: this experiment shows that the ablation has damaged the contacting membrane of both cells 3 and 4. (For interpretation of the references to color in this figure legend, the reader is referred to the Web version of this chapter.)

time window of few milliseconds drives the system out of equilibrium. This produces a change in subcellular geometries (cell vertex displacement for instance) and/or the displacement of subcellular structures. By measuring speed of displacement it is possible to infer the ratio of forces: in simple words greater will be the speed of displacement greater will be the acting force. The absolute value of cortical forces cannot be reliably estimated because the value of the drag coefficient is unknown.

We take now as example the ratio measurement of cortical forces acting at cell junctions in a tissue. If we consider a vertex from which three junctions (L1, L2, L3) radiate, we can model the system considering three forces (F1, F2, F3) acting on each of the three junctions' direction reflecting the contribution of the different contractile units (C1, C2, C3) acting at the junctional cortex (Fig. 4.15A). The resultant of these forces can be approximate to a null value reflecting the state of quasi-equilibrium of the system. If C1 is ablated, F1 is considered null (Fig. 4.15A). Thus, the system is locally and transiently driven out of equilibrium and the resultant of forces can be written as:

$$\vec{F}_{\text{tot}} = \vec{F}_2 + \vec{F}_3 + \mu \vec{v},$$

where μ is the drag coefficient (considered constant for simplicity) and v is the speed of vertex displacement. Shortly, after ablation a maximum speed displacement is reached v_{max} (Fig. 4.15B) which is proportional to remaining forces just after ablation:

$$v\text{max} = \frac{\left|\vec{F}_2 + \vec{F}_3\right|(t = 0^+)}{\mu}.$$

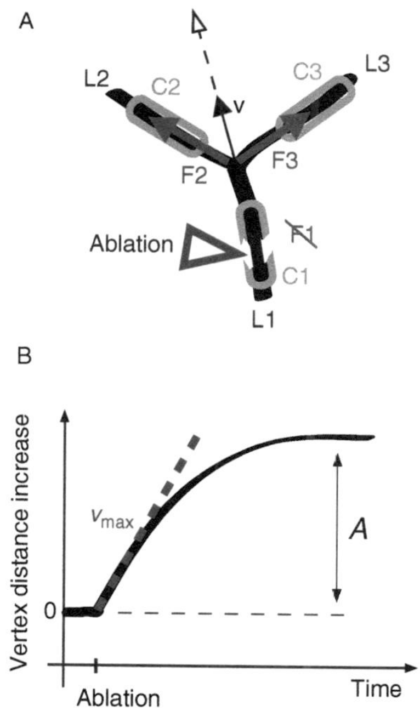

Figure 4.15 (A) Cartoon representing a three-way cell vertex. A contractile element (C, green) produces a cortical force along the junction (F, red). After ablating C1, F1 is null and the vertex moves at a certain speed v. (B) Graph representation showing the vertex distance increases after ablation in function of time. Distance increase reaches a maximal speed (v_{max}) shortly after ablation. At longer time scales, the vertex reaches an asymptotic distance (A). (For interpretation of the references to color in this figure legend, the reader is referred to the Web version of this chapter.)

If now we assume that $\left(\overrightarrow{F_2} + \overrightarrow{F_3}\right)(t = 0^-) = \overrightarrow{F_1}(t = 0^-)$, $v_{\max}$ reflects the tension at the targeted junction before ablation (Fernandez-Gonzalez *et al.*, 2009; Landsberg *et al.*, 2009; Rauzi *et al.*, 2008). Other studies model the contractile unit as a spring and a dashpot in parallel (Voigt model). Based on this model, the asymptotic distance retracted by the vertex after ablation (Fig. 4.15B) is proportional to the force acting on the ablated junction. Based on the same model, a characteristic time also can be measured reflecting the elastic and viscous properties of the system. This type of approach is better suited on living tissues showing long time scale dynamics compared to the time needed to reach asymptotic relaxations (e.g., imaginal wing disk of *Drosophila*) (Farhadifar *et al.*, 2007; Landsberg *et al.*, 2009).

5.5. Final methodological considerations

Laser ablation is an extraordinary tool, spatially and temporally specific. Many types of surgery systems can be conceived for different purposes and major analysis and controls must be done to characterize the laser-sample interaction. Some examples of such analysis and controls have been previously presented. Other examples can be found in the literature. For instance, Hutson et Ma developed a surgery system based on nano second UV laser photoablation (Hutson and Ma, 2007). By using a hydrophone needle, the authors were able to record the pressure transients associated with plasma formation, that is, shock wave mechanical propagation and cavitation bubbles collapse in the living *Drosophila* embryo. Such analysis is of primary importance especially when using nanosecond pulsed surgery because shock waves and bubbles are mechanical actors that could play a major role in tissue and certainly more in cell deformation. Many works use similar nanosecond laser ablation systems for subcellular studies without worrying to characterize the laser-sample interaction before using the tool. Laser surgery is a technique which is popular at the moment but it should be used cautiously: laser damage must be characterized accurately to avoid misleading results and misinterpretations. In addition, much work still has to be done to better characterize the "catalyzing" function of fluorescent proteins (used for *in vivo* imaging) that could play a major role in the process of laser ablation.

Other techniques which are spatially, temporally and protein specific have been conceived and are still under development. Chromophore-assisted light inactivation (CALI) is such an example (Bulina *et al.*, 2006a, b; Jay, 1988; Monier *et al.*, 2010; Rajfur *et al.*, 2002; Tour *et al.*, 2003). CALI is light mediated and it is used to selectively inactivate proteins within cells. This is achieved by tagging specific proteins with a chromophore and by irradiating the whole with high laser power (40 mW). Chromophores generate reacting oxygen species (ROS) capable to interact within 10 nm, thus denaturizing the protein to which they are associated with.

Major efforts are ongoing in order to optimize this technique by engineering more efficient photosensitizers that need shorter exposer time and lower laser power to produce ROS. The high degree of specificity of this technique makes it a very promising tool for *in vivo* studies.

6. Concluding Remarks

Cortical forces drive and support a large number of cell shape changes. We reviewed here the molecular origins of cortical forces and discuss how they build up and deploy in epithelial cells. To some extent, the steady-state distribution of cortical forces is sufficient to understand mechanics of cells in tissues and a large number of studies have been successful to predict cell shapes and cells shape changes under this simple assumption. However, the supramolecular structures underlying cortical forces, mainly the actomyosin networks, are active materials, which are able to reshape thereby changing their mechanical properties with time. These unique properties give rise to dynamic patterns of cortical forces that recent works have started to elucidate: pulsatile behavior, cortical flows, dynamic anchoring. How the mesoscopic dynamical organization of force generators and force transmitters cooperate to produce cell shape changes in tissues will be a key question for the future.

Acknowledgments

M. R. was supported by a Ph D fellowship from Région PACA and Amplitude Systems. P. -F. L. is supported by the CNRS (program ATIP), the ANR (program PCV), the Fondation Recherche Médicale, and a program grant from the HFSP.

References

Abe, K., and Takeichi, M. (2008). EPLIN mediates linkage of the cadherin catenin complex to F-actin and stabilizes the circumferential actin belt. *Proc. Natl. Acad. Sci. USA* **105,** 13–19.

Albiges-Rizo, C., Destaing, O., Fourcade, B., Planus, E., and Block, M. R. (2009). Actin machinery and mechanosensitivity in invadopodia, podosomes and focal adhesions. *J. Cell Sci.* **122,** 3037–3049.

Angres, B., Barth, A., and Nelson, W. J. (1996). Mechanism for transition from initial to stable cell-cell adhesion: Kinetic analysis of E-cadherin-mediated adhesion using a quantitative adhesion assay. *J. Cell Biol.* **134,** 549–557.

Bertet, C., Sulak, L., and Lecuit, T. (2004). Myosin-dependent junction remodelling controls planar cell intercalation and axis elongation. *Nature* **429,** 667–671.

Bertet, C., Rauzi, M., and Lecuit, T. (2009). Repression of Wasp by JAK/STAT signalling inhibits medial actomyosin network assembly and apical cell constriction in intercalating epithelial cells. *Development* **136,** 4199–4212.

Blanchard, G. B., Murugesu, S., Adams, R. J., Martinez-Arias, A., and Gorfinkiel, N. (2010). Cytoskeletal dynamics and supracellular organisation of cell shape fluctuations during dorsal closure. *Development* **137,** 2743–2752.

Blankenship, J. T., Backovic, S. T., Sanny, J. S., Weitz, O., and Zallen, J. A. (2006). Multicellular rosette formation links planar cell polarity to tissue morphogenesis. *Dev. Cell* **11,** 459–470.

Brodland, G. W., and Chen, H. H. (2000). The mechanics of heterotypic cell aggregates: Insights from computer simulations. *J. Biomech. Eng.* **122,** 402–407.

Bulina, M. E., Chudakov, D. M., Britanova, O. V., Yanushevich, Y. G., Staroverov, D. B., Chepurnykh, T. V., Merzlyak, E. M., Shkrob, M. A., Lukyanov, S., and Lukyanov, K. A. (2006a). A genetically encoded photosensitizer. *Nat. Biotechnol.* **24,** 95–99.

Bulina, M. E., Lukyanov, K. A., Britanova, O. V., Onichtchouk, D., Lukyanov, S., and Chudakov, D. M. (2006b). Chromophore-assisted light inactivation (CALI) using the phototoxic fluorescent protein KillerRed. *Nat. Protoc.* **1,** 947–953.

Buttery, S. M., Yoshida, S., and Pellman, D. (2007). Yeast formins Bni1 and Bnr1 utilize different modes of cortical interaction during the assembly of actin cables. *Mol. Biol. Cell* **18,** 1826–1838.

Cavey, M., Rauzi, M., Lenne, P. F., and Lecuit, T. (2008). A two-tiered mechanism for stabilization and immobilization of E-cadherin. *Nature* **453,** 751–756.

Chaudhuri, O., Parekh, S. H., and Fletcher, D. A. (2007). Reversible stress softening of actin networks. *Nature* **445,** 295–298.

Chesarone, M. A., DuPage, A. G., and Goode, B. L. (2010). Unleashing formins to remodel the actin and microtubule cytoskeletons. *Nat. Rev.* **11,** 62–74.

Chung, S. H., and Mazur, E. (2009). Femtosecond laser ablation of neurons in *C. elegans* for behavioral studies. *Appl. Phys. Mater. Sci. Process.* **96,** 335–341.

Clausi, D. A., and Brodland, G. W. (1993). Mechanical evaluation of theories of neurulation using computer-simulations. *Development* **118,** 1013–1023.

Conte, V., Munoz, J. J., and Miodownik, M. (2008). A 3D finite element model of ventral furrow invagination in the *Drosophila melanogaster* embryo. *J. Mech. Behav. Biomed. Mater.* **1,** 188–198.

Conte, V., Munoz, J. J., Baum, B., and Miodownik, M. (2009). Robust mechanisms of ventral furrow invagination require the combination of cellular shape changes. *Phys. Biol.* **6,** 16010.

David, D. J., Tishkina, A., and Harris, T. J. (2010). The PAR complex regulates pulsed actomyosin contractions during amnioserosa apical constriction in *Drosophila. Development* **137,** 1645–1655.

Dawes-Hoang, R. E., Parmar, K. M., Christiansen, A. E., Phelps, C. B., Brand, A. H., and Wieschaus, E. F. (2005). Folded gastrulation, cell shape change and the control of myosin localization. *Development* **132,** 4165–4178.

Dubash, A. D., Menold, M. M., Samson, T., Boulter, E., García-Mata, R., Doughman, R., and Burridge, K. (2009). Chapter 1. Focal adhesions: new angles on an old structure. *Int. Rev. Cell Mol. Biol.* **277,** 1–65.

del Rio, A., Perez-Jimenez, R., Liu, R., Roca-Cusachs, P., Fernandez, J. M., and Sheetz, M. P. (2009). Stretching single talin rod molecules activates vinculin binding. *Science* **323,** 638–641.

Desprat, N., Supatto, W., Pouille, P. A., Beaurepaire, E., and Farge, E. (2008). Tissue deformation modulates twist expression to determine anterior midgut differentiation in Drosophila embryos. *Dev. Cell* **15,** 470–477.

Drees, F., Pokutta, S., Yamada, S., Nelson, W. J., and Weis, W. I. (2005). Alpha-catenin is a molecular switch that binds E-cadherin-beta-catenin and regulates actin-filament assembly. *Cell* **123,** 903–915.

Ettensohn, C. A. (1985). Gastrulation in the sea urchin embryo is accompanied by the rearrangement of invaginating epithelial cells. *Dev. Biol.* **112,** 383–390.

Farhadifar, R., Roper, J. C., Aigouy, B., Eaton, S., and Julicher, F. (2007). The influence of cell mechanics, cell-cell interactions, and proliferation on epithelial packing. *Curr. Biol.* **17,** 2095–2104.

Fernandez-Gonzalez, R., Simoes Sde, M., Roper, J. C., Eaton, S., and Zallen, J. A. (2009). Myosin II dynamics are regulated by tension in intercalating cells. *Dev. Cell* **17,** 736–743.

Fletcher, D. A., and Mullins, R. D. (2010). Cell mechanics and the cytoskeleton. *Nature* **463,** 485–492.

Foe, V. E., and von Dassow, G. (2008). Stable and dynamic microtubules coordinately shape the myosin activation zone during cytokinetic furrow formation. *J. Cell Biol.* **183,** 457–470.

Franke, J. D., Montague, R. A., and Kiehart, D. P. (2005). Nonmuscle myosin II generates forces that transmit tension and drive contraction in multiple tissues during dorsal closure. *Curr. Biol.* **15,** 2208–2221.

Gardel, M. L., Shin, J. H., MacKintosh, F. C., Mahadevan, L., Matsudaira, P., and Weitz, D. A. (2004). Elastic behavior of cross-linked and bundled actin networks. *Science* **304,** 1301–1305.

Gardel, M. L., Nakamura, F., Hartwig, J. H., Crocker, J. C., Stossel, T. P., and Weitz, D. A. (2006). Prestressed F-actin networks cross-linked by hinged filamins replicate mechanical properties of cells. *Proc. Natl. Acad. Sci. USA* **103,** 1762–1767.

Geiger, B., Spatz, J. P., and Bershadsky, A. D. (2009). Environmental sensing through focal adhesions. *Nat. Rev. Mol. Cell Biol.* **10,** 21–33.

Gibson, M. C., Patel, A. B., Nagpal, R., and Perrimon, N. (2006). The emergence of geometric order in proliferating metazoan epithelia. *Nature* **442,** 1038–1041.

Glazier, J. A., and Graner, F. (1993). Simulation of the differential adhesion driven rearrangement of biological cells. *Phys. Rev. E Stat. Phys. Plasmas Fluids Relat. Interdiscip. Top.* **47,** 2128–2154.

Glezer, E. N., and Mazur, E. (1997). Ultrafast-laser driven micro-explosions in transparent materials. *Appl. Phys. Lett.* **71,** 882–884.

Goldstein, B., and Hird, S. N. (1996). Specification of the anteroposterior axis in *Caenorhabditis elegans*. *Development* **122,** 1467–1474.

Gorfinkiel, N., Blanchard, G. B., Adams, R. J., and Martinez Arias, A. (2009). Mechanical control of global cell behaviour during dorsal closure in Drosophila. *Development* **136,** 1889–1898.

Grill, S. W., Howard, J., Schaffer, E., Stelzer, E. H., and Hyman, A. A. (2003). The distribution of active force generators controls mitotic spindle position. *Science* **301,** 518–521.

Hardin, J., and Keller, R. (1988). The behaviour and function of bottle cells during gastrulation of *Xenopus laevis*. *Development* **103,** 211–230.

Harris, T. J., and Peifer, M. (2004). Adherens junction-dependent and -independent steps in the establishment of epithelial cell polarity in *Drosophila*. *J. Cell Biol.* **167,** 135–147.

Hayashi, T., and Carthew, R. W. (2004). Surface mechanics mediate pattern formation in the developing retina. *Nature* **431,** 647–652.

Helwani, F. M., Kovacs, E. M., Paterson, A. D., Verma, S., Ali, R. G., Fanning, A. S., Weed, S. A., and Yap, A. S. (2004). Cortactin is necessary for E-cadherin-mediated contact formation and actin reorganization. *J. Cell Biol.* **164,** 899–910.

Higashida, C., Miyoshi, T., Fujita, A., Oceguera-Yanez, F., Monypenny, J., Andou, Y., Narumiya, S., and Watanabe, N. (2004). Actin polymerization-driven molecular movement of mDia1 in living cells. *Science* **303,** 2007–2010.

Hird, S. N., and White, J. G. (1993). Cortical and cytoplasmic flow polarity in early embryonic cells of *Caenorhabditis elegans*. *J. Cell Biol.* **121,** 1343–1355.
Homem, C. C., and Peifer, M. (2009). Exploring the roles of diaphanous and enabled activity in shaping the balance between filopodia and lamellipodia. *Mol. Biol. Cell* **20,** 5138–5155.
Hotulainen, P., and Lappalainen, P. (2006). Stress fibers are generated by two distinct actin assembly mechanisms in motile cells. *J. Cell Biol.* **173,** 383–394.
Humphrey, D., Duggan, C., Saha, D., Smith, D., and Kas, J. (2002). Active fluidization of polymer networks through molecular motors. *Nature* **416,** 413–416.
Hutson, M. S., and Ma, X. Y. (2007). Plasma and cavitation dynamics during pulsed laser microsurgery in vivo. *Phys. Rev. Lett.* **99,** 158104.
Hutson, M. S., Tokutake, Y., Chang, M. S., Bloor, J. W., Venakides, S., Kiehart, D. P., and Edwards, G. S. (2003). Forces for morphogenesis investigated with laser microsurgery and quantitative modeling. *Science* **300,** 145–149.
Irvine, K. D., and Wieschaus, E. (1994). Cell intercalation during Drosophila germband extension and its regulation by pair-rule segmentation genes. *Development* **120,** 827–841.
Ivanov, A. I., McCall, I. C., Parkos, C. A., and Nusrat, A. (2004). Role for actin filament turnover and a myosin II motor in cytoskeleton-driven disassembly of the epithelial apical junctional complex. *Mol. Biol. Cell* **15,** 2639–2651.
Jacinto, A., Wood, W., Balayo, T., Turmaine, M., Martinez-Arias, A., and Martin, P. (2000). Dynamic actin-based epithelial adhesion and cell matching during *Drosophila* dorsal closure. *Curr. Biol.* **10,** 1420–1426.
Jacinto, A., Wood, W., Woolner, S., Hiley, C., Turner, L., Wilson, C., Martinez-Arias, A., and Martin, P. (2002). Dynamic analysis of actin cable function during *Drosophila* dorsal closure. *Curr. Biol.* **12,** 1245–1250.
Jacobson, A. G., and Gordon, R. (1976). Changes in the shape of the developing vertebrate nervous system analyzed experimentally, mathematically and by computer simulation. *J. Exp. Zool.* **197,** 191–246.
Jay, D. G. (1988). Selective destruction of protein function by chromophore-assisted laser inactivation. *Proc. Natl. Acad. Sci. USA* **85,** 5454–5458.
Kafer, J., Hogeweg, P., and Maree, A. F. (2006). Moving forward moving backward: Directional sorting of chemotactic cells due to size and adhesion differences. *PLoS Comput. Biol.* **2,** e56.
Kafer, J., Hayashi, T., Maree, A. F., Carthew, R. W., and Graner, F. (2007). Cell adhesion and cortex contractility determine cell patterning in the *Drosophila* retina. *Proc. Natl. Acad. Sci. USA* **104,** 18549–18554.
Kam, Z., Minden, J. S., Agard, D. A., Sedat, J. W., and Leptin, M. (1991). *Drosophila* gastrulation—analysis of cell-shape changes in living embryos by 3-dimensional fluorescence microscopy. *Development* **112,** 365–370.
Kametani, Y., and Takeichi, M. (2007). Basal-to-apical cadherin flow at cell junctions. *Nat. Cell Biol.* **9,** 92–98.
Kasza, K. E., Nakamura, F., Hu, S., Kollmannsberger, P., Bonakdar, N., Fabry, B., Stossel, T. P., Wang, N., and Weitz, D. A. (2009). Filamin A is essential for active cell stiffening but not passive stiffening under external force. *Biophys. J.* **96,** 4326–4335.
Keller, R. E. (1978). Time-lapse cinemicrographic analysis of superficial cell behavior during and prior to gastrulation in *Xenopus laevis*. *J. Morphol.* **157,** 223–247.
Keller, R., and Danilchik, M. (1988). Regional expression, pattern and timing of convergence and extension during gastrulation of *Xenopus laevis*. *Development* **103,** 193–209.
Keller, R., Cooper, M. S., Danilchik, M., Tibbetts, P., and Wilson, P. A. (1989). Cell intercalation during notochord development in *Xenopus laevis*. *J. Exp. Zool.* **251,** 134–154.

Keller, R., Davidson, L., Edlund, A., Elul, T., Ezin, M., Shook, D., and Skoglund, P. (2000). Mechanisms of convergence and extension by cell intercalation. *Philos. Trans. R. Soc. Lond. B Biol. Sci.* **355,** 897–922.

Keller, R., Davidson, L. A., and Shook, D. R. (2003). How we are shaped: The biomechanics of gastrulation. *Differentiation* **71,** 171–205.

Kiehart, D. P., Galbraith, C. G., Edwards, K. A., Rickoll, W. L., and Montague, R. A. (2000). Multiple forces contribute to cell sheet morphogenesis for dorsal closure in *Drosophila. J. Cell Biol.* **149,** 471–490.

Koenderink, G. H., Dogic, Z., Nakamura, F., Bendix, P. M., MacKintosh, F. C., Hartwig, J. H., Stossel, T. P., and Weitz, D. A. (2009). An active biopolymer network controlled by molecular motors. *Proc. Natl. Acad. Sci. USA* **106,** 15192–15197.

Kovacs, E. M., Goodwin, M., Ali, R. G., Paterson, A. D., and Yap, A. S. (2002). Cadherin-directed actin assembly: E-cadherin physically associates with the Arp2/3 complex to direct actin assembly in nascent adhesive contacts. *Curr. Biol.* **12,** 379–382.

Kumar, S., Maxwell, I. Z., Heisterkamp, A., Polte, T. R., Lele, T. P., Salanga, M., Mazur, E., and Ingber, D. E. (2006). Viscoelastic retraction of single living stress fibers and its impact on cell shape, cytoskeletal organization, and extracellular matrix mechanics. *Biophys. J.* **90,** 3762–3773.

Landsberg, K. P., Farhadifar, R., Ranft, J., Umetsu, D., Widmann, T. J., Bittig, T., Said, A., Julicher, F., and Dahmann, C. (2009). Increased cell bond tension governs cell sorting at the *Drosophila* anteroposterior compartment boundary. *Curr. Biol.* **19,** 1950–1955.

Le Goff, L., Amblard, F., and Furst, E. M. (2002). Motor-driven dynamics in actin-myosin networks. *Phys. Rev. Lett.* **88,** 018101.

Lecuit, T., and Lenne, P. F. (2007). Cell surface mechanics and the control of cell shape, tissue patterns and morphogenesis. *Nat. Rev. Mol. Cell Biol.* **8,** 633–644.

Leptin, M. (1995). *Drosophila* gastrulation: From pattern formation to morphogenesis. *Annu. Rev. Cell Dev. Biol.* **11,** 189–212.

Leptin, M. (2005). Gastrulation movements: The logic and the nuts and bolts. *Dev. Cell* **8,** 305–320.

Leptin, M., and Grunewald, B. (1990). Cell shape changes during gastrulation in *Drosophila. Development* **110,** 73–84.

Ma, X., Lynch, H. E., Scully, P. C., and Hutson, M. S. (2009). Probing embryonic tissue mechanics with laser hole drilling. *Phys. Biol.* **6,** 036004.

Major, R. J., and Irvine, K. D. (2005). Influence of Notch on dorsoventral compartmentalization and actin organization in the *Drosophila* wing. *Development* **132,** 3823–3833.

Major, R. J., and Irvine, K. D. (2006). Localization and requirement for Myosin II at the dorsal-ventral compartment boundary of the *Drosophila* wing. *Dev. Dyn.* **235,** 3051–3058.

Maree, A. F., and Hogeweg, P. (2001). How amoeboids self-organize into a fruiting body: Multicellular coordination in *Dictyostelium discoideum. Proc. Natl. Acad. Sci. USA* **98,** 3879–3883.

Martin, S. G., and Chang, F. (2006). Dynamics of the formin for3p in actin cable assembly. *Curr. Biol.* **16,** 1161–1170.

Martin, A. C., Kaschube, M., and Wieschaus, E. F. (2009). Pulsed contractions of an actin-myosin network drive apical constriction. *Nature* **457,** 495–499.

Martin, A. C., Gelbart, M., Fernandez-Gonzalez, R., Kaschube, M., and Wieschaus, E. F. (2010). Integration of contractile forces during tissue invagination. *J. Cell Biol.* **188,** 735–749.

Matzke, R., Jacobson, K., and Radmacher, M. (2001). Direct, high-resolution measurement of furrow stiffening during division of adherent cells. *Nat. Cell Biol.* **3,** 607–610.

Mayer, M., Depken, M., Bois, J. S., Julicher, F., and Grill, S. W. (2010). Anisotropies in cortical tension reveal the physical basis of polarizing cortical flows. *Nature* **467,** 617–621.

Millard, T. H., and Martin, P. (2008). Dynamic analysis of filopodial interactions during the zippering phase of *Drosophila* dorsal closure. *Development* **135,** 621–626.

Monier, B., Pelissier-Monier, A., Brand, A. H., and Sanson, B. (2010). An actomyosin-based barrier inhibits cell mixing at compartmental boundaries in *Drosophila* embryos. *Nat. Cell Biol.* **12,** 60–65.

Moore, S. W. (1994). A fiber optic system for measuring dynamic mechanical properties of embryonic tissues. *IEEE Trans. Biomed. Eng.* **41,** 45–50.

Moore, S. W., Keller, R. E., and Koehl, M. A. (1995). The dorsal involuting marginal zone stiffens anisotropically during its convergent extension in the gastrula of *Xenopus laevis*. *Development* **121,** 3131–3140.

Müller, H. A., and Wieschaus, E. (1996). Armadillo, bazooka, and stardust are critical for early stages in formation of the zonula adherens and maintenance of the polarized blastoderm epithelium in *Drosophila*. *J. Cell Biol.* **134,** 149–163.

Mullins, R. D., Heuser, J. A., and Pollard, T. D. (1998). The interaction of Arp2/3 complex with actin: Nucleation, high affinity pointed end capping, and formation of branching networks of filaments. *Proc. Natl. Acad. Sci. USA* **95,** 6181–6186.

Munro, E., Nance, J., and Priess, J. R. (2004). Cortical flows powered by asymmetrical contraction transport PAR proteins to establish and maintain anterior-posterior polarity in the early *C. elegans* embryo. *Dev. Cell* **7,** 413–424.

Niemz, M. H. (2004). Laser-Tissue Interactions: Fundamentals of Microscopy. Springer, Berlin.

Odell, G. M., Oster, G., Alberch, P., and Burnside, B. (1981). The mechanical basis of morphogenesis.I. Epithelial folding and invagination. *Dev. Biol.* **85,** 446–462.

Plateau, J. A. F. (1873). Statistique Experimentale et Theorique des Liquides Soumis aux Seules Forces Moleculaires. Gauthier-Villars, Paris.

Pollard, T. D. (2007). Regulation of actin filament assembly by Arp2/3 complex and formins. *Annu. Rev. Biophys. Biomol. Struct.* **36,** 451–477.

Pollard, T. D., and Borisy, G. G. (2003). Cellular motility driven by assembly and disassembly of actin filaments. *Cell* **112,** 453–465.

Pruyne, D., Evangelista, M., Yang, C., Bi, E., Zigmond, S., Bretscher, A., and Boone, C. (2002). Role of formins in actin assembly: Nucleation and barbed-end association. *Science* **297,** 612–615.

Rajfur, Z., Roy, P., Otey, C., Romer, L., and Jacobson, K. (2002). Dissecting the link between stress fibres and focal adhesions by CALI with EGFP fusion proteins. *Nat. Cell Biol.* **4,** 286–293.

Raucher, D., and Sheetz, M. P. (1999a). Enhanced endocytosis by local anesthetics requires cytoplasmic access and correlates with membrane tension. *Mol. Biol. Cell* **10,** 398a.

Raucher, D., and Sheetz, M. P. (1999b). Membrane expansion increases endocytosis rate during mitosis. *J. Cell Biol.* **144,** 497–506.

Rauzi, M., Verant, P., Lecuit, T., and Lenne, P. F. (2008). Nature and anisotropy of cortical forces orienting *Drosophila* tissue morphogenesis. *Nat. Cell Biol.* **10,** 1401–1410.

Rauzi, M., Lenne, P. F., and Lecuit, T. (2010). Planar polarized actomyosin contractile flows control epithelial junction remodelling. *Nature* **468,** 1110–1114.

Rose, R., Weyand, M., Lammers, M., Ishizaki, T., Ahmadian, M. R., and Wittinghofer, A. (2005). Structural and mechanistic insights into the interaction between Rho and mammalian Dia. *Nature* **435,** 513–518.

Sahai, E., and Marshall, C. J. (2002). ROCK and Dia have opposing effects on adherens junctions downstream of Rho. *Nat. Cell Biol.* **4,** 408–415.

Schaffer, C. B., Brodeur, A., Garcia, J. F., and Mazur, E. (2001). Micromachining bulk glass by use of femtosecond laser pulses with nanojoule energy. *Opt. Lett.* **26,** 93–95.

Schaffer, C. B., Garcia, J. F., and Mazur, E. (2003). Bulk heating of transparent materials using a high-repetition-rate femtosecond laser. *Appl. Phys. Mater. Sci. Process.* **76,** 351–354.

Schock, F., and Perrimon, N. (2002). Cellular processes associated with germ band retraction in Drosophila. *Dev. Biol.* **248,** 29–39.
Sheetz, M. P., and Dai, J. W. (1996). Modulation of membrane dynamics and cell motility by membrane tension. *Trends Cell Biol.* **6,** 85–89.
Sherrard, K., Robin, F., Lemaire, P., and Munro, E. (2010). Sequential activation of apical and basolateral contractility drives ascidian endoderm invagination. *Curr. Biol.* **20,** 1499–1510.
Shih, J., and Keller, R. (1992). Cell motility driving mediolateral intercalation in explants of *Xenopus laevis*. *Development* **116,** 901–914.
Skoglund, P., Rolo, A., Chen, X. J., Gumbiner, B. M., and Keller, R. (2008). Convergence and extension at gastrulation require a myosin IIB-dependent cortical actin network. *Development* **135,** 2435–2444.
Solon, J., Kaya-Copur, A., Colombelli, J., and Brunner, D. (2009). Pulsed forces timed by a ratchet-like mechanism drive directed tissue movement during dorsal closure. *Cell* **137,** 1331–1342.
Steinberg, M. S. (1963). Reconstruction of tissues by dissociated cells. Some morphogenetic tissue movements and the sorting out of embryonic cells may have a common explanation. *Science* **141,** 401–408.
Storm, C., Pastore, J. J., MacKintosh, F. C., Lubensky, T. C., and Janmey, P. A. (2005). Nonlinear elasticity in biological gels. *Nature* **435,** 191–194.
Svitkina, T. M., and Borisy, G. G. (1999). Arp2/3 complex and actin depolymerizing factor/cofilin in dendritic organization and treadmilling of actin filament array in lamellipodia. *J. Cell Biol.* **145,** 1009–1026.
Sweeton, D., Parks, S., Costa, M., and Wieschaus, E. (1991). Gastrulation in *Drosophila*: The formation of the ventral furrow and posterior midgut invaginations. *Development* **112,** 775–789.
Tepass, U., and Hartenstein, V. (1994). The development of cellular junctions in the *Drosophila* embryo. *Dev. Biol.* **161,** 563–596.
Tour, O., Meijer, R. M., Zacharias, D. A., Adams, S. R., and Tsien, R. Y. (2003). Genetically targeted chromophore-assisted light inactivation. *Nat. Biotechnol.* **21,** 1505–1508.
Toyama, Y., Peralta, X. G., Wells, A. R., Kiehart, D. P., and Edwards, G. S. (2008). Apoptotic force and tissue dynamics during *Drosophila* embryogenesis. *Science* **321,** 1683–1686.
Trepat, X., Grabulosa, M., Puig, F., Maksym, G. N., Navajas, D., and Farre, R. (2004). Viscoelasticity of human alveolar epithelial cells subjected to stretch. *Am. J. Physiol. Lung Cell. Mol. Physiol.* **287,** L1025–L1034.
Vasioukhin, V., Bauer, C., Yin, M., and Fuchs, E. (2000). Directed actin polymerization is the driving force for epithelial cell-cell adhesion. *Cell* **100,** 209–219.
Vavylonis, D., Wu, J. Q., Hao, S., O'Shaughnessy, B., and Pollard, T. D. (2008). Assembly mechanism of the contractile ring for cytokinesis by fission yeast. *Science* **319,** 97–100.
Verma, S., Shewan, A. M., Scott, J. A., Helwani, F. M., den Elzen, N. R., Miki, H., Takenawa, T., and Yap, A. S. (2004). Arp2/3 activity is necessary for efficient formation of E-cadherin adhesive contacts. *J. Biol. Chem.* **279,** 34062–34070.
Vicente-Manzanares, M., Ma, X., Adelstein, R. S., and Horwitz, A. R. (2009). Non-muscle myosin II takes centre stage in cell adhesion and migration. *Nat. Rev.* **10,** 778–790.
Vogel, A., and Venugopalan, V. (2003). Mechanisms of pulsed laser ablation of biological tissues. *Chem. Rev.* **103,** 577–644.
Vogel, A., Noack, J., Huttman, G., and Paltauf, G. (2005). Mechanisms of femtosecond laser nanosurgery of cells and tissues. *Appl. Phys. B: Lasers Opt.* **81,** 1015–1047.
Wachsstock, D. H., Schwarz, W. H., and Pollard, T. D. (1994). Cross-linker dynamics determine the mechanical properties of actin gels. *Biophys. J.* **66,** 801–809.

Wagner, B., Tharmann, R., Haase, I., Fischer, M., and Bausch, A. R. (2006). Cytoskeletal polymer networks: The molecular structure of cross-linkers determines macroscopic properties. *Proc. Natl. Acad. Sci. USA* **103,** 13974–13978.

Wang, N., Tolic-Norrelykke, I. M., Chen, J., Mijailovich, S. M., Butler, J. P., Fredberg, J. J., and Stamenovic, D. (2002). Cell prestress. I. Stiffness and prestress are closely associated in adherent contractile cells. *Am. J. Physiol. Cell Physiol.* **282,** C606–C616.

Warga, R. M., and Kimmel, C. B. (1990). Cell movements during epiboly and gastrulation in zebrafish. *Development* **108,** 569–580.

Wu, J. Q., Sirotkin, V., Kovar, D. R., Lord, M., Beltzner, C. C., Kuhn, J. R., and Pollard, T. D. (2006). Assembly of the cytokinetic contractile ring from a broad band of nodes in fission yeast. *J. Cell Biol.* **174,** 391–402.

Yamada, S., Pokutta, S., Drees, F., Weis, W. I., and Nelson, W. J. (2005). Deconstructing the cadherin-catenin-actin complex. *Cell* **123,** 889–901.

Yonemura, S., Wada, Y., Watanabe, T., Nagafuchi, A., and Shibata, M. (2010). Alpha-catenin as a tension transducer that induces adherens junction development. *Nat. Cell Biol.* **12,** 533–542.

Young, P. E., Pesacreta, T. C., and Kiehart, D. P. (1991). Dynamic changes in the distribution of cytoplasmic myosin during *Drosophila* embryogenesis. *Development* **111,** 1–14.

Zallen, J. A., and Wieschaus, E. (2004). Patterned gene expression directs bipolar planar polarity in Drosophila. *Dev. Cell* **6,** 343–355.

Zang, J. H., and Spudich, J. A. (1998). Myosin II localization during cytokinesis occurs by a mechanism that does not require its motor domain. *Proc. Natl. Acad. Sci. USA* **95,** 13652–13657.

Zhou, J., Kim, H. Y., and Davidson, L. A. (2009). Actomyosin stiffens the vertebrate embryo during crucial stages of elongation and neural tube closure. *Development* **136,** 677–688.

CHAPTER FIVE

Tension and Epithelial Morphogenesis in *Drosophila* Early Embryos

Claire M. Lye *and* Bénédicte Sanson

Contents

Department of Physiology, Development, and Neuroscience, University of Cambridge, Cambridge, United Kingdom

Current Topics in Developmental Biology, Volume 95
ISSN 0070-2153, DOI: 10.1016/B978-0-12-385065-2.00005-0

Abstract

During morphogenesis, tissues are shaped by cell behaviors such as apical cell constriction and cell intercalation, which are the result of cell intrinsic forces, but are also shaped passively by forces acting on the cells. The latter extrinsic forces can be produced either within the deforming tissue by the tissue-scale integration of intrinsic forces, or outside the tissue by other tissue movements or by fluid flows. Here we review the intrinsic and extrinsic forces that sculpt the epithelium of early *Drosophila* embryos, focusing on three conserved morphogenetic processes: tissue internalization, axis extension, and segment boundary formation. Finally, we look at how the actomyosin cytoskeleton forms force-generating structures that power these three morphogenetic events at the cell and the tissue scales.

1. Introduction

In her compelling review on epithelial morphogenesis, Fristom (1988) wrote "Tissue morphogenesis results from an interplay of intrinsic and extrinsic forces. It is important to consider, not only the structural and molecular processes that accompany morphogenesis, but the nature and origin of mechanical forces that develop within the tissue." In the context of morphogenesis, intrinsic forces refer to the forces generated within a cell that alter its shape or movement. These are distinguished from extrinsic forces produced outside the cell, which can also alter the cell's behavior. This distinction is similar to the distinction between cell-autonomous and non-cell-autonomous properties used in developmental genetics, for example during clonal analysis, to describe mutations which have effects confined to the mutated patch of cells, by contrast to those mutations which have an effect within the patch but also beyond.

In the field of developmental biology, the bulk of recent studies on morphogenesis have focused on the mechanisms underlying cell-autonomous behaviors (Lecuit and Lenne, 2007; Montell, 2008). Cell-autonomous behaviors such as polarized cell intercalation or cell apical constriction are at the heart of common morphogenetic movements such as convergence and extension (Keller, 2002) or tissue internalization (Leptin, 2005; Sawyer *et al.*, 2010), respectively. These cell-autonomous behaviors are reused in different tissues and animals, implying that mechanisms of morphogenesis are conserved throughout development and evolution (Martin and Parkhurst, 2004; Solnica-Krezel, 2005). The molecular signals and mechanisms governing these cell behaviors are being successfully deciphered, and we are starting to understand how molecular programs generate intrinsic forces that deform or move cells.

However, cells are not only actively producing morphogenetic forces, they also respond passively to extrinsic forces (Hardin and Walston, 2004; Keller *et al.*, 2008b; Zhang *et al.*, 2010). Extrinsic forces can come from the deformation of neighboring tissues (Glickman *et al.*, 2003) or from flows of fluids at the surface of tissues (Hahn and Schwartz, 2009). So, in addition to intrinsic forces, we need to understand how extrinsic forces can shape tissues. This is a much less-explored area and is one focus of this review. In particular, we need to understand how extrinsic forces impact on cell behaviors. Computer modeling suggests that extrinsic forces could cause passive cell elongation and also cell intercalation (Chen and Brodland, 2000). In support of the latter, external forces have been shown to promote cell intercalation in epithelial tubes during tracheal development (Caussinus *et al.*, 2008) and in *Xenopus* ectoderm when this tissue is artificially stretched (Beloussov *et al.*, 2000). Additionally, we need to understand how forces feedback on molecular processes inside cells. Mechano-transduction mechanisms have already been explored in specific contexts (Brierley, 2010), and it is likely that these will have a widespread role in morphogenesis (Brouzes and Farge, 2004; Fernandez-Gonzalez and Zallen, 2009; Gjorevski and Nelson, 2010a).

Our aim here is to review what is known about epithelial morphogenesis in the early *Drosophila* embryo, considering both intrinsic and extrinsic forces. We cover the morphogenetic events shaping the developing epithelium from gastrulation to parasegment formation (Developmental stages 5–11, Fig. 5.1). These early stages include three key morphogenetic processes, which are conserved during evolution: tissue internalization, convergence and extension of the main body axis and segmentation.

2. Anatomy and Cell Biology of the Early Embryo: An Overview

The anatomy of the early *Drosophila* embryo has been reviewed extensively and the reader should refer to these earlier works for an in-depth description (Costa *et al.*, 1993; Foe *et al.*, 1993; Hartenstein and Campos-Ortega, 1985, Campos-Ortega and Hartenstein, 1997; Martinez-Arias, 1993; Nussein-Volhard and Wieschaus, 1990). Equally, many aspects of the cell biology of the early embryo have been covered in excellent reviews recently and the reader should refer to these for details of the molecular mechanisms (Harris *et al.*, 2009 and references therein). Note that the morphogenesis of internal tissues in early embryogenesis such as mesoderm migration (Wilson and Leptin, 2000), tracheal tube formation (Affolter and

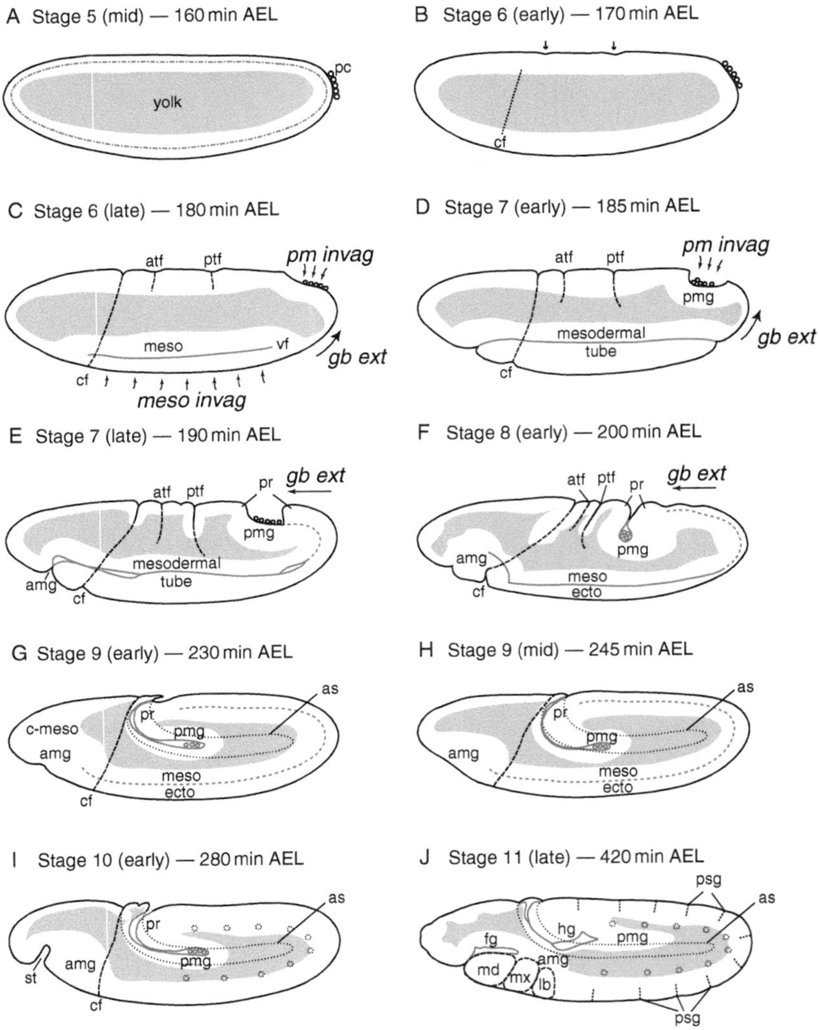

Figure 5.1 Anatomy of *Drosophila* embryos from stage 5 to 11. Camera lucida drawings of sagittal sections of fixed whole mount embryos (Campos-Ortega and Hartenstein, 1997, Chapter 2) are shown for each embryonic stage. Further anatomical details were added based on observations of fixed embryos from (Larsen *et al.*, 2008; Martinez-Arias, 1993) and our own observations of fixed and live embryos. Approximate times of development after egg laying (AEL) are shown for embryos maintained at 25 °C. Left is anterior, right posterior, top dorsal, and bottom ventral. Arrows indicate the movement of tissues unless otherwise indicated, solid circles represent pole cells. *External anatomy*: Black solid lines show the external sagittal outline of the embryo. Black dashed lines show the position of folds or grooves at the surface of the embryo; dots are used when these are subtle. Small black dots show the approximate location of the amnioserosa (as), which forms at stage 8 and is folded between the two halves of the germ-band at stages 9–11. Black dotted circles from stage 10 onward indicate the approximate position of tracheal pits. *Internal anatomy*: Gray lines show the internal structures apparent in sagittal sections. Dashed gray lines are used when the exact position of these structures is difficult to ascertain. Gray shading represents the yolk.

Abbreviations: pc, pole cells; cf, cephalic furrow; atf, anterior transverse furrow; ptf, posterior transverse furrow; pm invag, posterior midgut invagination; meso invag, mesoderm invagination; gb ext, germ-band extension; meso, mesoderm; vf, ventral furrow; pmg, posterior midgut; pr, proctodeum (note: will become hindgut); amg, anterior midgut; ecto, ectoderm; c-meso, cephalic mesoderm; as, amnioserosa; st, stomodeal invagination; fg, foregut; hg, hindgut; psg, parasegment groove; md, mandibular bud; mx, maxillary bud; lb, labial bud. (A) *Stage 5 (mid)*: Mid cellularization. The in-growing membranes (cellularization front: gray dotted/dashed line) are approximately halfway through the peripheral zone of "cortical" cytoplasm (white) and are further advanced on the ventral side. (B) *Stage 6 (early)*: Cellularization is complete ventrally. Irregularities appear in the cells on the dorsal side of the embryo (very mild indentations can be seen, arrows). The posterior ectoderm is flattening and angled toward the dorsal side and the pole cells are shifted in position compared to stage 5. The cephalic furrow is beginning to form laterally. Mesoderm invagination is also beginning along the ventral side of the embryo (but no evidence of it can yet be seen in this view of the embryo). (C) *Stage 6 (late)*: The cephalic furrow is clearly visible and extends circumferentially around the embryo. The transverse furrows have formed (at the positions of the mild indentations on the dorsal side of early stage 6 embryos). The mesoderm is almost fully invaginated, and a deep ventral furrow has formed. Invagination of the posterior midgut begins. Germ-band extension is starting and the posterior midgut invagination and pole cells shift onto the dorsal side of the embryo. (D) *Stage 7 (early)*: The transverse furrows deepen and extend laterally. Mesoderm invagination is complete resulting in the formation of a "mesodermal tube." Posterior midgut invagination continues. Germ-band extension continues and the posterior midgut and pole cells shift further onto the dorsal side of the embryo. (E) *Stage 7 (late)*: The transverse furrows deepen and extend laterally further. At the anterior of the "mesoderm tube" the anterior midgut starts to develop, and posteriorly the lumen of the mesodermal tube becomes continuous with the posterior midgut. The posterior midgut invagination deepens further and is flanked on the surface by cells of the proctodeum. Germ-band extension continues: the posterior midgut and pole cells become displaced further toward the anterior of the embryo on the dorsal side, and the cephalic furrow is displaced toward the anterior on the ventral side. (F) *Stage 8 (early)*: The transverse furrows are closer together than at previous stages. The "mesodermal tube" has collapsed and the mesodermal cells are dividing and start migrating (the demarcation between the mesodermal and ectodermal cell layers is marked by a gray solid/dotted line). The posterior midgut (containing pole cells) is internalized, and the proctodeum begins to be internalized. Germ-band extension continues: the posterior gut opening is displaced further toward the anterior of the embryo on the dorsal side, while the cephalic furrow is displaced further anteriorly on the ventral side. (G) *Stage 9 (early)*: The transverse furrows have disappeared and the amnioserosa (a squamous epithelium) has formed on the dorsal side of the embryo and extends laterally and between the ventrally and dorsally situated parts of the germ-band. Neuroblast delamination starts (not depicted). The posterior midgut has changed shape and the proctodeum has internalized further. Germ-band extension is complete and the posterior gut opening now lies immediately posterior to the dorsal side of the head. (H) *Stage 9 (mid)*: The indentation associated with the anterior midgut has disappeared, and a slight flattening can be seen on ventral side of head which marks the position of the cells of the presumptive stomodeum. (I) *Stage 10 (early)*: The stomodeum has invaginated while the cells of the anterior midgut have shifted posteriorly. The tracheal placodes appear in segments T2–A8 (approximate position indicated by dotted circles). (J) *Stage 11 (mid)*: The foregut and hindgut have formed and the anterior and posterior midguts consist of masses of cells growing toward the posterior. The parasegmental grooves have formed and some can be recognized as subtle indentations in sagittal sections. The tracheal pits have now invaginated and the salivary glands have started to form (not depicted). The segments of the gnathocephalon (mandibular bud, maxillary bud, and labial bud) have formed.

Caussinus, 2008), and salivary glands formation (Myat, 2005) are not discussed here, nor are morphogenetic events shaping the epithelium later in development, such as dorsal closure (Heisenberg, 2009) or head morphogenesis (VanHook and Letsou, 2008).

Here we summarize how the embryonic epithelium forms, and what morphogenetic events then shape it. In the next sections, we consider the intrinsic and extrinsic forces underlying these morphogenetic events. Figure 5.1 illustrates the main anatomical features in embryos from cellularization until just before germ-band retraction (i.e. from embryonic stages 5 to 11).

2.1. Formation of the cellularized embryo

The egg of *Drosophila melanogaster* is an ovoid about 500 μm long and 180 μm wide with a curved ventral side and a flat dorsal side. Two protective membranes, an external air-filled chorion and an internal vitelline membrane, containing perivitelline fluid, surround the embryo. The latter is essential for viability and needs to be kept when imaging living embryos (Nussein-Volhard and Wieschaus, 1990). The *Drosophila* embryo has a syncytial mode of development at first (Foe *et al.*, 1993). Starting from the zygotic nucleus located in the middle of the yolk cell, rapid and synchronous divisions quickly increase the number of nuclei. During the first nine divisions, the majority of nuclei migrate incrementally from their central position toward the embryo's surface. After the ninth division, bulges start to be visible at the surface of the embryo, indicating that the nuclei have reached a cortical position. During this syncytial blastoderm stage, the nuclei divide synchronously four more times. This period corresponds to the period of the mid-blastula transition, when development becomes controlled by the zygote's transcriptional program (De Renzis *et al.*, 2007).

This syncytial blastoderm stage is accompanied by the expansion of a cortical cytoplasmic layer which segregates from the yolk components (Foe *et al.*, 1993). Although there are no cells yet, recent work indicates that the syncytial blastoderm membrane above each nucleus behaves as a "cell-like" compartment: neither integral nor peripheral membrane proteins are exchanged between adjacent cell-like units during the remaining four syncytial divisions (Mavrakis *et al.*, 2009). Moreover, the membrane above each nucleus already has an epithelial-like organization with apical and basolateral-like domains. Microvilli are found apically, which will unfold during the fast phase cellularization, and reform later (Fig 5.2). The centrosomes, microtubules, and secretory apparatus are also already organized and polarized at this early stage (Harris *et al.*, 2009).

Once the 13th and last syncytial division is completed, cellularization starts (Mazumdar and Mazumdar, 2002). Membranes invaginate first slowly around each nuclei until they reach their basal side, then proceed quickly

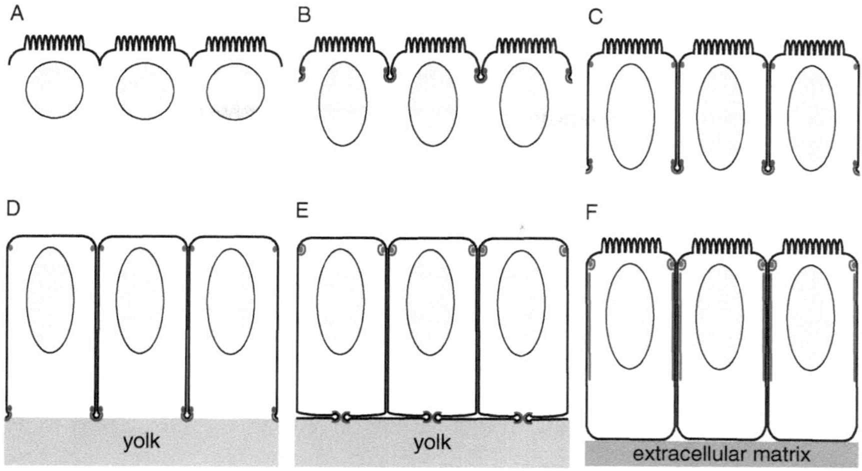

Figure 5.2 Formation of the embryonic epithelium in *Drosophila*. (A) *Syncytial blastoderm*: The plasma membrane proximal to each nucleus (yellow) is polarized with an apical domain, which forms microvilli, immediately above the nucleus and a basolateral domain situated more laterally. (B) *Cellularization begins (slow phase)*: The furrow canals form and the nuclei start elongating. The furrow canals have enrichments of Myosin II (red) and basal adherens junctions (blue dots) associated with them. (C) *Cellularization continues (slow phase ends, fast phase begins)*: During the slow phase of cellularization, nuclei elongate and plasma membranes lengthen between them as the furrow canals, and their associated Myosin II and basal adherens junctions, move basally. Apical adherens junctions (green) appear at mid-cellularization, but at this stage the majority of Myosin II remain associated with the basal adherens junctions. When the nuclei have fully elongated and the furrow canals reach the level of the bases of the elongated nuclei, the fast phase of cellularization begins. (D) *Cellularization continues (fast phase)*: During the fast phase of cellularization, the furrow canals, and their associated Myosin II and basal adherens junctions, continue to move basally. The apical microvilli are lost during this phase, and might constitute a source of membrane for the fast phase of cellularization. (E) *Cellularization ends*: Cellularization is complete when an actomyosin-driven cytokinesis-like event severs the ectodermal cells from the yolk. (F) *Mature epithelium*: Later in embryonic development, the extracellular matrix (orange) is deposited, and mature epithelial cells are attached to it on their basal side. Additionally the septate junctions (purple; equivalent to vertebrate tight junctions) form laterally. (For interpretation of the references to color in this figure legend, the reader is referred to the Web version of this chapter.)

until columnar-shaped cells are formed, about 35–40 μm in length (Fig. 5.2B–D). When membranes start growing, junctions start to form as clusters termed spot adherens junctions (Muller and Wieschaus, 1996; Tepass *et al.*, 2001). These clusters form at two distinct sites: in the apical region, but also at the basal ends of the invaginating membranes (Fig. 5.2C). At the level of these basal junctions, an accumulation of actomyosin forms a ring the width of each cell, which reduces in diameter in the last steps of cellularization (Fig. 5.2D and E). The basal junctions disappear at the end of cellularization, and the material composing it is thought to fuse with the maturing apical adherens junctions. At this stage, actomyosin becomes strongly enriched at the apical junctions (Fig. 5.2E).

At the end of stage 5, the approximately 6000 cells that compose the embryo are still open on the basal side, connecting to the yolk through a canal formed by the actomyosin ring (Fig. 5.2E). During gastrulation, contraction of the actomyosin ring severs this canal in a cytokinesis-like event which detaches the epithelial cells from the yolk. This severing occurs first in regions undergoing morphogenetic movements such as in the ventral furrow, cephalic furrow, and posterior midgut invagination (N. Lawrence and B. Sanson, unpublished; Blankenship and Wieschaus, 2001). Perhaps because of these lingering connections and importantly for experimental design, drugs, double strand RNA, and dyes can be injected in the yolk and be taken up by cells throughout cellularization and during early germ-band extension. Despite their clear epithelial nature, the blastoderm cells lack two key features of mature epithelia (Tepass *et al.*, 2001; Fig. 5.2E and F). First, the cells of the early embryo lack a basal extracellular matrix, which will be deposited at later stages. Second, the functional equivalent of vertebrate tight junctions, the septate junctions, is not formed yet: these are recognizable by electron microscopy much later, in stages 14–15 embryos (Banerjee *et al.*, 2006).

2.2. Morphogenetic movements of gastrulation

Just before gastrulation starts, the columnar blastoderm cells are virtually indistinguishable from each other, with the exception of the pole cells, sitting at the surface of the epithelial layer at the posterior end of the embryo (Fig. 5.1A and B). The first morphogenetic movement is the invagination of the mesodermal precursors (Costa *et al.*, 1993), which begins 5 min after the completion of cellularization on the ventral side of the embryo (Witzberger *et al.*, 2008; Fig. 5.1B and C). The cephalic furrow, which separates the head from the trunk tissue, forms around the same time. Invagination of mesodermal precursors through the ventral furrow leads to the formation of an mesodermal tube transiently inside the embryo (Fig. 5.1D and E; see also Fig. 5.4H). This tube collapses later and the mesodermal cells start dividing and initiate migration onto the basal surface of the ectodermal cells (Fig. 5.1F; McMahon *et al.*, 2010).

Partially overlapping with mesoderm invagination, the embryo's trunk, known as the germ-band, undergoes a substantial elongation at gastrulation (Costa *et al.*, 1993; Fig. 5.1C–G). Germ-band extension is a typical movement of convergence and extension, where the tissue elongates in one axis while narrowing in the perpendicular axis (here anteroposterior and dorsoventral, respectively; Keller, 2002). During this movement, the germ-band more than doubles in length while narrowing by a roughly equivalent factor. The germ-band elongates mainly in the posterior direction (and only moderately in the anterior direction) (Fig. 5.1D–F). At the posterior, because the extraembryonic membranes limit the embryo's volume, the germ-band elongates by folding on itself, with the end of the germ-band extending toward the head region on the dorsal side of the embryo (Fig. 5.1E–G).

Temporally overlapping with the beginning of germ-band extension, the posterior end of the embryo invaginates to form the posterior midgut invagination. The pole cells located at the posterior are transported by these tissue movements onto the dorsal side and then disappear into the pit formed by the posterior midgut invagination (Fig. 5.1C–F). The anterior midgut invagination forms in the head region (Fig. 5.1E). The anterior and posterior midgut will later join up to form the embryonic gut. While germ-band extension is progressing, deep folds, the anterior and posterior transverse furrows, form on the dorsal side of the embryo between the head region and the tip of the advancing germ-band (Fig. 5.1C–F). Around this time the dorsal tissue flattens, forming the extraembryonic amnioserosa tissue, which then folds between the two halves of the extended germ-band (Fig. 5.1G; Schock and Perrimon, 2002). While the beginning of gastrulation occurs without cell division, these resume during germ-band extension, with a complex spatiotemporal pattern which is a manifestation of the zygotic morphogenetic program (Foe *et al.*, 1993).

2.3. The beginning of metamerization

Metamerization in the epidermis is first evident in the patterns of cell division and neuroblast delamination. During the extended germ-band phase (stages 9–11), cell division is active, with each cell in the epidermis undergoing two to three rounds of divisions, depending of their positions (Martinez-Arias, 1993). Mitotic activity diminishes at stage 11, ceasing first dorsally then ventrally in the epidermis (Martinez-Arias, 1993). During the proliferative phase, the cells do not grow significantly and thus become smaller with each division. Note that the early epithelium is not competent for programmed cell death until mid-stage 11, when apoptosis occurs in a few cells in the amnioserosa, then in larger subsets of cells in the epidermis from stage 12 onward (Abrams *et al.*, 1993). Although the ectoderm does not lose significant cell numbers through apoptosis, cells delaminate heavily from the ventral ectoderm (the neuroectoderm) to form the neurons. Several waves of delamination occur successively from the end of the fast phase of germ-band extension to the end of stage 11 (Martinez-Arias, 1993).

In addition to the repeated patterns of cell division and neuron formation, regularly spaced grooves appears at the surface of the germ-band at mid-stage 10 (Fig. 5.1J). These are the parasegmental grooves that form at the interface between *wingless* and *engrailed*-expressing cells (Hatini and

DiNardo, 2001; Sanson, 2001). These grooves are transient and will disappear at germ-band retraction. Another overt sign of metamerization in the epidermis at the germ-band extended stage is the formation of the tracheal pits at stage 10 (Fig. 5.1I). These form from discs of about 10 cells in the trunk parasegments, which adopt a more basal position, with the more central cells invaginating to form a tubular structure. The tubes join up and fuse at later stages to form the tracheal system (Affolter and Caussinus, 2008). At the end of stage 11, the germ-band retracts, restoring a dorsal and a ventral side to the embryo (Schock and Perrimon, 2002, 2003). While the parasegmental grooves disappear at the anterior edge of the Engrailed stripes, the segmental folds form at their posterior edge: these will persist during the rest of the embryo and larval life. Germ-band retraction unfolds the amnioserosa, which now covers the whole of the dorsal side of the trunk. The contraction of the amnioserosa later contributes to bring the two halves of the dorsal epithelium toward each other in the morphogenetic movement of dorsal closure (Heisenberg, 2009).

3. Cell-Autonomous Behaviors and Intrinsic Forces in the Gastrulating Embryo

During *Drosophila* gastrulation, specific cell-autonomous behaviors have been identified that shape the embryo (Fig. 5.3). Much is known about the genetic pathways controlling these cell-autonomous behaviors, and we are starting to understand the dynamics of the cytoskeleton that produce them. Below is an overview of the current knowledge on these topics.

3.1. Invaginations and folds

3.1.1. Mesoderm invagination

The earliest morphogenetic movement in *Drosophila* gastrulation, mesoderm invagination (Fig. 5.1C), is controlled by two transcription factors, Twist and Snail (Leptin, 2005). The nuclear gradient of the maternal protein Dorsal triggers the expression of these two genes in a stripe of cells along the ventral side of the embryo (Fig. 5.3A). Twist and Snail are both required for fating these cells to become mesoderm, and for controlling the shape changes in these cells which lead to their movement inside the embryo (Fig. 5.4). The ventral cells first flatten their apical cell surfaces (Fig. 5.4A), and then constrict their apex (Fig. 5.4B–D), first stochastically, then in an apparently coordinated fashion. The cells elongate in the apicobasal axis (Fig. 5.4C and D) and a ventral furrow forms (Fig. 5.4E). The deepening of the furrow and then formation of an epithelial tube is accompanied by a

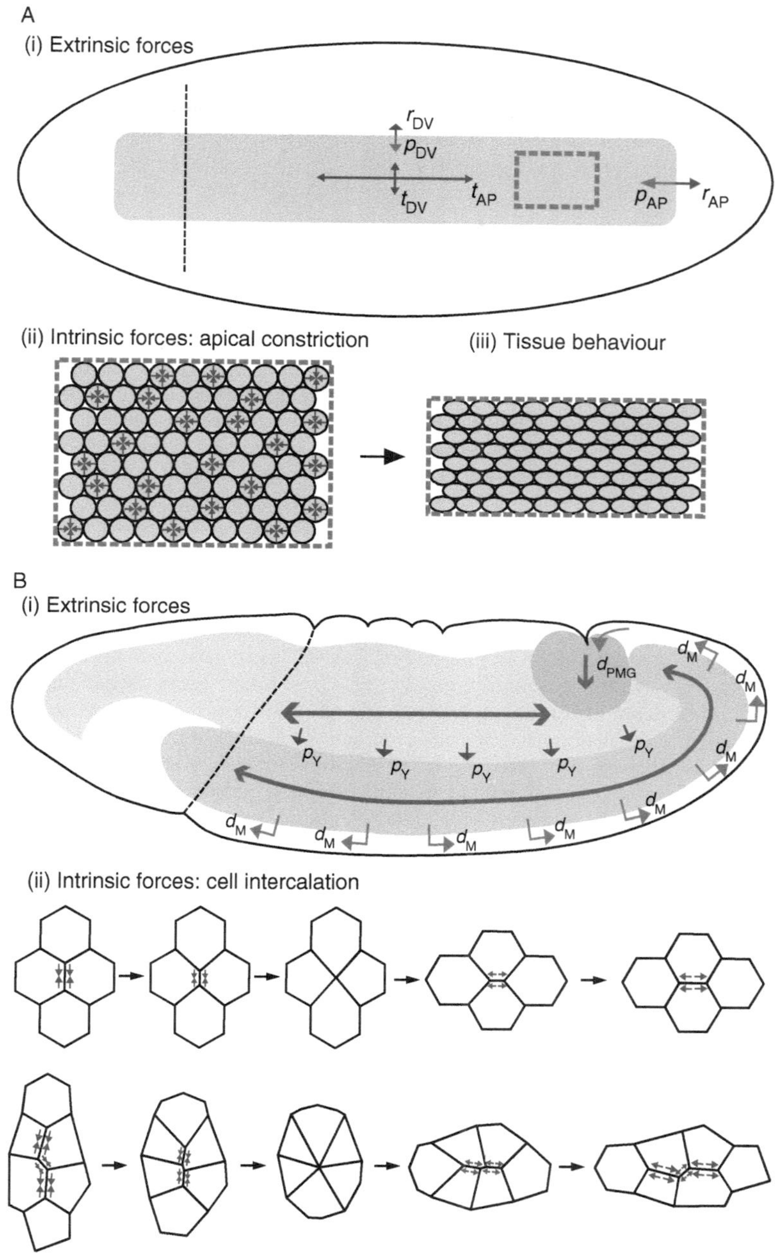

Figure 5.3 Forces acting during mesoderm invagination and germ-band extension. *Anatomy key*: Ectoderm, white; mesoderm, pink; posterior midgut, purple; yolk, gray; position of cephalic furrow, black dashed line; gray arrows (in B), main direction of movement of yolk, mesoderm, and endoderm. *Forces*: Intrinsic forces, blue arrows;

shortening of the cells (Fig. 5.4F and G). In *snail* mutants, there are neither apical constrictions nor ventral furrow. In the absence of Twist, there are still some apical contractions due to residual activity of Snail, which can lead to some ventral furrowing in *twist* mutant embryos, but these furrows are of variable depth and tend to regress after a while (Seher *et al.*, 2007). In the absence of *both* Twist and Snail, the cells do not contract their apices and there is no ventral invagination.

Twist and Snail being transcription factors and thus located in the nucleus, other proteins must translate the transcriptional information into

---

extrinsic forces acting on the mesoderm, red arrows; extrinsic forces acting on the ectoderm, green arrows. (A) Forces acting in mesoderm invagination. (i) Schematic view of the ventral side of an embryo showing the putative extrinsic forces acting early in mesoderm invagination. A tensile force acting in both anterior–posterior (t_{AP}) and medial-lateral (t_{DV}) directions may result from the integration of intrinsic contractile forces throughout the mesodermal tissue. Because of the shape of the presumptive mesoderm, there are on average a larger number of cells undergoing apical contraction in the AP axis than in the DV axis at any time, and thus $t_{AP} > t_{DV}$. Pulling forces (p_{AP} and p_{DV}) will act on the ectoderm, as the mesoderm pulls on it as it invaginates. Again the shape of the presumptive mesoderm means that $p_{AP} > p_{DV}$. Resistive forces (r_{AP} and r_{ML}) from the ectoderm will act to prevent invagination; however invagination will occur, so $p_{AP} > r_{AP}$ and $p_{DV} > r_{DV}$. Because the embryo is a long ovoid, there are a larger number of ectodermal cells in total resisting the invagination along the AP (long) axis than the DV (short) axis, and thus $r_{AP} > r_{DV}$. The boxed (gray dashed line) area is shown in more detail in (ii) and, at a later stage of mesoderm invagination, in (iii). (ii) Intrinsic forces acting during mesoderm invagination. Apical contraction (blue arrows), driven by pulsed actomyosin activity, occurs in a subset of cells at any one time. Cortical tension (not depicted) is maintained in cells between pulsed contractions allowing forces to be transmitted between cells. The integrated contractile forces manifest as a global "extrinsic" tensile force (t_{AP} and t_{DV} in (i)). (iii) Tissue behavior in the invaginating mesoderm. As a result of the intrinsic and extrinsic forces described above, mesodermal cells undergo anisotropic apical constriction, eventually resulting in the invagination of the mesodermal tissue to form an internal tube. (B) Forces acting in germ-band extension. (i) Schematic lateral cross-section of an embryo showing the putative extrinsic forces acting during germ-band extension. Mesoderm invagination causes a flow of yolk (gray), primarily in the anterior–posterior direction (double headed gray arrow) and increased pressure in the yolk (p_Y) which pushes down on the basal sides of cells (including those of the mesodermal tube, pink). The mesodermal tube may undergo convergent extension (gray arrows). This would be bidirectional, but because of the limited possible displacement of the head tissue, would be primarily in the posterior direction. The postulated mesodermal convergent extension could produce a drag force (d_M) on the overlying ectoderm. The posterior midgut (purple) may also produce a drag force (d_{PMG}) as it invaginates, which may pull the posterior end of the germ-band. (ii) Intrinsic forces during germ-band extension. Autonomous cell intercalation behaviors act to elongate the germ-band. Actomyosin-driven junction shrinkage and subsequent junction expansion in the perpendicular axis produce forces leading to convergence in the dorsoventral axis and extension in the anterior–posterior axis (blue arrows), in pair of cells (top panel) or columns of cells forming rosettes (bottom panel). (For interpretation of the references to color in this figure legend, the reader is referred to the Web version of this chapter.)

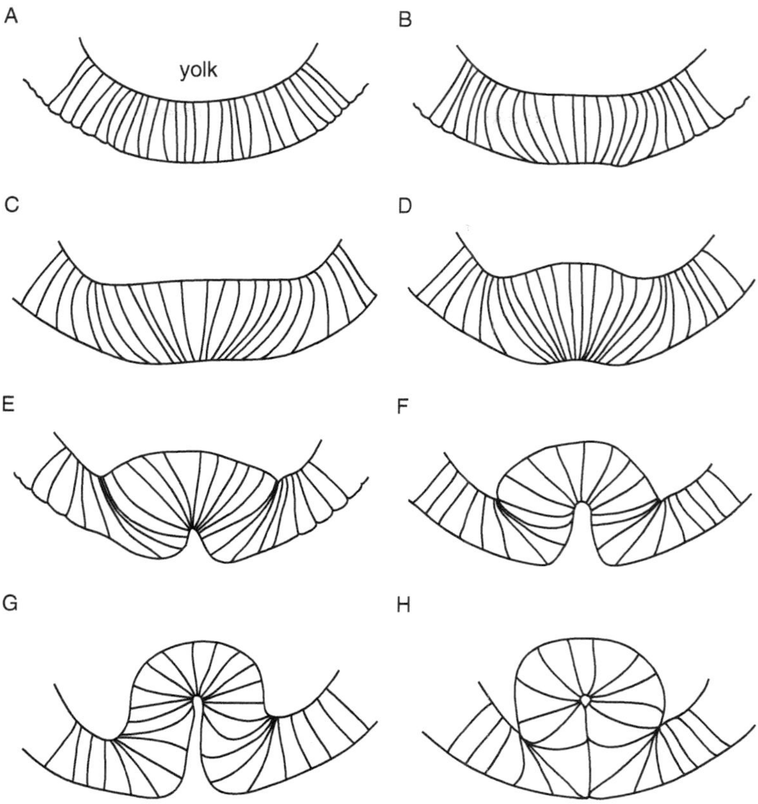

Figure 5.4 Cell shape changes during mesoderm invagination. Camera lucida drawings of scanning electron micrographs of cross-sections through *Drosophila* embryos from Sweeton *et al.* (1991). The sections show the ventral part of the embryo, with the apical side of the cells facing outside (bottom) and the basal side contacting the yolk (top). (A) The ventral-most cells (presumptive mesoderm) have flattened apically (compare the flat apices of these cells to the more rounded apices of the more lateral cells) and mild apical constriction can be noted in some cells. (B–D) Further apical constriction occurs in the most ventrally situated cells, while the slightly more lateral apically flattened cells expand their apices. The ventral cells elongate in the apicobasal axis, and a shallow groove is formed. (E and F) The apically constricted cells shorten in length as invagination occurs (and the ventral furrow forms). (G) Invagination completes resulting in the complete internalization of the mesodermal cells through the ventral furrow. (H) The ventral furrow closes leaving a tube of mesodermal cells that remains attached to the ectoderm.

the cell and tissue behaviors of invagination. The first protein discovered with this function is encoded by *folded gastrulation (fog)*, a zygotic gene which is expressed under the control of Twist in the mesodermal cells (Costa *et al.*, 1994; Sweeton *et al.*, 1991). In the absence of *fog*, there is a clear defect in the dynamics of apical constriction in the mesodermal cells (Costa *et al.*,

1994; Oda and Tsukita, 2001; Sweeton *et al.*, 1991). However, mesoderm invagination still occurs with only a few minutes delay. Fog is a putative secreted ligand for a G protein-coupled receptor, and acts through *concertina* (*cta*), a maternally contributed gene, which codes for a Galpha subunit. The putative G protein receptor has not been identified yet. *cta* mutants also show a delay in ventral furrow formation, which is a little more severe than in *fog* mutants. Although the loss of function phenotypes of *fog* and *cta* mutants are mild in term of mesoderm invagination, overexpression of *fog*, and a gain of function mutation in *cta*, both trigger ectopic apical flattening in all cells in the embryo (Morize *et al.*, 1998). This suggests that both *fog* and *cta* have a significant function in mesoderm invagination, but that these genes act redundantly with other pathways.

A gene identified as acting in a parallel pathway is *T48*, which encodes a transmembrane protein expressed in the mesodermal cells under the control of Twist (Kolsch *et al.*, 2007). Similar to *fog* or *cta*, loss of *T48* on its own has a subtle mesoderm invagination defect, whereas double mutants of *cta* and *T48* exhibit a severe mesoderm invagination defect, with loss of apical constrictions and no ventral furrowing (Kolsch *et al.*, 2007). Both Fog/Cta and T48 are thought to act through RhoGEF2, based on genetic evidence (Dawes-Hoang *et al.*, 2005; Nikolaidou and Barrett, 2004), and interaction data. The mammalian homologue of Cta and RhoGEF2 interact (Hart *et al.*, 1998), while T48 and RhoGEF2 interact in *Drosophila* (Kolsch *et al.*, 2007). RhoGEF2 is in turn required for apical constriction of mesodermal cells, and the ventral furrow does not form in *rhoGEF2* mutant embryos (Barrett *et al.*, 1997; Hacker and Perrimon, 1998).

RhoGEF2 is required for the apical localization of Myosin II in mesodermal cells (Dawes-Hoang *et al.*, 2005; Nikolaidou and Barrett, 2004). This apical localization occurs a few minutes after the completion of cellularization on the basal side of ventral cells (which also requires Myosin II; Witzberger *et al.*, 2008). RhoGEF2 activates Rho by stimulating GDP release from inactive Rho. Rho-kinase, which is thought to be activated by Rho (Somlyo and Somlyo, 2003), has also been shown to be required for apical localization of Myosin II in mesodermal cells (Dawes-Hoang *et al.*, 2005). During apical constriction, reorganization of actin at the cell apex is important (Fox and Peifer, 2007), as is the apparent disassembly then reassembly of adherens junctions from a subapical position to an apical position, which depends upon the activity of Snail (Kolsch *et al.*, 2007).

Remarkably, the mesh of actomyosin present in the apical cap of the mesodermal cells (called the medial web) has been shown recently to constrict in a pulsatile fashion, with actin and Myosin II concentrating in visible foci (Martin *et al.*, 2009). Snail is required to initiate this pulsatile behavior. The pathway downstream of Snail that induces this pulsatile behavior is not known. In wild-type, the pulsed contractions of the actomyosin network appear to pull on the apical plasma membrane via spot adherens junctions, leading to incremental apical surface reduction (Martin *et al.*, 2009, 2010). Therefore, the redistribution of the adherens junctions to these more apical spots might be important for productive apical constriction. Twist is required to stabilize the reduction in area after each contraction, in a way analogous to a ratchet mechanism. In absence of Twist, the mutant cells often fail to maintain their apical constriction, as they are pulled open by the contractions of adjacent cells.

3.1.2. Posterior midgut invagination

The posterior midgut invagination (Fig. 5.1C and D) forms from a disc of cells that contract their apices, elongate their apicobasal axis, and invaginate, with a sequence of cell shape changes very similar to those occurring at the ventral furrow. As for mesoderm invagination, apical cell constrictions require *rhoGEF2* (Barrett *et al.*, 1997). Although the presumptive posterior midgut cells do not express *twist* or *snail*, they do express *fog* (Costa *et al.*, 1994). Here, expression of *fog* is under the control of the genes *huckebein* and *tailless*, which function in the cascade controlling the terminal patterning in early embryos (St Johnston and Nusslein-Volhard, 1992). In absence of *fog* (or *cta*), the posterior midgut invagination does not form at all, in contrast to the mere delay observed for mesoderm invagination. In these mutants, the posterior end of the germ-band does not move away in front of the extending germ-band (see below). Because of this, the germ-band buckles and forms large folds which give the embryo a concertinaed appearance, hence the gene names (*concertina* and *folded gastrulation*).

3.1.3. Transient ectodermal folds

In addition to the ventral and posterior invaginations, the gastrulating embryo exhibits some stereotypical, although transient folds, which are also thought to occur as a consequence of cell-autonomous behaviors (Costa *et al.*, 1993). The most prominent and long-lived fold, the cephalic furrow, separates the head region from the trunk of the embryo (Fig. 5.1B–I). Cephalic furrow formation starts by the apicobasal shortening of a single row of cells on the lateral sides of the embryo. This cell shape change requires the activity of the gap gene *buttonhead* and the pair-rule transcription factor *even-skipped* (Vincent *et al.*, 1997). It is not clear if the subsequent deepening of the cephalic furrow, and its extension into the dorsal and ventral parts of the embryo, is caused by cell-autonomous behaviors, or by pushing from the extending germ-band. Two other folds, the anterior and posterior transverse folds, form in the trunk, in front of the advancing germ-band (Costa *et al.*, 1993; Fig. 5.1C–F). These folds initiate from the dorsal side and extend onto the lateral side. The position at which they form is reproducible between embryos. They could be caused by cell-autonomous behaviors similar to those initiating cephalic furrow formation, but at present the gene activities controlling these are unknown. The cell shape changes initiating cephalic and transverse folds do not appear to be accompanied by a constriction of the apical cell surface. Consistent with this, the activity of RhoGEF2 is not required for cephalic furrow formation (Barrett *et al.*, 1997).

3.1.4. Intrinsic forces in invagination and folding

As reviewed above, specific behaviors of cells have been identified that correlate with the invaginations and folds of gastrulation. Cephalic and transverse folds are associated with a cell shortening event without apical constriction. By contrast, for the mesoderm and posterior midgut invaginations, apical redistribution of actomyosin and subsequent constriction of the cell apex is thought to be the active mechanism generating force (Sawyer *et al.*, 2010). In the case of mesoderm invagination, computer simulations suggest that all the other behaviors, namely initial flattening of the cell surface, elongation then shortening of the mesodermal cells and nuclear movement, are passive behaviors which happen as a consequence of Myosin II-dependent apical constriction (Pouille and Farge, 2008).

3.2. Tissue elongation: Germ-band extension and amnioserosa flattening

3.2.1. Germ-band extension

Halfway through mesoderm invagination, the germ-band commences convergence and extension. Cell-autonomous behaviors are crucial for this process: cells intercalate in a polarized way in the ventro-lateral ectoderm, causing an extension in the anteroposterior axis and a convergence in the dorsoventral axis (Hartenstein and Campos-Ortega, 1985; Irvine and Wieschaus, 1994). Polarized cell intercalation requires the anteroposterior patterning system (but not the dorsoventral patterning system; Irvine and Wieschaus, 1994). Mutants in either maternal determinants such as *bicoid* and *nanos*, or gap genes such as *Kruppel* or *knirps*, or pair-rule genes such as *eve* or *runt*, all exhibit incomplete germ-band extension associated with a decrease of polarized cell intercalation (Bertet *et al.*, 2004; Irvine and Wieschaus, 1994; Zallen and Wieschaus, 2004). However, segment polarity genes, which are downstream of pair-rule genes in the segmentation cascade, and are starting to be expressed at gastrulation, do not exhibit detectable germ-band extension defects (Irvine and Wieschaus, 1994).

The extent of the germ-band extension defects in maternal and gap mutants are proportional to the size of the trunk region where AP positional information is lost. For example, in a gap gene mutant such as *knirps*, the region of the trunk where AP patterning appears normal (based on *eve* expression) undergoes polarized intercalation, whereas in the region with the patterning defect, no polarized intercalation is observed (Butler *et al.*, 2009; Irvine and Wieschaus, 1994). The pair-rule mutants have phenotypes of varying severity, with *eve* and *runt* having the strongest defects (Irvine and Wieschaus, 1994; Zallen and Wieschaus, 2004). The following evidence suggests that it is the correct expression in stripes of the pair-rule genes, rather than their expression *per se*, that is critical for germ-band extension. First, uniform expression of either *eve* or *runt* gives a stronger germ-band extension defect than the loss of function of either gene. Second, in embryos with extra copies of *bicoid*, in which the trunk is shorter due to an expansion of the head domain, the narrower stripes of pair-rule expression are associated with a faster rate of germ-band extension.

Irvine and Wieschaus (1994) postulated that pair-rule genes may control the expression in stripes of cell adhesion molecules, and that these differences in adhesion at cell–cell interfaces could promote cell intercalation. Although a role for differential adhesion has not been proven, recent work has examined cell–cell interface properties, focusing on the role of actomyosin. Strikingly, anteroposterior patterning is necessary and sufficient for the planar polarized enrichment of the apical actomyosin cytoskeleton at the cellular interfaces perpendicular to the anteroposterior axis (called "vertical" interfaces; Bertet *et al.*, 2004; Zallen and Wieschaus, 2004; Fig. 5.3B). This enrichment of actomyosin is thought to promote the remodeling of junctions at these interfaces, causing oriented cell intercalation (Bertet *et al.*, 2004; Blankenship *et al.*, 2006). Two distinct modes of Myosin II-dependent cell intercalation have been described in the ventrolateral ectoderm, one that is predominant at the beginning of GBE, and another that occurs later in extension, when the arrangement of cells has become more disordered.

The first type of cell intercalation is the so-called T1–T2–T3 transition involving four epithelial cells (Fig. 5.3B). In the germ-band, a pair of cells in T1 figures share one "vertical" cell–cell interface enriched in Myosin II, and this interface shrinks to a point-like vertex, forming a T2 figure (Bertet *et al.*, 2004). This four-cell vertex grows a new interface oriented in the perpendicular ("horizontal") direction, to form a T3 figure where cells have exchanged neighbors. The second pattern of intercalation is the formation of rosettes which can contain 5–12 cells rearranging as a group (Fig. 5.3B). Rosettes are formed from DV-oriented columns of cells sharing "vertical" interfaces enriched in Myosin II and linked together (Blankenship *et al.*, 2006; Fernandez-Gonzalez *et al.*, 2009). These linked interfaces shrink progressively to a single vertex, with all the cells in the group forming a rosette-like structure around this vertex. Then a new series of interfaces grow again in the perpendicular direction, resolving the rosette in an AP extended patch of tissue.

It has been shown that disrupting Myosin II polarization in the vertical interfaces, by either perturbing AP patterning, removing some of the maternal contribution of Myosin II or injecting the embryonic yolk with the Rho-kinase inhibitor Y27632, leads to an inhibition of polarized intercalation (Bertet *et al.*, 2004; Zallen and Wieschaus, 2004). Bertet *et al.* (2004) postulated that "A Myosin-dependent polarized contractile force might direct cell intercalation by forcing E-cadherin contact remodeling through the underlying actin–myosin cytoskeleton." Evidence for such an intrinsic force has been recently provided by nano-ablation of the cell cortex at vertical junctions enriched in Myosin II (Rauzi *et al.*, 2008). The interface's vertices pull away from the ablation site, and the speed of this "relaxation" was used to obtain a relative measure of the elastic force at a given interface prior to ablation. Vertical interfaces have a faster recoil than horizontal interfaces, indicating that the former are under more tension. This tension is dependent upon Myosin II activity since injection of Y27632 decreases it (Rauzi *et al.*, 2008). Computer simulations show that anisotropy of tension between vertical and horizontal interfaces is sufficient to elongate *in silico* models of the germ-band (Honda *et al.*, 2008; Rauzi *et al.*, 2008).

During cell intercalation, Bazooka/Par-3 (Baz/Par3) and adherens junction components such as E-Cadherin and Armadillo/Beta-catenin are depleted in the shrinking "vertical" junctions and enriched in the growing "horizontal" junctions (Blankenship *et al.*, 2006; Zallen and Wieschaus, 2004; Fig. 5.3B(ii)). Baz/Par3 is required for the establishment of adherens junctions at cellularization (Harris and Peifer, 2004; McGill *et al.*, 2009; Muller and Wieschaus, 1996), and for the correct localization and polarization of Armadillo/Beta-catenin in germ-band cells (Simoes Sde *et al.*, 2010). Thus it is likely that in the germ-band, low concentration of Baz/Par3 in the vertical junctions might be important for the destabilization of adherens junctions there, while high concentration in the horizontal junctions might conversely stabilize the new contacts. A precise mechanism for these possible roles remains to be elucidated.

While active cell intercalation in the germ-band is clearly dependent upon the planar polarized distribution of several proteins, we still know very little about the signals that set up this polarization (reviewed in Bertet and Lecuit, 2009; Zallen and Blankenship, 2008). Recently, Rho-kinase was found to be enriched at vertical junctions like Myosin II, where it excludes Baz/Par3 from the cortex by phosphorylation of its C-terminal coiled-coil domain (Simoes Sde *et al.*, 2010). In turn, Baz/Par3 is required for the planar polarization of both Myosin II and adherens junction components. However, how Rho-kinase itself becomes planar polarized in the first place remains unknown. Clearly, the striped expression of pair-rule genes is crucial to establish this planar polarity. Since the two pair-rule genes with the strongest germ-band extension defects, *eve* and *runt*, are transcription factors, downstream genes must exist that translate the transcriptional information to a change at the cortex. Intriguingly, a G protein-coupled receptor, 5-HT2, a homologue of the mammalian serotonin receptors, is expressed in a pair-rule pattern, and has been reported to disrupt germ-band extension in loss and gain of function mutants (Schaerlinger *et al.*, 2007). Genes such as this one are potential candidates to be part of a signaling pathway translating AP patterning into planar polarization.

3.2.1.1. Amnioserosa cell elongation While the germ-band is extending, another tissue elongation occurs on the dorsal side of the embryo. Here the cell-autonomous behaviors are of a completely different nature: the dorsal cells flatten and elongate dramatically to form the amnioserosa, changing their shape from columnar to squamous in about 20 min of development, while retaining the same cell neighbors (Pope and Harris, 2008; Schock and Perrimon, 2002, 2003). These cell shape changes require the transcription factor Zen, which is positioned on the dorsal side of the embryo through repression elsewhere by the Dorsal nuclear gradient. In absence of Zen, the dorsal ectoderm remains columnar. The elongation of the dorsal cells involves a 90° rotation of the microtubule cytoskeleton, from an apicobasal orientation to a planar orientation (Pope and Harris, 2008). The dorsal cells elongates primarily in the DV axis as cell junctions parallel to the DV axis increase in length more than those oriented in the perpendicular direction. Once elongated, these cells form the amnioserosa tissue which folds between the two halves of the extended germ-band (Fig. 5.1G–J).

4. Passive Cell Behaviors and Extrinsic Forces in the Gastrulating *Drosophila* Embryo

Whereas spectacular progress has been made in understanding the mechanisms underlying cell-autonomous behaviors during *Drosophila* gastrulation, especially for mesoderm invagination and germ-band extension, less attention has been given to the extrinsic forces that could influence these morphogenetic movements. In epithelial cells, the apical adhesive belt is thought to be linked to the cortical actomyosin cytoskeleton (although the molecular details of this connection are not fully elucidated; Drees *et al.*, 2005; Yamada *et al.*, 2005). Because of this linkage, one can reasonably assume that epithelial embryonic cells are mechanically coupled to each other, and thus a deformation in one place of the embryo could have an effect somewhere else. In this section, we review the evidence for such extrinsic forces playing a role in the morphogenesis of the *Drosophila* gastrula.

4.1. Extrinsic forces in mesoderm invagination

In the previous section, we described the intrinsic forces that govern the apical constriction of mesodermal cells necessary for their invagination. Recent work suggests that extrinsic forces contribute to this morphogenetic event by introducing an anisotropy in the shape of the constricting cells (Martin *et al.*, 2010). Cells in the ventral furrow do not have an isodiametric shape: their apical surface becomes on average twice as long in AP compared to DV (Fig. 5.3A(iii)). This anisotropy requires an integration of contractile forces through a "supracellular actomyosin network" made up of each cell's actomyosin apical medial web connected to each other by spot adherens junctions (Martin *et al.*, 2010). Disrupting this network by either knocking down E-Cadherin or Armadillo/Beta-catenin to abolish spot adherens junctions, or by laser ablating the actomyosin medial web, results in tissue-wide tears with a geometry suggesting an anteroposterior tensile force (Martin *et al.*, 2010). The still constricting mesodermal cells on each side of the tears lose their AP elongated shape, suggesting that this putative anteroposterior tensile force causes the cells' anisotropy. The activity of the genes *twist* and *snail* are important for the production of this tensile force, since tissue-wide tears do not occur in embryos in which either of these genes have been knocked down in combination with adherens junction components (Martin *et al.*, 2010). *twist* is required to stabilize actomyosin at the medial web in between *snail*-induced pulsed contractions (Martin *et al.*, 2009), and the cells' anisotropy was observed to increase between, as well as during, contractions (Martin *et al.*, 2010). These results emphasize the importance of the stabilization phase for the anteroposterior force to be transmitted across the supracellular actomyosin network.

One possible simple explanation for the AP elongated shape of the constricting mesodermal cells is that the geometry of the stripe of cells undergoing pulsatile contraction determines this anisotropy (Fig. 5.3A(i)). The mesoderm primordium is about 100 cells long and about 18 cells wide. Because of this elongated shape, there are about five times more cells connected to each other in the AP axis compared to the DV axis. Force generated by each contracting mesodermal cell pulling on its immediate neighbors is transmitted across the supracellular actomyosin network to the rest of the tissue. These pulling forces might sum into tissue-scale forces that are greater in the AP axis than the DV axis, simply because of the sheer difference in cell numbers in AP compared to DV (more cells along the AP axis, than along the DV axis, will be contracting at any given time). Passive deformation in response to these tissue scale forces would in turn explain the AP elongated shape of the constricting cells. Thus in this case, the integration of forces generated by individual cells could in turn influence the shape of every cell in the tissue. The result of this integration can be considered as an extrinsic force, to which the cells respond passively by elongating. Another factor to consider is the resistance of the ectoderm to the mesoderm pulling forces. An alternative, though not mutually exclusive, explanation for the AP elongated shape of constricting mesodermal cells is that the shape of the embryo determines the anisotropy. The embryo is about two and half times as long as it is wide, and since all the ectodermal cells around the circumference of the embryo are likely to resist the pulling from the mesoderm (if they are mechanically coupled together through the adherens junctions and the actomyosin cytoskeleton), there must be a larger number of ectodermal cells resisting the invagination along the AP (long) axis than the DV (short) axis. Thus the shape of the embryo might cause the resistive forces to be greater in the AP than the DV axis. This could also cause the anisotropy observed in the constricting mesodermal cells. We think that it is likely that it is the balance between contractile forces in the mesoderm and resistive forces in the ectoderm that determines the anisotropic shape of the mesodermal cells (Fig. 5.3A).

Intriguingly, the contractile mesoderm primordium also seems able to integrate externally applied forces. Pouille *et al.* (2009) showed that applying a mechanical deformation to the mesoderm can rescue apical redistribution of Myosin II and mesoderm invagination in Snail mutants. This rescue depends upon *fog*, suggesting a role for this gene in the response to mechanical forces during mesoderm invagination. The molecular mechanism behind this interaction is unknown, but it has been suggested that endocytosis of Fog plays a part (Pouille *et al.*, 2009). The rescue also requires *twist*, since mechanical deformation of double mutant embryos for *twist* and *snail* embryos do not produce Myosin II redistribution and ventral furrow formation. Together with the findings from Martin *et al.* (2010), this suggests that integration of mechanical forces by the supracellular actomyosin network of the mesoderm primordium might be important for invagination.

Are other forces contributing to mesoderm invagination in the wild-type embryo? Apical constriction is thought to be the main source of intrinsic forces during mesoderm invagination (see Section 2; Fig. 5.3A (ii)), and as found by Martin *et al.* (2010), integration of these intrinsic forces at the tissue scale is likely to be important. In addition, other forces originating outside the mesoderm primordium might be necessary for mesoderm invagination. This question has been approached using computer simulations of the embryo (Allena *et al.*, 2010; Conte *et al.*, 2008, 2009; Pouille and Farge, 2008). Whereas apical contraction seems necessary and sufficient for ventral furrow formation in these *in silico* embryos, it does not appear to be enough to completely invaginate a mesodermal tube. For a normal invagination, *in silico* embryos need a somewhat pressurized system, with a vitelline membrane and perivitelline fluid surrounding the embryo (Pouille and Farge, 2008). Conte *et al.* (2009, 2008) also suggest that pushing from the ectodermal tissue on either side of the ventral furrow might be important to seal it. These studies open the possibility that forces external to the mesoderm primordium contribute to normal mesoderm invagination, but this is not yet supported by experimental evidence.

4.2. Apical surface deformation as a signature of extrinsic forces

Changes in cell surface area can be interpreted as a signature of forces passively deforming the tissue, by analogy with the behavior of a passive cellular material such as a foam, in which shape reflects stress (Weaire and

Hutzler, 2001). Accordingly, if a tissue experiences a pull in the plane of the epithelium, the cells will increase their apical surface area. Conversely, if the tissue is subjected to a compression, the cells surface area is expected to decrease. This simple analogy can be used to look for extrinsic forces that deform the embryo (Butler *et al.*, 2009). We have already seen that the anteroposterior elongation of the surface of the constricting mesodermal cells may be a passive response to extrinsic forces (Martin *et al.*, 2010). In that case, the extrinsic force may have its source within the mesodermal tissue itself. But extrinsic forces can also have their source outside the tissue they deform. For example, mesoderm invagination deforms the adjacent ectoderm: ectodermal cells on either side of the invaginating mesoderm elongate in the dorsoventral axis (Butler *et al.*, 2009; Kam *et al.*, 1991; Oda and Tsukita, 2001). Quantifications show that this elongation is accompanied by an increase in surface area, consistent with the ectoderm being pulled by the invaginating mesoderm (Butler *et al.*, 2009; see Fig. 5.3A(ii)). Quantification of the rate of tissue deformation (strain rate) shows that this dorsoventral ectodermal stretch is maximum when the mesoderm is rapidly involuting. These ectodermal cell shape changes are absent in *twist* mutants, confirming that they are a consequence of mesoderm invagination, rather than caused by active cell behaviors in the ectoderm.

The maximum rate of ectoderm stretching in the dorsoventral direction coincides with the start of germ-band band extension (Butler *et al.*, 2009). Germ-band extension is associated not only with polarized cell intercalation (see Section 2; Fig. 5.3B(ii)), but also with cell shape changes, namely, an elongation of the cell apical surface in the anteroposterior direction (Butler *et al.*, 2009). In wild-type embryos, this anteroposterior elongation rectifies almost perfectly the earlier dorsoventral elongation caused by mesoderm invagination, and cell apical areas return to an almost isodiametric shape 10 min after the beginning of germ-band extension. In anteroposterior patterning mutants, where polarized cell intercalation is disrupted, cells return briefly to an isometric shape at about 10 min, but then elongate visibly in the anteroposterior axis, in a gradient increasing from the middle of the embryo to its posterior. These elongations are accompanied by an increase in surface area, indicating that an extrinsic force stretches the germ-band tissue.

4.3. Extrinsic forces influence germ-band extension

Where does the extrinsic force deforming the germ-band come from? We have found that in *twist* mutant embryos, the cell shape elongations observed in a gradient from the middle of the embryo toward its posterior are significantly reduced (Butler *et al.*, 2009). This suggests a simple explanation, that mesoderm invagination is required for the production of an extrinsic tensile force that deforms the germ-band in the anteroposterior direction. A role for mesoderm invagination is strikingly temporally consistent: in wild-type embryos, there is a tight correlation between the point of rapid mesoderm involution and the onset of germ-band extension. Although a connection between extrinsic force contributing to germ-band extension and mesoderm invagination is the simplest and most parsimonious explanation based on the above, imagining a mechanism for how such AP force could be generated by mesoderm invagination is much more challenging.

There might be a relationship between the extrinsic force deforming the germ-band and the AP-oriented tensile force detected in the mesoderm primordium (Martin *et al.*, 2010; Zhang *et al.*, 2010). However, we think that this latter force is likely to be the result of a balance between contractile forces in the mesoderm and resistive forces in the ectoderm (Fig. 5.3A). If this is correct, then the mesoderm should be mainly producing contractile forces, which seem at odds with the idea of an extrinsic force contributing to *extend* the germ-band. Also, the period when the tensile forces in the mesoderm are observed correspond to a narrow window of 5–10 min just before the start of germ-band extension, whereas the extrinsic force deform the germ-band for the next 20–25 min. If the two forces are related, then mechanisms must exist that transform a brief contractile force into a more lasting pulling force.

A possibility is that the epithelial mesodermal tube (Fig. 5.4H) that forms inside the embryo upon rapid mesoderm invagination extends in the anteroposterior direction, and that this tube extension in turn drags the attached ectoderm (Fig. 5.3B). Cell shape changes contributing to AP extension are detectable for about the first 20–25 min of axis extension, which is consistent with the period when the mesoderm is thought to have a tubular shape (McMahon *et al.*, 2010; Murray and Saint, 2007). At the end of this period, the tube collapses and the mesodermal cells dissociate and migrate inside the embryo (Leptin, 2005). A drag force from the invaginated mesoderm is also consistent with our findings that the germ-band extends fastest close to the ventral midline, suggesting an axial pull (Butler *et al.*, 2009). Moreover, this axial bias ceases after 25 min of extension, which is again temporally consistent with mesodermal tube collapse and the end of cell shape changes in the germ-band. Why would the mesodermal tube extend? When the mesoderm invaginates, it has to push against the yolk fluid inside the embryo and displace it (Fig. 5.3B). Since the embryo is itself contained in the perivitelline fluid and encased by the extraembryonic membranes, this fluid displacement will have to be accommodated within a finite volume, and could generate an increase of pressure inside the embryo (Fig. 5.3B). Increasing pressure at the surface of the invaginating mesodermal tube could conceivably force (by mechanical constraints) cell rearrangement in the tube and reduce its diameter. In support of this idea, mechanical load has been shown to cause cell rearrangement which leads to the elongation of tracheal tubes (Caussinus *et al.*, 2008). Mesoderm elongation has not been reported explicitly, but careful transverse sectioning of the mesodermal tube revealed a loss of two cells in average out of about 20 during mesodermal tube formation (Sweeton *et al.*, 1991). A possible interpretation for this observation is that the mesodermal tube is reducing its diameter by convergence and extension. The resulting extension could in turn drag the ectoderm and explain the cell shape changes that we are observing (Butler *et al.*, 2009; Fig. 5.3B).

Yolk displacement itself could be an alternative explanation for generation of an extrinsic AP force deforming the ectoderm. When observed in sagittal and transverse sections, the invaginating mesoderm can be estimated to displace a third of the yolk volume (Leptin, 2005), which could generate a significant flow. Because the embryo is a long ovoid, it is likely that the bulk of yolk displacement occurs in the anteroposterior axis, for example, that the yolk is displaced toward both ends of the embryo when the mesoderm invaginates (Fig. 5.3B(i)). Simulations of the ventral invagination in *in silico* embryos shows a middle to end surface deformation of the embryo, which could potentially describe the impact of a yolk flow inside the embryo (Allena *et al.*, 2010; Conte *et al.*, 2008). Moreover, sections through embryos fixed during mesoderm invagination show thinning of cells at the anterior and posterior of the embryo (see Fig. 5.1C and D), providing some supporting evidence for an anteroposterior yolk flow. Such a flow could deform the germ-band cells from their basal sides up.

Alternatively, an extrinsic force could also be a composite of influences, with other morphogenetic movements playing a role. It is important to note that the cell shape changes contributing to germ-band extension are not completely abolished in *twist* mutants (Butler *et al.*, 2009). This could be because the cells are still able to contract and partially internalize in *twist* mutants due to some remaining *snail* expression, and that this is sufficient to produce some residual tensile force in the ectoderm, or it could indicate a role for other morphogenetic movements in contributing to this force. If the former explanation is correct, the cell shape changes contributing to axis extension should disappear in *snail* mutants. Alternatively, other morphogenetic events in the embryo could provide either compressive DV forces or AP pulling forces. For example, posterior midgut invagination could be pulling the end of the germ-band round the posterior of the embryo (See Figs. 5.3B(i) and 5.1). Quantitative analysis of a *fog* mutant, in which posterior midgut invagination is blocked, shows an average rate of cell shape change similar to wild-type (Butler *et al.*, 2009). This data suggests that posterior midgut invagination does not contribute to cell shape change in the extending germ-band. However, the presence of folds in the extending germ-band of *fog* mutant complicates the analysis of strain, and a clearer experiment would be to look at cell shape changes in mutants where both *fog* and polarized cell intercalation have been eliminated. Another potential contributor to extrinsic forces in the germ-band is the flattening of the amnioserosa cells onto the dorsal side. Amnioserosa cells takes 20 min to elongate by a factor 8 (Pope and Harris, 2008). This massive change of shape could conceivably generate a pushing force compressing the tissue from dorsal to ventral. It is not clear, however, if amnioserosa elongation begins early enough in development to explain the cell shape changes contributing to the fast phase of axis extension. Also, in the regions of stretched tissue in anteroposterior patterning mutants such as *Kruppel*, the apical cell areas are increased, rather than decreased, suggesting a pulling rather than a compressive force (Butler *et al.*, 2009).

4.4. Interaction between intrinsic and extrinsic forces in the embryo

Active cell behaviors can relieve stress produced by extrinsic forces. Our quantitative analyses show that the presence of an extrinsic force is almost "invisible" in wild-type germ-band extension, as its tissue deforming capabilities are balanced by polarized cell intercalation (Butler *et al.*, 2009). It is only when polarized cell intercalation is defective that the germ-band tissue becomes visibly stretched. Thus, polarized intercalation can be seen as a cell behavior "fluidifying" the germ-band tissue to relieve the stress imposed by this extrinsic force. Interestingly, positing an extrinsic force provides an explanation for why there is some residual germ-band extension in all anteroposterior patterning mutants analyzed so far, including those in which no polarized cell intercalation should be remaining (Irvine and Wieschaus, 1994). We have confirmed this quantitatively by analyzing *bicoid nanos* mutants and *knirps hunchback* mutants (Butler, 2009; Butler *et al.*, 2009). In these mutants, we cannot detect polarized cell intercalation in the trunk contributing to anteroposterior axis extension, but there is still some tissue extension which is caused exclusively by cell shape changes. Another cell behavior, oriented cell divisions, can potentially relieve stress from extrinsic forces. Oriented cell divisions have been reported on the posterior side of the extending germ-band (da Silva and Vincent, 2007): they could serve a similar function to polarized cell intercalation in this other region, releasing stress induced, for example, by the nearby posterior midgut invagination.

Another potential interaction between extrinsic and intrinsic forces is that extrinsic forces could facilitate active cell behaviors. We have found that cell intercalation strain rates are decreased by 30% in *twist* mutants (Butler *et al.*, 2009). An intriguing possibility is that the anteroposterior extrinsic force increases the rate of cell-autonomous polarized intercalation in the germ-band.

4.5. Conclusion: The embryo as a system

When looking for the source of any passive cell behavior, it becomes essential to think about the embryo as a system or biological machine. In the previous paragraphs, we have considered the different morphogenetic movements and folds occurring in the embryo that could generate passive cell behaviors. We have also discussed yolk flows and the constraints imposed by embryonic membranes. Another factor we have considered is the geometry of the embryo (its ovoid shape), which might also influence the balance of forces in the embryo. To understand how these different parts and factors influence each other, *in silico* modeling contributes by formally describing the *Drosophila* embryo as a system (Allena *et al.*, 2010; Conte *et al.*, 2008, 2009; Pouille and Farge, 2008). Work remains to be done to include a more complete description of morphogenetic movements into these simulations, including their dynamics.

5. Tension and Supracellular Actomyosin Structures in the Early Embryo

In the previous section, we have discussed how tension can be generated in an embryo and how this in turn influences cell and tissue morphogenesis. In molecular terms, it has become increasingly apparent that a key factor in epithelial morphogenesis is the actomyosin cytoskeleton (Conti and Adelstein, 2008; Corrigall *et al.*, 2007; Escudero *et al.*, 2007; Franke *et al.*, 2005; Gally *et al.*, 2009; Lee *et al.*, 2006; Quintin *et al.*, 2008; Vicente-Manzanares *et al.*, 2009). The early *Drosophila* embryo is an excellent illustration of this, since many of the morphogenetic cell and tissue behaviors have been shown to depend upon the activity of the apical actomyosin cytoskeleton. This activity is important at the cell-autonomous level, such as in constricting mesodermal cells (Martin *et al.*, 2009) or in intercalating germ-band cells (Bertet *et al.*, 2004). There is also increasing evidence for a role for supracellular actomyosin structures in morphogenesis (Blankenship *et al.*, 2006; Fernandez-Gonzalez *et al.*, 2009; Franke *et al.*, 2005; Martin *et al.*, 2010; Monier *et al.*, 2010).

5.1. Actomyosin structures and invaginations

We have seen that the contractility of an apical medial web of actomyosin is at the heart of mesoderm invagination (Barrett *et al.*, 1997; Dawes-Hoang *et al.*, 2005; Martin *et al.*, 2009, 2010; Nikolaidou and Barrett, 2004). Actomyosin has also be shown to be important for other invaginations in the early embryo such as posterior midgut invagination during gastrulation (Nikolaidou and Barrett, 2004), and later for the formation of tracheal pits at the onset of metamerization (Nishimura *et al.*, 2007). It will be interesting to see whether in these other examples, a contraction of the apical cell surface is achieved through pulsatile contractions of an apical medial web. An intriguing possibility is that pulsatile contractions linked to a ratchet mechanism represent a widespread strategy to contract the apex of epithelial cells, since amnioserosa cells also use this mechanism to contract during dorsal closure (Blanchard *et al.*, 2010; David *et al.*, 2010; Gorfinkiel *et al.*, 2009; Solon *et al.*, 2009). Factors like Snail are required to promote pulsatile behavior in the mesodermal cells (Martin *et al.*, 2009), whereas recent work shows that the formation of a medial web of actin and subsequent apical constriction is inhibited in the germ-band cells by inactivation of Wasp by JAK/STAT signaling (Bertet *et al.*, 2009). Future research is needed to find the factors that unleash the pulsatile behavior of the actomyosin apical cytoskeleton at the right time and location, in order to promote apical constriction. Another question is how the actomyosin network's pulsation in an individual cell synergizes with those of its neighbors to give tissue-scale behaviors (Blanchard *et al.*, 2010). As discussed in the previous section, an example of such tissue-scale behavior is the anteroposterior tensile force in the contracting mesoderm, which is thought to emerge from a supracellular behavior of the interconnected actomyosin networks (Martin *et al.*, 2010).

5.2. Actomyosin structures and axis extension

We have also seen that polarized cell intercalation during germ-band extension is thought to be a direct consequence of the planar polarized enrichment of actomyosin at the "vertical" (perpendicular to AP axis) junctions of the germ-band cells (Bertet *et al.*, 2004; Blankenship *et al.*, 2006; Zallen and Wieschaus, 2004). Planar polarization of apical actin and Myosin II can be first detected at stage 6, when the cells still have a relatively hexagonal shape and a regular organization just prior to the start of intercalation (Blankenship *et al.*, 2006). The cells then lose their regular arrangement with the progress of polarized intercalation and the cell junctions enriched in Myosin II are more frequently found linked to one another (Fernandez-Gonzalez *et al.*, 2009; Fig. 5.5). These linked vertical junctions can shrink together to a point, forming rosettes (Blankenship *et al.*, 2006; Fig. 5.6B). The linked junctions are found to be under more tension than the single Myosin II-enriched interfaces (Fernandez-Gonzalez *et al.*, 2009). This was shown by using a laser to ablate the cortical actomyosin at cell–cell interfaces between two vertices, comparing single interfaces with linked interfaces. The speed of recoil of the severed interfaces was measured and used to deduce a relative estimate of tension at both types of interface. This experiment showed that linked interfaces were approximately 1.7 times more tensile than single interfaces.

A possible explanation for this result is that the tension in a chain of connected interfaces enriched in actomyosin sums up across these "supracellular" actomyosin cables, and thus the longer the chain of connected interfaces, the higher the tensile force in a given interface. Another (non exclusive) possibility is that the enriched interfaces recruit more Myosin II

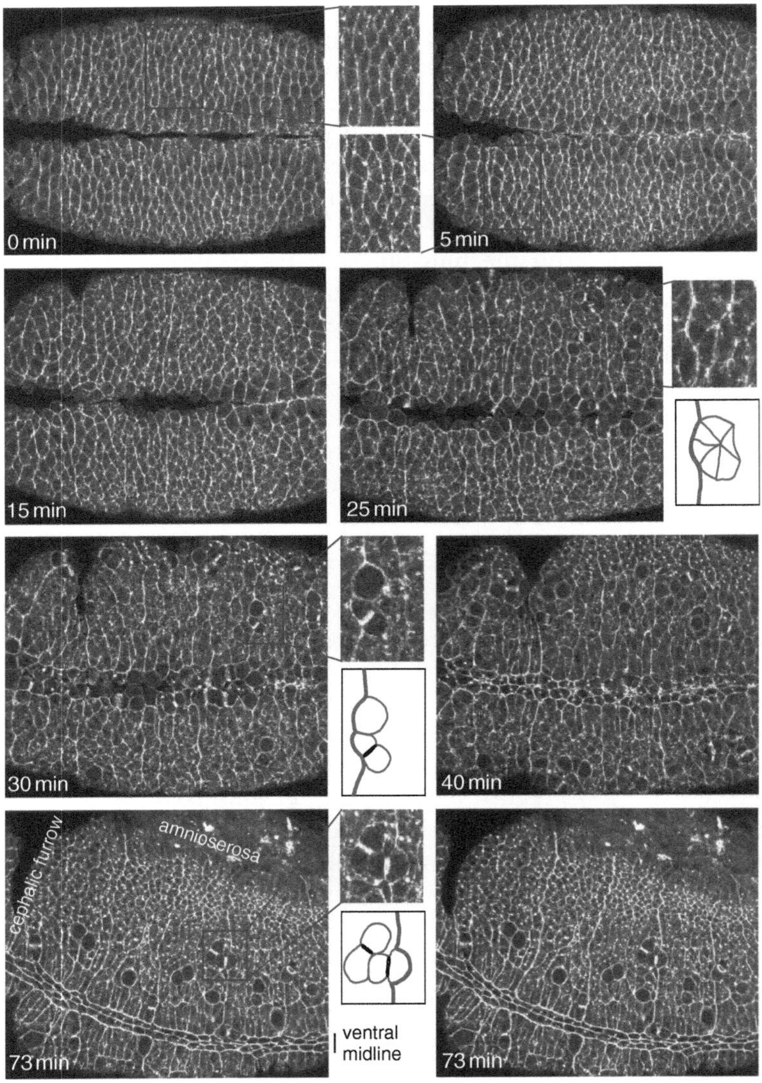

Figure 5.5 Planar polarized enrichments of Myosin II from germ-band extension to parasegmentation. Stills from a time-lapse movie of an embryo labeled with RMLC-GFP imaged ventrally from early in germ-band extension (Monier *et al.*, 2010). Note that this field of view captures approximately 60% of the ventro-lateral ectoderm on each side of the midline at the beginning of the movie. Later in the movie, the embryo rolls, so in the last timepoint the view includes the ventro-lateral and dorsal ectoderm, as well as some amnioserosa. Time from the beginning of the movie is shown on each frame. Key for line drawings: thin gray line—cell outlines; thick gray lines—parasegment boundary; black lines—cytokinesis rings. *At 0 and 5 min* Myosin II is enriched at both single vertical interfaces (colored in magenta in close-ups), and linked vertical interfaces (colored in cyan in close-ups). *By 5 min* single vertical interfaces enriched in

in response to the tension generated by the linked interfaces pulling on each other. Increase in Myosin II recruitment would in turn translate into an increase in tension at each interface. Supporting this notion, linked enriched vertical interfaces have a more abundant and stable pool of cortical Myosin II than single enriched vertical interfaces, as indicated by FRAP experiments (Fernandez-Gonzalez *et al.*, 2009). Moreover, exerting a pulling force at the surface of the embryonic ectoderm (by suction using a pipette), recruits Myosin II at the apical cortex (Fernandez-Gonzalez *et al.*, 2009). This suggests that mechanical pulling of ectodermal cells can result in Myosin II recruitment to the cell cortex (Fernandez-Gonzalez and Zallen, 2009). This is reminiscent of the recruitment of Myosin II on the apical side of ventral cells of *snail* mutants, in response to artificial mechanical deformation (Pouille *et al.*, 2009). Another observation which indirectly supports the idea of Myosin II recruitment in response to tension is the transient enrichment of Myosin II in supracellular cables in the germ-band of DIAP1 mutant embryos (Chandraratna *et al.*, 2007). In these embryos, the premature activation of the apoptosis cascade at gastrulation is accompanied with huge contractions of the ectoderm. In regions where the ectoderm is pulled strongly, actomyosin cables thicken visibly in a way suggestive of an immediate recruitment in response to tension. During morphogenesis, swift recruitment of Myosin II to the cortex in response to tension could be important in two ways: to resist pulling forces in order to preserve tissue integrity, and to increase the contractility of supracellular actomyosin structures, which could in turn increase the efficiency of a given morphogenetic process. The molecular mechanism behind this apparent mechano-sensitive recruitment of Myosin II in the early *Drosophila* embryo remains to be determined (reviewed by Fernandez-Gonzalez and Zallen, 2009).

Myosin II have become more rare than at the beginning of the movie. *By 15 min* planar polarization in every ectodermal cell has decreased, while enrichment of Myosin II can be discerned at most parasegmental boundaries (red arrows). *At 25 min* cell division has begun (both at the midline and in clusters of cells in the ventro-lateral epidermis). When cells adjacent to a parasegment boundary divide or intercalate they can transiently deform it. The close-up shows by a rosette deforming the parasegment boundary. *At 30 min*, Myosin II enrichment is present at all of parasegment boundaries in the frame of view (red arrows). The close-up shows an example of parasegment boundary being deformed by a dividing cell and by a cell that has just rounded up prior to division. *At 40 min*, the cell divisions at the midline are nearly finished, but cells are still dividing in the ventro-lateral ectoderm. The embryo is beginning to roll, bringing into view some cells of the dorsal ectoderm (top of frame). *At 73 min*, the embryo has rolled further revealing the whole width of the ectoderm and some amnioserosa. Myosin II is enriched along the whole length of parasegment boundaries, from the ventral midline to the amnioserosa (green in the right-hand panel). Cell divisions are still continuing in the ventro-lateral epidermis. The close-up shows another example of a dividing cell deforming the parasegment boundary. (For interpretation of the references to color in this figure legend, the reader is referred to the Web version of this chapter.)

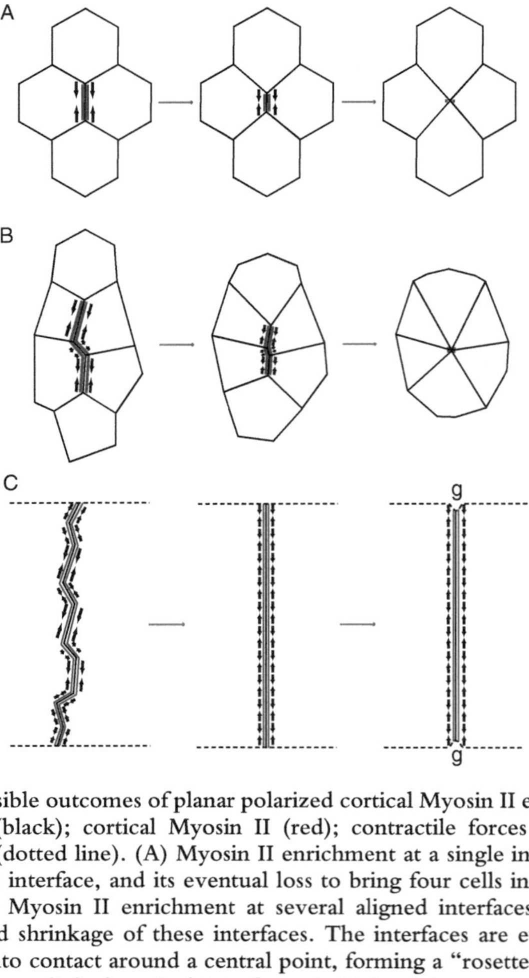

Figure 5.6 Possible outcomes of planar polarized cortical Myosin II enrichment. Key: Cell interfaces (black); cortical Myosin II (red); contractile forces (black arrows); adjacent tissues (dotted line). (A) Myosin II enrichment at a single interface results in shrinkage of this interface, and its eventual loss to bring four cells into contact at the same point. (B) Myosin II enrichment at several aligned interfaces results in both straightening and shrinkage of these interfaces. The interfaces are eventually lost to bring the cells into contact around a central point, forming a "rosette". (C) Myosin II enrichment at many linked vertical interfaces, (which might span a whole tissue, as shown here) results in both straightening and shrinkage of these interfaces. Once the linked interfaces have formed a straight line, further Myosin II-shrinkage could result in groove formation (g). (For interpretation of the references to color in this figure legend, the reader is referred to the Web version of this chapter.)

5.3. Actomyosin structures and compartment boundaries

During the rapid phase of germ-band extension, "vertical" interfaces in the ventro-lateral ectoderm cells are enriched in actomyosin, either as isolated interfaces or as interfaces linked in supracellular cables (Bertet *et al.*, 2004; Blankenship *et al.*, 2006; Zallen and Wieschaus, 2004; Figs. 5.5 and 5.6A and B). Later on, actomyosin enrichment at vertical interfaces diminishes, except in segmentally repeated cables which end up spanning the full dorsoventral width of the ectoderm (Monier *et al.*, 2010; Fig. 5.5). There is a temporal overlap between the formation of these segmentally repeated cables and the global planar polarization of Myosin II, but the relationship between these two different enrichments is unknown. At stage 9/10, the segmentally repeated cables have been shown to coincide with the interface between each stripe of Wingless-expressing cells and Engrailed-expressing cells (Monier *et al.*, 2010). This corresponds to the position of the parasegmental boundaries which are anatomically apparent as shallow grooves at stage 11 (Larsen *et al.*, 2008). The formation of the parasegmental actomyosin cables precedes the formation of these grooves, and the cables disappear when the grooves disappear, at the beginning of germ-band retraction (Monier *et al.*, 2010). The parasegmental boundary has been demonstrated to behave as a boundary of lineage restriction, that is, cells do not cross it in either direction (Vincent and O'Farrell, 1992). Another characteristic of the parasegmental boundary interfaces is that they align to form a straight line (Monier *et al.*, 2010). Linked interfaces that do not coincide with parasegmental boundaries do not align. Also, parasegmental alignments and actomyosin enrichments are lost in anteroposterior mutants that disrupt compartment formation. Thus, there is a correlation between the enrichment of actin and Myosin II at the apical cortex of the boundary cells, and the straightness of the compartmental boundary.

We have shown that Myosin II activity is required for compartmental lineage restriction at the parasegmental boundary (Monier *et al.*, 2010). When Myosin II function is disrupted in mutant or drug-treated embryos, the parasegmental boundary loses its straightness and cells can be displaced into the wrong compartment. When Myosin II is specifically inactivated by Chromophore-Assisted Laser Inactivation (CALI) at the supracellular cable, the same cell sorting phenotypes are observed. This shows that it is the specialized enrichment of Myosin II along the length of the boundary that maintains a lineage restriction. The results of CALI experiments indicate that the parasegmental actomyosin cables are required to stop cell crossing only in the presence of a challenge to boundary integrity. One challenge is the intercalation of cells close to the boundary, which can cause local boundary deformation (Fig. 5.5). This suggests that parasegmental cables prevent inappropriate cell mixing during cell intercalation in the extending germ-band. However, intercalation figures become rarer as parasegmentation progresses, due to the completion of germ-band extension. The more frequently observed challenge to parasegmental boundaries is from proliferative cells (Martinez-Arias, 1993). The proliferative period (between stages 8 and 11) corresponds to the period when the actomyosin cable is detected at parasegmental boundaries (Monier *et al.*, 2010). When Myosin II is inactivated at PS cables by CALI, compartmental cell crossing is observed when boundary cells are dividing (Monier *et al.*, 2010). Strikingly, in wild-type embryos, the enrichment of actomyosin is maintained at the cortex of boundary cells undergoing division (Fig. 5.5). Despite this, the ballooning of the dividing cells transiently deforms the actomyosin-enriched cortex, suggesting that the cell shape change of cell division imposes a stronger force than the contractile force of the actomyosin-enriched cortex. Once cells have divided, the Myosin II-enriched interfaces, which are part of the parasegmental boundary, contract back and straighten, pushing the daughter cells back into their compartment of origin. This simple tension-based mechanism is remarkably effective at keeping dividing boundary cells from two distinct compartments separate, even if many boundary cells divide at the same time (Monier *et al.*, 2010; Fig. 5.5).

Direct evidence for the importance of tension to prevent cell sorting at compartment boundaries comes from work on the *Drosophila* wing disc. Here too, a correlation between actomyosin enrichment and lineage restriction has been reported at both the AP and DV compartmental boundaries (Landsberg *et al.*, 2009; Major and Irvine, 2005, 2006). Both boundaries show enrichment in actin (and actin regulators) and, to a lesser extent, in Myosin II at boundary cell–cell interfaces. Moreover, hypomorphic *zipper* mutants (*zipper* encodes for Myosin Heavy Chain) have irregular compartmental boundaries. Although the requirement for Myosin II at these specific actomyosin enrichments has not been directly tested, experiments designed to estimate tensile forces were performed for the AP boundary (Landsberg *et al.*, 2009). Laser severing of single cell interfaces at the AP boundary compared to cell interfaces elsewhere in the tissue, and analysis of the initial velocity of recoil suggests a 2.5-fold increase in tensile force at cell interfaces at the AP lineage restriction boundary. In the presence of the Rho-kinase inhibitor Y27632, this ratio was reduced to 1.5, suggesting that increased tensile forces at the AP compartment interfaces requires Myosin II activity. The authors also ran computer simulations that suggest that the increase in tension is sufficient to maintain straight interfaces between growing cell populations. Interestingly, in the wing disc as well as in the embryo, there is a correlation between proliferative activity in the epidermis at the vicinity of the boundary and formation of a transient enrichment of actomyosin at a compartmental boundary interface (see discussion in Monier *et al.*, 2010).

5.4. Actomyosin supracellular structures as versatile factors of morphogenesis

Tension at the parasegmental boundary is not only effective at keeping dividing cells in the correct compartment, but it also aligns cellular interfaces at the boundary (Landsberg *et al.*, 2009; Monier *et al.*, 2010). This could be important for the precise patterning of each compartment, as diffusion of morphogens from a straight line of cells will give a regular pattern. Straightening or smoothing of linked cellular interfaces enriched in Myosin II has been observed in several tissues in *Drosophila* (Corrigall *et al.*, 2007;

Escudero *et al.*, 2007; Mulinari *et al.*, 2008; Nishimura *et al.*, 2007; Simone and DiNardo, 2010), and is often followed by the formation of a furrow or a fold. Similarly, it is likely that the formation of parasegmental grooves is a consequence of the increased tension at parasegmental interfaces. As we have seen earlier, another outcome of linked cortical actomyosin enrichments is rosette formation during germ-band extension.

It is remarkable that the same subcellular specialization, namely the enrichment of Myosin II in a portion of the apical cortex, can generate so many diverse tissue-scale behaviors in the early *Drosophila* embryo. From the beginning of germ-band extension through to the end of parasegmentation, a sequence of behaviors is observed that involve an increasing number of linked interfaces (Fig. 5.6). Single Myosin II-enriched interfaces generate a T1–T3 tetrad transition (Bertet *et al.*, 2004); 2–6 linked interfaces produce rosettes (Blankenship *et al.*, 2006); more than 10 linked interfaces act as a boundary which is both straight and sorts cells, and evolves further to form a groove (Monier *et al.*, 2010). A key question is whether genetic regulations are required for the transition between these distinct tissue behaviors? It is intriguing to think that, outside positional information, the emergent mechanical properties of the linked interfaces could explain these behaviors. For example, the transition between rosette formation and parasegmental boundary formation could only require that the chain of linked contractile interfaces becomes anchored to a dorsal and a ventral edge of the ectoderm (which they do, at the dorsal amnioserosa and the ventral midline, respectively; Fig. 5.5). Alternatively, above a threshold number of linked contractile interfaces, rosette formation might become impossible, and so a boundary could form instead. The shrinkage of each interface by contractile forces could be balanced by the pulling forces exerted on it by the adjacent ventral and dorsal interfaces. If the contractile force of each interface is equivalent to the pulling force exerted on it at both its ends by the other contractile interfaces, then the length of each of the connected interfaces could stabilize and the column of connected interfaces could straighten. If however the contractile forces at each interface are stronger than the forces exerted on their ends, then each interface could shrink to a similar degree, and this could generate an indentation or groove in the tissue (Fig. 5.6C).

Interestingly, the morphologies associated with a chain of actomyosin-enriched interfaces vary between tissues. For example, whereas a groove is present at embryonic anteroposterior boundaries, there is no anatomical manifestation associated with the anteroposterior boundary in the wing disc. It is interesting to note that overexpression of RhoGEF2 produces much deeper grooves at the parasegmental boundary in the embryo (Mulinari *et al.*, 2008). Even more intriguing, loss of function of the transcription factor Optomotor-blind causes the formation of a groove at the AP compartmental boundary of the wing disc, where normally none is present (Shen *et al.*, 2008; Umemori *et al.*, 2007). This suggests that genetic factors can control the degree of shrinkage associated with chains of actomyosin-rich interfaces. One could envisage that very strong contractile forces in a chain of connected interfaces could produce very deep grooves and even sever a tissue, a version of cytokinesis at the tissue scale. Thus chains of actomyosin-rich interfaces could conceivably act as a very versatile supracellular structure to shape developing embryos and organs.

In addition to the chains of enriched interfaces, actomyosin also has a supracellular organization when present as an apical medial web connected through adhesion contacts. As discussed earlier, this supracellular organization could explain the anisotropy of shape of the mesodermal cells (AP elongated, Fig. 5.3A(iii)), but could also help a bending inward of the stripe of interconnected mesodermal cells, and formation of a hollow tube (Fig. 5.2). This might facilitate internalization of the mesoderm tube and explain why invagination happens rather suddenly once the mesodermal cells have reached their AP elongated shape.

In all the examples discussed above, (invagination of the mesoderm, convergence and extension of the germ-band, cell sorting and groove formation at boundaries), the tissue-scale behavior can be hypothesized to be dependent upon the balance of forces between contractile forces at each interface (or apex in the case of the mesodermal cells) and pulling forces at the vertexes or edges of connected cell–cell interfaces. These forces will be influenced by factors that control the contractility and density of cortical actomyosin as well as the adhesive properties of cell–cell interfaces (reviewed by Lecuit and Lenne, 2007). Finding how these factors regulate supracellular actomyosin structures will be important to understand the genetic control of morphogenesis. In addition, it is becoming increasingly likely that forces might signal directly through mechano-transduction pathways, which will need to be characterized in the context of each morphogenetic process (Desprat *et al.*, 2008; Fernandez-Gonzalez and Zallen, 2009; Fernandez-Gonzalez *et al.*, 2009; Pouille *et al.*, 2009).

6. Final Remarks

6.1. A widespread role for Myosin II activity in morphogenesis

Myosin II-dependent apical cell constriction has been found to be important for tissue internalization, but also for tissue bending and tissue thickening in a variety of organisms (Sawyer *et al.*, 2010). In vertebrates, Shroom3, an actin-binding protein, is required for apical cell constriction during neural tube closure (Haigo *et al.*, 2003; Lee *et al.*, 2009; Nishimura and Takeichi, 2008), lens placode invagination (Plageman *et al.*, 2010), and gut development (Chung *et al.*, 2010). The activity of Shroom3 in apical constriction is upstream of Myosin II and Rho-kinase (Hildebrand, 2005). In cochlear hair cells, remodeling of the apical cell circumference has been linked to Myosin II activity (Etournay *et al.*, 2010). These examples suggest that Myosin II-dependent regulation of apical cortical tension is important in the morphogenesis of vertebrate epithelia. It will be interesting to see if the pulsatile behavior of actomyosin uncovered in the *Drosophila* mesodermal (Martin *et al.*, 2009) and amnioserosa cells (Blanchard *et al.*, 2010; Solon *et al.*, 2009), also powers apical cell constriction or remodeling in vertebrate tissues.

Although apical cell constriction is clearly required for many cases of tissue internalization (Sawyer *et al.*, 2010), it might not be always sufficient. During tissue internalization, apical cell constriction and concurrent cell elongation *before* invagination, is often followed by cell shortening *during* tissue invagination. Sherrard *et al.* (2010) have shown that in ascidians, cell shortening is an active cell behavior which requires Myosin II enrichment at the lateral cell cortex in a Rho/Rho-kinase independent pathway. Moreover, they show that *both* apical cell constriction and lateral cell shortening are required for normal tissue internalization (here endoderm). This raises the question of the role of similar cell shortening behaviors in other tissue invaginations, including in *Drosophila* mesoderm invagination, and whether these are active or passive.

In vertebrates, Myosin II activity has been linked to the intercalatory behaviors of mesodermal and neural cells during convergence and extension in gastrulation and neural tube formation, respectively (Rolo *et al.*, 2009; Skoglund *et al.*, 2008; Weiser *et al.*, 2009). Although vertebrate mesenchymal cells have a distinct shape and cytoskeletal organization from *Drosophila* epithelial cells, in both cases the planar polarization of the actomyosin cytoskeleton is important for the intercalatory cell behavior (Skoglund *et al.*, 2008). Related to this, in vertebrate gastrulation, cell intercalation is regulated by the Wnt/PCP pathway (Roszko *et al.*, 2009). There are also hints that Myosin II could be important for boundary formation in vertebrates: Myosin II and actin accumulate at rhombomere boundaries in the zebrafish mutant for the myosin phosphatase regulator Mypt (Gutzman and Sive, 2010).

In conclusion, understanding vertebrate morphogenesis will require elucidating how patterning systems and signals control the activity of the actomyosin cytoskeleton, and how this cytoskeleton produces the intrinsic forces underlying specific cell behaviors. The mechanisms currently being deciphered in the invertebrate models *Caenorhabditis elegans* (Gally *et al.*, 2009; Zhang *et al.*, 2010) and *Drosophila* (Martin, 2010) will help this quest.

6.2. Extrinsic forces in epithelial morphogenesis: Future directions

In addition to the intrinsic forces produced by actomyosin activity, the integration of these forces at the tissue scale needs to be further understood (Blanchard *et al.*, 2010; Martin *et al.*, 2010). This will require knowing more details about how cells are coupled mechanically and what the material properties of a given tissue are (Gjorevski and Nelson, 2010b; Kabla *et al.*, 2010). It will also require finding out how actomyosin contractility and/or enrichment can increase in response to tensile forces (Fernandez-Gonzalez and Zallen, 2009). Moreover, forces extrinsic to the tissue can modify active cell behaviors (Gjorevski and Nelson, 2010a; Zhang *et al.*, 2010). We have evidence for a decrease in cell intercalation when extrinsic forces diminish in germ-band extension (Butler *et al.*, 2009). In *Xenopus*, convergent extension of the deep cells appears to tow the overlying ectoderm and help cell intercalation there (Keller *et al.*, 2008b). This points to a cooperation between extrinsic and intrinsic forces to shape tissues, the molecular mechanisms of which remains to be explored. Measuring forces in tissue morphogenesis is another challenge. Measuring tissue deformation (strain) is an important start (Blanchard *et al.*, 2009; Butler *et al.*, 2009; England *et al.*, 2006; Keller *et al.*, 2008a; McMahon *et al.*, 2008; Sherrard *et al.*, 2010), and combined with measurements of material properties of tissues, this has the potential to be developed into a noninvasive description of forces in living organisms (Kabla *et al.*, 2010). Here again, *C. elegans* and *Drosophila* embryos (Zhang *et al.*, 2010 and this review), which are easily accessible to live imaging and genetic manipulation, provide an opportunity to describe intrinsic and extrinsic forces and to understand how these interact to sculpt developing tissues.

ACKNOWLEDGMENTS

We thank Bruno Monier, Huw Naylor, Daniel St Johnston, and Robert Tetley for comments on the chapter and acknowledge support for our research from the Human Science Frontier Science Program and the Wellcome Trust.

REFERENCES

Abrams, J. M., White, K., Fessler, L. I., and Steller, H. (1993). Programmed cell death during *Drosophila* embryogenesis. *Development* **117,** 29–43.

Affolter, M., and Caussinus, E. (2008). Tracheal branching morphogenesis in *Drosophila*: New insights into cell behaviour and organ architecture. *Development* **135,** 2055–2064.

Allena, R., Mouronval, A. S., and Aubry, D. (2010). Simulation of multiple morphogenetic movements in the *Drosophila* embryo by a single 3D finite element model. *J. Mech. Behav. Biomed. Mater.* **3,** 313–323.

Banerjee, S., Sousa, A. D., and Bhat, M. A. (2006). Organization and function of septate junctions: An evolutionary perspective. *Cell Biochem. Biophys.* **46,** 65–77.

Barrett, K., Leptin, M., and Settleman, J. (1997). The Rho GTPase and a putative RhoGEF mediate a signaling pathway for the cell shape changes in *Drosophila* gastrulation. *Cell* **91,** 905–915.

Beloussov, L. V., Louchinskaia, N. N., and Stein, A. A. (2000). Tension-dependent collective cell movements in the early gastrula ectoderm of Xenopus laevis embryos. *Dev. Genes Evol.* **210,** 92–104.

Bertet, C., and Lecuit, T. (2009). Planar polarity and short-range polarization in *Drosophila* embryos. *Semin. Cell Dev. Biol.* **20,** 1006–1013.
Bertet, C., Sulak, L., and Lecuit, T. (2004). Myosin-dependent junction remodelling controls planar cell intercalation and axis elongation. *Nature* **429,** 667–671.
Bertet, C., Rauzi, M., and Lecuit, T. (2009). Repression of Wasp by JAK/STAT signalling inhibits medial actomyosin network assembly and apical cell constriction in intercalating epithelial cells. *Development* **136,** 4199–4212.
Blanchard, G. B., Kabla, A. J., Schultz, N. L., Butler, L. C., Sanson, B., Gorfinkiel, N., Mahadevan, L., and Adams, R. J. (2009). Tissue tectonics: Morphogenetic strain rates, cell shape change and intercalation. *Nat. Methods* **6,** 458–464.
Blanchard, G. B., Murugesu, S., Adams, R. J., Martinez-Arias, A., and Gorfinkiel, N. (2010). Cytoskeletal dynamics and supracellular organisation of cell shape fluctuations during dorsal closure. *Development* **137,** 2743–2752.
Blankenship, J. T., and Wieschaus, E. (2001). Two new roles for the *Drosophila* AP patterning system in early morphogenesis. *Development* **128,** 5129–5138.
Blankenship, J. T., Backovic, S. T., Sanny, J. S., Weitz, O., and Zallen, J. A. (2006). Multicellular rosette formation links planar cell polarity to tissue morphogenesis. *Dev. Cell* **11,** 459–470.
Brierley, S. M. (2010). Molecular basis of mechanosensitivity. *Auton. Neurosci.* **153,** 58–68.
Brouzes, E., and Farge, E. (2004). Interplay of mechanical deformation and patterned gene expression in developing embryos. *Curr. Opin. Genet. Dev.* **14,** 367–374.
Butler, L. C. (2009). *Analysis of Cell Behaviours underlying Germband Extension in the* Drosophila *Embryo*. University of Cambridge, Cambridge, UK, PhD Thesis, Department of Physiology, Development and Neuroscience.
Butler, L. C., Blanchard, G. B., Kabla, A. J., Lawrence, N. J., Welchman, D. P., Mahadevan, L., Adams, R. J., and Sanson, B. (2009). Cell shape changes indicate a role for extrinsic tensile forces in *Drosophila* germ-band extension. *Nat. Cell Biol.* **11,** 859–864.
Campos-Ortega, J. A., and Hartenstein, V. (1997). The Embryonic Development of *Drosophila melanogaster*. 2nd edn. Springer-Verlag, Berlin Heidelberg.
Caussinus, E., Colombelli, J., and Affolter, M. (2008). Tip-cell migration controls stalk-cell intercalation during *Drosophila* tracheal tube elongation. *Curr. Biol.* **18,** 1727–1734.
Chandraratna, D., Lawrence, N., Welchman, D. P., and Sanson, B. (2007). An in vivo model of apoptosis: Linking cell behaviours and caspase substrates in embryos lacking DIAP1. *J. Cell Sci.* **120,** 2594–2608.
Chen, H. H., and Brodland, G. W. (2000). Cell-level finite element studies of viscous cells in planar aggregates. *J. Biomech. Eng.* **122,** 394–401.
Chung, M. I., Nascone-Yoder, N. M., Grover, S. A., Drysdale, T. A., and Wallingford, J. B. (2010). Direct activation of Shroom3 transcription by Pitx proteins drives epithelial morphogenesis in the developing gut. *Development* **137,** 1339–1349.
Conte, V., Munoz, J. J., and Miodownik, M. (2008). A 3D finite element model of ventral furrow invagination in the *Drosophila* melanogaster embryo. *J. Mech. Behav. Biomed. Mater.* **1,** 188–198.
Conte, V., Munoz, J. J., Baum, B., and Miodownik, M. (2009). Robust mechanisms of ventral furrow invagination require the combination of cellular shape changes. *Phys. Biol.* **6,** 016010.
Conti, M. A., and Adelstein, R. S. (2008). Nonmuscle myosin II moves in new directions. *J. Cell Sci.* **121,** 11–18.
Corrigall, D., Walther, R. F., Rodriguez, L., Fichelson, P., and Pichaud, F. (2007). Hedgehog signaling is a principal inducer of Myosin-II-driven cell ingression in *Drosophila* epithelia. *Dev. Cell* **13,** 730–742.
Costa, M., Sweeton, D., and Wieschaus, E. (1993). Gastrulation in *Drosophila*: Cellular mechanisms of morphogenetic movements. *In* "The Development of *Drosophila*

melanogaster," (M. Bate and A. Martinez-Arias, eds.), Cold Spring Harbor Laboratory Press, New York.

Costa, M., Wilson, E. T., and Wieschaus, E. (1994). A putative cell signal encoded by the folded gastrulation gene coordinates cell shape changes during *Drosophila* gastrulation. *Cell* **76,** 1075–1089.

da Silva, S. M., and Vincent, J. P. (2007). Oriented cell divisions in the extending germband of *Drosophila. Development* **134,** 3049–3054.

David, D. J., Tishkina, A., and Harris, T. J. (2010). The PAR complex regulates pulsed actomyosin contractions during amnioserosa apical constriction in *Drosophila. Development* **137,** 1645–1655.

Dawes-Hoang, R. E., Parmar, K. M., Christiansen, A. E., Phelps, C. B., Brand, A. H., and Wieschaus, E. F. (2005). Folded gastrulation, cell shape change and the control of myosin localization. *Development* **132,** 4165–4178.

De Renzis, S., Elemento, O., Tavazoie, S., and Wieschaus, E. F. (2007). Unmasking activation of the zygotic genome using chromosomal deletions in the *Drosophila* embryo. *PLoS Biol.* **5,** e117.

Desprat, N., Supatto, W., Pouille, P. A., Beaurepaire, E., and Farge, E. (2008). Tissue deformation modulates twist expression to determine anterior midgut differentiation in *Drosophila* embryos. *Dev. Cell* **15,** 470–477.

Drees, F., Pokutta, S., Yamada, S., Nelson, W. J., and Weis, W. I. (2005). Alpha-catenin is a molecular switch that binds E-cadherin-beta-catenin and regulates actin-filament assembly. *Cell* **123,** 903–915.

England, S. J., Blanchard, G. B., Mahadevan, L., and Adams, R. J. (2006). A dynamic fate map of the forebrain shows how vertebrate eyes form and explains two causes of cyclopia. *Development* **133,** 4613–4617.

Escudero, L. M., Bischoff, M., and Freeman, M. (2007). Myosin II regulates complex cellular arrangement and epithelial architecture in *Drosophila. Dev. Cell* **13,** 717–729.

Etournay, R., Lepelletier, L., Boutet de Monvel, J., Michel, V., Cayet, N., Leibovici, M., Weil, D., Foucher, I., Hardelin, J. P., and Petit, C. (2010). Cochlear outer hair cells undergo an apical circumference remodeling constrained by the hair bundle shape. *Development* **137,** 1373–1383.

Fernandez-Gonzalez, R., and Zallen, J. A. (2009). Cell mechanics and feedback regulation of actomyosin networks. *Sci. Signal.* **2,** 78.

Fernandez-Gonzalez, R., Simoes Sde, M., Roper, J. C., Eaton, S., and Zallen, J. A. (2009). Myosin II dynamics are regulated by tension in intercalating cells. *Dev. Cell* **17,** 736–743.

Foe, V., Odell, G. M., and Edgar, B. A. (1993). Mitosis and Morphogenesis in the *Drosophila* Embryo: Point and Counterpoint. *In* "The Development of *Drosophila* Melanogaster," (M. Bate and A. Martinez-Arias, eds.), .Cold Spring Harbor Laboratory Press.

Fox, D. T., and Peifer, M. (2007). Abelson kinase (Abl) and RhoGEF2 regulate actin organization during cell constriction in *Drosophila. Development* **134,** 567–578.

Franke, J. D., Montague, R. A., and Kiehart, D. P. (2005). Nonmuscle myosin II generates forces that transmit tension and drive contraction in multiple tissues during dorsal closure. *Curr. Biol.* **15,** 2208–2221.

Fristom, D. (1988). The cellular basis of epithelial morphogenesis. A review. *Tissue Cell* **20,** 645–690.

Gally, C., Wissler, F., Zahreddine, H., Quintin, S., Landmann, F., and Labouesse, M. (2009). Myosin II regulation during C. elegans embryonic elongation: LET-502/ROCK, MRCK-1 and PAK-1, three kinases with different roles. *Development* **136,** 3109–3119.

Gjorevski, N., and Nelson, C. M. (2010a). Endogenous patterns of mechanical stress are required for branching morphogenesis. *Integr. Biol. (Camb)* **2,** 424–434.

Gjorevski, N., and Nelson, C. M. (2010b). The mechanics of development: Models and methods for tissue morphogenesis. *Birth Defects Res. C Embryo Today* **90,** 193–202.

Glickman, N. S., Kimmel, C. B., Jones, M. A., and Adams, R. J. (2003). Shaping the zebrafish notochord. *Development* **130,** 873–887.

Gorfinkiel, N., Blanchard, G. B., Adams, R. J., and Martinez Arias, A. (2009). Mechanical control of global cell behaviour during dorsal closure in *Drosophila. Development* **136,** 1889–1898.

Gutzman, J. H., and Sive, H. (2010). Epithelial relaxation mediated by the myosin phosphatase regulator Mypt1 is required for brain ventricle lumen expansion and hindbrain morphogenesis. *Development* **137,** 795–804.

Hacker, U., and Perrimon, N. (1998). DRhoGEF2 encodes a member of the Dbl family of oncogenes and controls cell shape changes during gastrulation in *Drosophila. Genes Dev.* **12,** 274–284.

Hahn, C., and Schwartz, M. A. (2009). Mechanotransduction in vascular physiology and atherogenesis. *Nat. Rev. Mol. Cell Biol.* **10,** 53–62.

Haigo, S. L., Hildebrand, J. D., Harland, R. M., and Wallingford, J. B. (2003). Shroom induces apical constriction and is required for hingepoint formation during neural tube closure. *Curr. Biol.* **13,** 2125–2137.

Hardin, J., and Walston, T. (2004). Models of morphogenesis: The mechanisms and mechanics of cell rearrangement. *Curr. Opin. Genet. Dev.* **14,** 399–406.

Harris, T. J., and Peifer, M. (2004). Adherens junction-dependent and -independent steps in the establishment of epithelial cell polarity in *Drosophila. J. Cell Biol.* **167,** 135–147.

Harris, T. J., Sawyer, J. K., and Peifer, M. (2009). How the cytoskeleton helps build the embryonic body plan: Models of morphogenesis from *Drosophila. Curr. Top. Dev. Biol.* **89,** 55–85.

Hart, M. J., Jiang, X., Kozasa, T., Roscoe, W., Singer, W. D., Gilman, A. G., Sternweis, P. C., and Bollag, G. (1998). Direct stimulation of the guanine nucleotide exchange activity of p115 RhoGEF by Galpha13. *Science* **280,** 2112–2114.

Hartenstein, V., and Campos-Ortega, J. A. (1985). Fate-mapping in wild-type *Drosophila* melanogaster. 1. The spatio-temporal pattern of embryonic cell divisions. Roux's. *Arch. Dev. Biol.* **194,** 181–195.

Hatini, V., and DiNardo, S. (2001). Divide and conquer: Pattern formation in *Drosophila* embryonic epidermis. *Trends Genet.* **17,** 574–579.

Heisenberg, C. P. (2009). Dorsal closure in *Drosophila*: Cells cannot get out of the tight spot. *Bioessays* **31,** 1284–1287.

Hildebrand, J. D. (2005). Shroom regulates epithelial cell shape via the apical positioning of an actomyosin network. *J. Cell Sci.* **118,** 5191–5203.

Honda, H., Nagai, T., and Tanemura, M. (2008). Two different mechanisms of planar cell intercalation leading to tissue elongation. *Dev. Dyn.* **237,** 1826–1836.

Irvine, K. D., and Wieschaus, E. (1994). Cell intercalation during *Drosophila* germband extension and its regulation by pair-rule segmentation genes. *Development* **120,** 827–841.

Kabla, A. J., Blanchard, G. B., Adams, R. J., and Mahadevan, L. (2010). Bridging Cell and Tissue Behavior in Embryo Development. *In* "Cell Mechanics: From Single Scale-Based Models to Multiscale Modeling," (A. Chauvière, L. Preziosi, and C. Verdier, eds.), Taylor and Francis. Chapman & Hall/CRC Mathematical & Computational Biology.

Kam, Z., Minden, J. S., Agard, D. A., Sedat, J. W., and Leptin, M. (1991). *Drosophila* gastrulation: Analysis of cell shape changes in living embryos by three-dimensional fluorescence microscopy. *Development* **112,** 365–370.

Keller, R. (2002). Shaping the vertebrate body plan by polarized embryonic cell movements. *Science* **298,** 1950–1954.

Keller, P. J., Schmidt, A. D., Wittbrodt, J., and Stelzer, E. H. (2008a). Reconstruction of zebrafish early embryonic development by scanned light sheet microscopy. *Science* **322,** 1065–1069.

Keller, R., Shook, D., and Skoglund, P. (2008b). The forces that shape embryos: Physical aspects of convergent extension by cell intercalation. *Phys. Biol.* **5,** 15007.

Kolsch, V., Seher, T., Fernandez-Ballester, G. J., Serrano, L., and Leptin, M. (2007). Control of *Drosophila* gastrulation by apical localization of adherens junctions and RhoGEF2. *Science* **315,** 384–386.

Landsberg, K. P., Farhadifar, R., Ranft, J., Umetsu, D., Widmann, T. J., Bittig, T., Said, A., Julicher, F., and Dahmann, C. (2009). Increased cell bond tension governs cell sorting at the *Drosophila* anteroposterior compartment boundary. *Curr. Biol.* **19,** 1950–1955.

Larsen, C., Bardet, P. L., Vincent, J. P., and Alexandre, C. (2008). Specification and positioning of parasegment grooves in *Drosophila*. *Dev. Biol.* **321,** 310–318.

Lecuit, T., and Lenne, P. F. (2007). Cell surface mechanics and the control of cell shape, tissue patterns and morphogenesis. *Nat. Rev. Mol. Cell Biol.* **8,** 633–644.

Lee, J. Y., Marston, D. J., Walston, T., Hardin, J., Halberstadt, A., and Goldstein, B. (2006). Wnt/Frizzled signaling controls C. elegans gastrulation by activating actomyosin contractility. *Curr. Biol.* **16,** 1986–1997.

Lee, C., Le, M. P., and Wallingford, J. B. (2009). The shroom family proteins play broad roles in the morphogenesis of thickened epithelial sheets. *Dev. Dyn.* **238,** 1480–1491.

Leptin, M. (2005). Gastrulation movements: The logic and the nuts and bolts. *Dev. Cell* **8,** 305–320.

Major, R. J., and Irvine, K. D. (2005). Influence of Notch on dorsoventral compartmentalization and actin organization in the *Drosophila* wing. *Development* **132,** 3823–3833.

Major, R. J., and Irvine, K. D. (2006). Localization and requirement for Myosin II at the dorsal-ventral compartment boundary of the *Drosophila* wing. *Dev. Dyn.* **235,** 3051–3058.

Martin, A. C. (2010). Pulsation and stabilization: Contractile forces that underlie morphogenesis. *Dev. Biol.* **341,** 114–125.

Martin, P., and Parkhurst, S. M. (2004). Parallels between tissue repair and embryo morphogenesis. *Development* **131,** 3021–3034.

Martin, A. C., Kaschube, M., and Wieschaus, E. F. (2009). Pulsed contractions of an actin-myosin network drive apical constriction. *Nature* **457,** 495–499.

Martin, A. C., Gelbart, M., Fernandez-Gonzalez, R., Kaschube, M., and Wieschaus, E. F. (2010). Integration of contractile forces during tissue invagination. *J. Cell Biol.* **188,** 735–749.

Martinez-Arias, A. (1993). Development and Patterning of the Larval Epidermis of *Drosophila*. *In* "The Development of *Drosophila* Melanogaster," (M. Bate and A. Martinez-Arias, eds.), Cold Spring Harbor Laboratory Press.

Mavrakis, M., Rikhy, R., and Lippincott-Schwartz, J. (2009). Plasma membrane polarity and compartmentalization are established before cellularization in the fly embryo. *Dev. Cell* **16,** 93–104.

Mazumdar, A., and Mazumdar, M. (2002). How one becomes many: Blastoderm cellularization in *Drosophila* melanogaster. *Bioessays* **24,** 1012–1022.

McGill, M. A., McKinley, R. F., and Harris, T. J. (2009). Independent cadherin-catenin and Bazooka clusters interact to assemble adherens junctions. *J. Cell Biol.* **185,** 787–796.

McMahon, A., Supatto, W., Fraser, S. E., and Stathopoulos, A. (2008). Dynamic analyses of *Drosophila* gastrulation provide insights into collective cell migration. *Science* **322,** 1546–1550.

McMahon, A., Reeves, G. T., Supatto, W., and Stathopoulos, A. (2010). Mesoderm migration in *Drosophila* is a multi-step process requiring FGF signaling and integrin activity. *Development* **137,** 2167–2175.

Monier, B., Pelissier-Monier, A., Brand, A. H., and Sanson, B. (2010). An actomyosin-based barrier inhibits cell mixing at compartmental boundaries in *Drosophila* embryos. *Nat. Cell Biol.* **12,** 60–65.

Montell, D. J. (2008). Morphogenetic cell movements: Diversity from modular mechanical properties. *Science* **322,** 1502–1505.
Morize, P., Christiansen, A. E., Costa, M., Parks, S., and Wieschaus, E. (1998). Hyperactivation of the folded gastrulation pathway induces specific cell shape changes. *Development* **125,** 589–597.
Mulinari, S., Barmchi, M. P., and Hacker, U. (2008). DRhoGEF2 and diaphanous regulate contractile force during segmental groove morphogenesis in the *Drosophila* embryo. *Mol. Biol. Cell* **19,** 1883–1892.
Muller, H. A., and Wieschaus, E. (1996). armadillo, bazooka, and stardust are critical for early stages in formation of the zonula adherens and maintenance of the polarized blastoderm epithelium in *Drosophila*. *J. Cell Biol.* **134,** 149–163.
Murray, M. J., and Saint, R. (2007). Photoactivatable GFP resolves *Drosophila* mesoderm migration behaviour. *Development* **134,** 3975–3983.
Myat, M. M. (2005). Making tubes in the *Drosophila* embryo. *Dev. Dyn.* **232,** 617–632.
Nikolaidou, K. K., and Barrett, K. (2004). A Rho GTPase signaling pathway is used reiteratively in epithelial folding and potentially selects the outcome of Rho activation. *Curr. Biol.* **14,** 1822–1826.
Nishimura, T., and Takeichi, M. (2008). Shroom3-mediated recruitment of Rho kinases to the apical cell junctions regulates epithelial and neuroepithelial planar remodeling. *Development* **135,** 1493–1502.
Nishimura, M., Inoue, Y., and Hayashi, S. (2007). A wave of EGFR signaling determines cell alignment and intercalation in the *Drosophila* tracheal placode. *Development* **134,** 4273–4282.
Nussein-Volhard, C., and Wieschaus, E. (1990). Looking at Embryos. *In* "*Drosophila*, a Practical Approach," (D. B. Roberts, ed.), Oxford University Press, Oxford.
Oda, H., and Tsukita, S. (2001). Real-time imaging of cell-cell adherens junctions reveals that *Drosophila* mesoderm invagination begins with two phases of apical constriction of cells. *J. Cell Sci.* **114,** 493–501.
Plageman, T. F., Jr., Chung, M. I., Lou, M., Smith, A. N., Hildebrand, J. D., Wallingford, J. B., and Lang, R. A. (2010). Pax6-dependent Shroom3 expression regulates apical constriction during lens placode invagination. *Development* **137,** 405–415.
Pope, K. L., and Harris, T. J. (2008). Control of cell flattening and junctional remodeling during squamous epithelial morphogenesis in *Drosophila*. *Development* **135,** 2227–2238.
Pouille, P. A., and Farge, E. (2008). Hydrodynamic simulation of multicellular embryo invagination. *Phys. Biol.* **5,** 15005.
Pouille, P. A., Ahmadi, P., Brunet, A. C., and Farge, E. (2009). Mechanical signals trigger Myosin II redistribution and mesoderm invagination in *Drosophila* embryos. *Sci. Signal.* **2,** 16.
Quintin, S., Gally, C., and Labouesse, M. (2008). Epithelial morphogenesis in embryos: Asymmetries, motors and brakes. *Trends Genet.* **24,** 221–230.
Rauzi, M., Verant, P., Lecuit, T., and Lenne, P. F. (2008). Nature and anisotropy of cortical forces orienting *Drosophila* tissue morphogenesis. *Nat. Cell Biol.* **10,** 1401–1410.
Rolo, A., Skoglund, P., and Keller, R. (2009). Morphogenetic movements driving neural tube closure in Xenopus require myosin IIB. *Dev. Biol.* **327,** 327–338.
Roszko, I., Sawada, A., and Solnica-Krezel, L. (2009). Regulation of convergence and extension movements during vertebrate gastrulation by the Wnt/PCP pathway. *Semin. Cell Dev. Biol.* **20,** 986–997.
Sanson, B. (2001). Generating patterns from fields of cells: Examples from *Drosophila* segmentation. *EMBO Rpts* **2,** 1083–1088.
Sawyer, J. M., Harrell, J. R., Shemer, G., Sullivan-Brown, J., Roh-Johnson, M., and Goldstein, B. (2010). Apical constriction: A cell shape change that can drive morphogenesis. *Dev. Biol.* **341,** 5–19.

Schaerlinger, B., Launay, J. M., Vonesch, J. L., and Maroteaux, L. (2007). Gain of affinity point mutation in the serotonin receptor gene 5-HT2Dro accelerates germband extension movements during *Drosophila* gastrulation. *Dev. Dyn.* **236,** 991–999.

Schock, F., and Perrimon, N. (2002). Cellular processes associated with germ band retraction in *Drosophila. Dev. Biol.* **248,** 29–39.

Schock, F., and Perrimon, N. (2003). Retraction of the *Drosophila* germ band requires cell-matrix interaction. *Genes Dev.* **17,** 597–602.

Seher, T. C., Narasimha, M., Vogelsang, E., and Leptin, M. (2007). Analysis and reconstitution of the genetic cascade controlling early mesoderm morphogenesis in the *Drosophila* embryo. *Mech. Dev.* **124,** 167–179.

Shen, J., Dorner, C., Bahlo, A., and Pflugfelder, G. O. (2008). optomotor-blind suppresses instability at the A/P compartment boundary of the *Drosophila* wing. *Mech. Dev.* **125,** 233–246.

Sherrard, K., Robin, F., Lemaire, P., and Munro, E. (2010). Sequential activation of apical and basolateral contractility drives ascidian endoderm invagination. *Curr. Biol.* **20,** 1499–1510.

Simoes Sde, M., Blankenship, J. T., Weitz, O., Farrell, D. L., Tamada, M., Fernandez-Gonzalez, R., and Zallen, J. A. (2010). Rho-kinase directs Bazooka/Par-3 planar polarity during *Drosophila* axis elongation. *Dev. Cell* **19,** 377–388.

Simone, R. P., and DiNardo, S. (2010). Actomyosin contractility and Discs large contribute to junctional conversion in guiding cell alignment within the *Drosophila* embryonic epithelium. *Development* **137,** 1385–1394.

Skoglund, P., Rolo, A., Chen, X., Gumbiner, B. M., and Keller, R. (2008). Convergence and extension at gastrulation require a myosin IIB-dependent cortical actin network. *Development* **135,** 2435–2444.

Solnica-Krezel, L. (2005). Conserved patterns of cell movements during vertebrate gastrulation. *Curr. Biol.* **15,** R213–R228.

Solon, J., Kaya-Copur, A., Colombelli, J., and Brunner, D. (2009). Pulsed forces timed by a ratchet-like mechanism drive directed tissue movement during dorsal closure. *Cell* **137,** 1331–1342.

Somlyo, A. P., and Somlyo, A. V. (2003). Ca2+ sensitivity of smooth muscle and nonmuscle myosin II: modulated by G proteins, kinases, and myosin phosphatase. *Physiol. Rev.* **83,** 1325–1358.

St Johnston, D., and Nusslein-Volhard, C. (1992). The origin of pattern and polarity in the *Drosophila* embryo. *Cell* **68,** 201–219.

Sweeton, D., Parks, S., Costa, M., and Wieschaus, E. (1991). Gastrulation in *Drosophila*: The formation of the ventral furrow and posterior midgut invaginations. *Development* **112,** 775–789.

Tepass, U., Tanentzapf, G., Ward, R., and Fehon, R. (2001). Epithelial cell polarity and cell junctions in *Drosophila. Annu. Rev. Genet.* **35,** 747–784.

Umemori, M., Takemura, M., Maeda, K., Ohba, K., and Adachi-Yamada, T. (2007). *Drosophila* T-box transcription factor Optomotor-blind prevents pathological folding and local overgrowth in wing epithelium through confining Hh signal. *Dev. Biol.* **308,** 68–81.

VanHook, A., and Letsou, A. (2008). Head involution in *Drosophila*: Genetic and morphogenetic connections to dorsal closure. *Dev. Dyn.* **237,** 28–38.

Vicente-Manzanares, M., Ma, X., Adelstein, R. S., and Horwitz, A. R. (2009). Non-muscle myosin II takes centre stage in cell adhesion and migration. *Nat. Rev. Mol. Cell Biol.* **10,** 778–790.

Vincent, J. P., and O'Farrell, P. H. (1992). The state of engrailed expression is not clonally transmitted during early *Drosophila* development. *Cell* **68,** 923–931.

Vincent, A., Blankenship, J. T., and Wieschaus, E. (1997). Integration of the head and trunk segmentation systems controls cephalic furrow formation in *Drosophila. Development* **124,** 3747–3754.

Weaire, D., and Hutzler, S. (2001). *The Physics of Foams.* Oxford University Press, Oxford.
Weiser, D. C., Row, R. H., and Kimelman, D. (2009). Rho-regulated myosin phosphatase establishes the level of protrusive activity required for cell movements during zebrafish gastrulation. *Development* **136,** 2375–2384.
Wilson, R., and Leptin, M. (2000). Fibroblast growth factor receptor-dependent morphogenesis of the *Drosophila* mesoderm. *Philos. Trans. R. Soc. Lond. B Biol. Sci.* **355,** 891–895.
Witzberger, M. M., Fitzpatrick, J. A., Crowley, J. C., and Minden, J. S. (2008). End-on imaging: A new perspective on dorsoventral development in *Drosophila* embryos. *Dev. Dyn.* **237,** 3252–3259.
Yamada, S., Pokutta, S., Drees, F., Weis, W. I., and Nelson, W. J. (2005). Deconstructing the cadherin-catenin-actin complex. *Cell* **123,** 889–901.
Zallen, J. A., and Blankenship, J. T. (2008). Multicellular dynamics during epithelial elongation. *Semin. Cell Dev. Biol.* **19,** 263–270.
Zallen, J. A., and Wieschaus, E. (2004). Patterned gene expression directs bipolar planar polarity in *Drosophila. Dev. Cell* **6,** 343–355.
Zhang, H., Gally, C., and Labouesse, M. (2010). Tissue morphogenesis: How multiple cells cooperate to generate a tissue. *Curr. Opin. Cell Biol.* **22**(5), 575–582.

CHAPTER SIX

Cell Sorting in Development

S. F. Gabby Krens *and* Carl-Philipp Heisenberg

Contents

Abstract

During the development of multicellular organisms, cell fate specification is followed by the sorting of different cell types into distinct domains from where the different tissues and organs are formed. Cell sorting involves both the segregation of a mixed population of cells with different fates and properties into distinct domains, and the active maintenance of their segregated state. Because of its biological importance and apparent resemblance to fluid segregation in physics, cell sorting was extensively studied by both biologists and physicists over the last decades. Different theories were developed that try to explain cell sorting on the basis of the physical properties of the constituent cells. However, only recently the molecular and cellular mechanisms that control the physical properties driving cell sorting, have begun to be unraveled. In this review, we will provide an overview of different cell-sorting processes

Institute of Science and Technology Austria, Klosterneuburg, Austria

Current Topics in Developmental Biology, Volume 95
ISSN 0070-2153, DOI: 10.1016/B978-0-12-385065-2.00006-2

in development and discuss how these processes can be explained by the different sorting theories, and how these theories in turn can be connected to the molecular and cellular mechanisms driving these processes.

1. Introduction

The development of multicellular organisms relies on the specification of different cell types and their sorting and assembly into distinct tissues and organs. Generally, cell sorting describes both the segregation of a mixed population of cells with different properties into distinct domains, and the active maintenance of their segregated state. The process of cell sorting appears to be conserved between vertebrates and invertebrates, and was postulated to occur in embryonic, postnatal, adult, and diseased tissues. Cell sorting and assembly is controlled by various molecular interactions among the constituent cells, but is ultimately governed by differences between the physical properties of the specific cell types. The earliest notion that physical cell properties are important for cell sorting was provided by Holtfreter and colleagues (Townes and Holtfreter, 1955), who showed that dissociated cells from gastrulating amphibian embryos are able to aggregate into their distinct germ layers. This sorting behavior was initially attributed to differential cell type/tissue specific "affinity" or "adhesion," although the molecular and cellular definition of these terms remained unclear. Here, we will give an overview of the principles underlying cell sorting in development, and describe exemplary developmental processes in which cell sorting is observed.

2. Principles of Cell Sorting

Cell sorting is the process by which a heterotypic mix of cells segregates into distinct homotypic domains. The outcome of cell-sorting events occurring *in vivo* is frequently reproduced *ex vivo* in heterotypic cocultures of primary cells, although the sorting dynamics *in vitro* often differs considerably from the situation *in vivo* (Armstrong, 1989). In order to explain cell sorting on the basis of the physical properties of the constituent cells, different theories were proposed, of which some were also experimentally addressed. Most of these theories assume that the constituent cells do not change cell fate during the sorting process, although position-dependent cell fate changes were observed in long-term *in vitro* experiments and are often associated with *in vivo* cell-sorting processes (Armstrong, 1989; Rossant *et al.*, 2003). In the following section, we will summarize some of the most prevalent theories explaining cell sorting in situations in which cell fate changes are not observed.

2.1. Liquid behavior of tissues

In an attempt to explain cell sorting, Steinberg and colleagues first proposed that cell sorting in heterotypic cell aggregates resembles the segregation of Newtonian fluids in dispersion (Steinberg, 1962a). The segregation behavior of Newtonian fluids is controlled by their relative surface and interfacial tensions (Rowlinson and Widom, 1989), determining both the dynamics of the segregation process and the geometrical organization in the segregated state. Consequently, cell sorting was proposed to be driven by differences in the surface tension of the tissues formed during the segregation process. Liquids surface tension is defined by intermolecular forces that drive the reorganization of the system to a geometric organization that maximizes intermolecular attraction and minimizes free surface energy. When two liquids are mixed, the geometric organization of the segregated state is such that the liquid molecules with stronger intermolecular forces form a central mass, which is surrounded by the liquid characterized by weaker intermolecular forces. To understand how differences in tissue surface tension (TST) drive cell sorting, one therefore needs to identify which molecular and cellular processes in multicellular aggregates are analogous to intermolecular forces in liquids.

2.2. Differential adhesion hypothesis

The most widely accepted theoretical explanation of cell sorting is the "differential adhesion hypothesis" (DAH). The DAH interprets cell sorting as the tendency of cells to maximize their intercellular adhesion in the same way that liquids maximize their intermolecular attraction and concurrently minimize the free surface energy. Consequently, the DAH regards the cells of an aggregate as being equivalent to the molecules of a liquid, and forces of adhesion between cells as being equivalent to intermolecular forces between liquid molecules (Foty *et al.*, 1994, 1996; Heintzelman *et al.*, 1978). Cell adhesion can be thought of as the energy released by a cell upon binding of a cell to another cell or matrix. In a heterotypic aggregate of cell types A and B, we define the (reversible) work (W) of homotypic cell–cell adhesion (cohesion) for each cell type as $W_{\text{A–A}}$ and $W_{\text{B–B}}$, and the work of heterotypic adhesion between the two cell types as $W_{\text{A–B}}$. The relative values of homotypic and heterotypic work of adhesion for cell types A and B will determine the final configuration of a heterotypical aggregate after sorting. Complete engulfment of tissue A by tissue B, for example, is achieved under the following conditions: $0 < W_{\text{B–B}} < W_{\text{A–B}} < (W_{\text{A–A}} + W_{\text{B–B}}/2) < W_{\text{A–A}}$ (Borghi and Nelson, 2009; Foty and Steinberg, 2005; Foty *et al.*, 1996; Steinberg, 1962b). The work associated with adhesion, and consequently TST, was shown to be correlated with the specific amount of adhesion molecules expressed in a cell type (Foty and Steinberg, 2005), although the expression level of adhesion molecules is only one of many cell properties determining TST (see also Section 2.3).

2.3. Differential surface contraction hypothesis

In addition to cell adhesion, cell contraction was proposed to be critical for cell sorting. The 'differential surface contraction hypothesis' (DSCH), first put forward by Harris (1976), proposes that TST is not solely dependent on cell adhesion, but also on the ability of tissues to assemble a contractile actin–myosin network at the tissue surface (cell-to-medium interface). Differential cell contractility leads to differences in TST and consequently determines the order of cell sorting, when the differences in contractility between cell/tissue types are interface specific. High cell contractility at the cell-to-medium interface increases TST, whereas high cell contractility at the cell-to-cell interface will decrease cell–cell adhesion and, as a result TST. Only when contractility between cell/tissue types is different at one, but not the other of these interfaces, will differential contractility lead to differences in TST and thus influence the outcome of cell sorting (Harris, 1976). Although the DSCH was criticized for being incompatible with the notion of tissues behaving like liquids, as liquid molecules are not contractile (Armstrong, 1989; Steinberg, 1996), it is in line with observation that cell aggregates assemble contractile elements at their surface. Recent studies in zebrafish have provided direct supporting evidence in favor of the DSCH. In gastrulating zebrafish embryos, ectoderm progenitor cells have a higher contractility than both mesoderm and endoderm cells. Since contractile elements are localized at the surface of germ layer progenitor cell aggregates, a higher contractility of ectoderm progenitors coincides with segregation to the center of heterotypic aggregates with mesoderm and/or endoderm cells (Krieg *et al.*, 2008).

A recent theory aims at explaining TST by the combined activities of cell adhesion and contractility. This theory adds to the DAH, as it predicts that there are regimes where the DAH does not correctly forecast TST, and that changes in TST will be accompanied by changes in the shape of surface cells as a function of the ratio between adhesion and contractility (Manning *et al.*, 2010). The experimental evidence for this theory is still very limited and thus needs further evaluation of the actual contributions of cell adhesion and contractility to TST.

2.4. Differential interfacial tension hypothesis

The 'differential interfacial tension hypothesis' (DITH; Brodland, 2002) combines elements from both the DAH and DSCH theories. It is based on analytic mechanics showing that differences in adhesion alone are not sufficient to drive cell sorting, and that cell sorting is instead driven by the combined effect of several force generators, which generate cell interfacial tension (Brodland and Chen, 2000). The DITH does not treat cells as liquid molecules, but rather proposes that cells generate mechanical tension at their cell-to-cell and cell-to-medium interfaces (interfacial tension), which depends largely on cell contraction and adhesion. The membrane and the underlying cortical cytoskeletal components are thought to generate contractile forces (F_{mem} and F_{mf}, respectively), which reduce the contact area, while cell adhesion creates an adhesive force (F_{adh}), which acts in the opposite direction and thus increases the contact area. Consequently, the interfacial tension of cell A at the cell-to-medium interface can be expressed as $\gamma = F_{mem}{}^{A} + F_{mf}{}^{A} - F_{adh}{}^{AM} + F_{other}{}^{AM}$, where F_{other} is the force resulting from the contraction of noncortical elements and viscous properties of the cell and its constituent organelles, reducing the interfacial area. The interfacial tension at the interface between cells A and B can be expressed as $\gamma = F_{mem}{}^{A} + F_{mem}{}^{B} + F_{mf}{}^{A} + F_{mf}{}^{B} - F_{adh}{}^{AB} + F_{other}{}^{AB}$.

In a heterotypic aggregate, cells of type B tend to envelop cells of type A, when $\gamma^{AM} > \gamma^{BM}$, and fully engulf the latter when $\gamma^{AM} > \gamma^{AB} + \gamma^{BM}$. The DITH differs from the DAH and the DSCH theories by combining cell contraction and adhesion and taking into account the possibility of additional properties of the cell (e.g., cytoplasm viscosity, density of the cell content) that are likely to influence interfacial tensions and, consequently, drive cell sorting.

2.5. Other factors contributing to cell sorting

Various other mechanisms were proposed to explain cell sorting, such as differences in cell motility (Jones *et al.*, 1989), specific adhesion (Curtis, 1960; Moscona, 1962; Townes and Holtfreter, 1955), chemotaxis, directed migration (Townes and Holtfreter, 1955), and the timing of sorting initiation (Armstrong, 1989; Curtis, 1961; Steinberg, 1996). Most of these mechanisms are likely to be insufficient to explain cell sorting, as the experimental patterns of cell sorting generally do not correspond to the patterns predicted by those mechanisms (Steinberg, 1996). Several other cell properties, such as cell attraction and repulsion, membrane fluctuations, and cell polarity, were also implicated to influence cell sorting, and were used in mathematical models simulating cell sorting (Armstrong, 1989; Brodland and Chen, 2000; Glazier and Graner, 1993; Maeda *et al.*, 2007; Mombach, 1999; Palsson, 2001; Shinbrot *et al.*, 2009; Steinberg, 1996; Voß-Böhme and Deutsch, 2009). However, convincing experimental evidence that any of those factors plays a critical role is still missing.

3. Measuring Tissue Surface Tension (TST)

While the molecular and cellular basis of TST continues to be intensely debated, there is ample evidence that differences in TST influence the spatial organization of tissues during sorting. Studies analyzing the function of TST in cell sorting used several different methods to measure

TST (Fig. 6.1). Surface tension is the energy required to increase the surface area of a liquid by a unit amount. Assuming that individual cells in an aggregate possess unconstrained motile properties, like for molecules in a liquid, any deformation of the spherical aggregate will induce translocation of particles (cells) from the interior to the exterior of the aggregate. Consequently, any increase in aggregate surface area is accompanied by a loss of cell-to-cell interface area. Based on these assumptions, TST directly relates to cell's interfacial tensions and can be expressed as $\sigma_{cm} = \gamma_{cm} - (\gamma_{cc}/2)$, where σ_{cm} is the bulk surface tension, and γ_{cc} and γ_{cm} are the cell-to-cell and cell-to-medium interfacial tensions, respectively (Brodland *et al.*, 2009). Currently at least three methods to measure TST are used, all of which are based on determining the forces required to deform the tissue surface.

3.1. Parallel plate compression apparatus (PPCA)

The surface tension of tissues can be measured by deforming an aggregate under a known force using a tissue surface tensiometer, or PPCA (Fig. 6.1A; Brodland *et al.*, 2009; Foty *et al.*, 1994; Mgharbel *et al.*, 2009; Narotte *et al.*, 2008). The aggregate is placed between two parallel plates, which are moved together to compress the tissue to a certain degree, thereby deforming the tissue by applying a constant strain. Both the compression force and the shape change of the aggregate are simultaneously recorded during the time that the aggregate takes to reach a new force–equilibrium under fixed compression. Typically, cell aggregates display a biphasic behavior under compression: an initial viscoelastic response on a short time scale, followed by a fluid like viscous behavior on longer time scales. The initial viscoelastic behavior is attributed to the rapid deformation of the cells within the aggregates (solid-like behavior), whereas the long-term viscous behavior is likely to be associated to cellular rearrangements and neighbor exchanges. The resistance force at equilibrium is the force necessary to keep the aggregate in its deformed configuration. In analogy with the surface tension of a liquid drop compressed between two plates (Adamson, 1967), the surface tension of the aggregate can be calculated by applying the Young–Laplace equation (Davis, 1984). The measured compression force at equilibrium (F_{eq}) results from the pressure difference (ΔP) between the aggregates interior and the external medium upon deformation. The force difference is distributed over the contact area of the aggregate with the outside medium as a function of its surface tension (σ). The TST can thus be calculated from the compression force and radii of curvature of the compressed aggregate at its interface with the outside medium using the Young–Laplace equation with $F_{eq} = \pi R_3^2 \sigma [(1/R_1) + (1/R_2)]$ (Adamson, 1967; Foty *et al.*, 1996), where R_1 and R_2 are the principal radii of curvature, and R_3 is the area of contact between the aggregate and either of the parallel compression plates (Fig. 6.1A). However, recent publications have shown

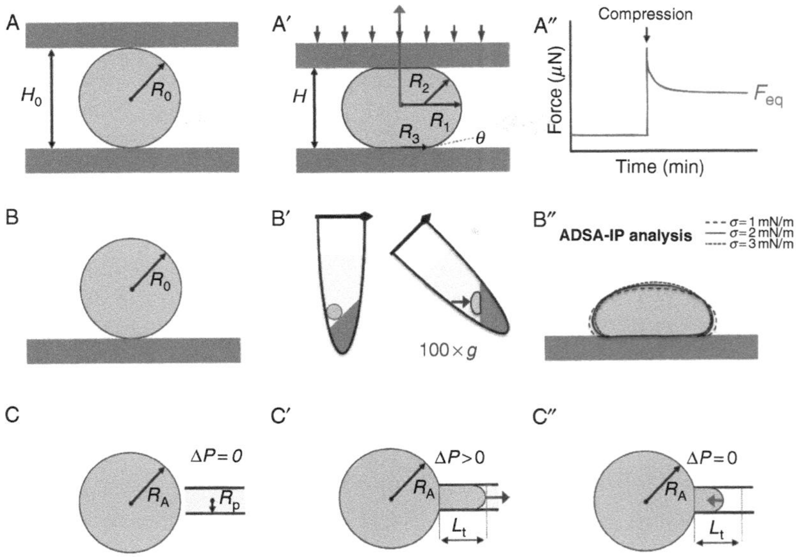

Figure 6.1 Principles of tissue surface tensiometry. (A) PPCA. A spherical aggregate with radius R_0 is placed between two nonadhesive compression plates which are moved together until $H_0 = 2 \times R_0$. The aggregate deformation is initiated by moving the plates together until height H is reached, thereby exerting a compressive force on the aggregate (blue arrows in (A′)), while the force exerted by the aggregate upon the upper compression plate (red arrow in (A′)) is monitored as a function of time (red line in (A″)). Over time, a new shape equilibrium is reached with principle radii of curvature R_1 at the aggregates equator, R_2 in a plane through its axis of symmetry and R_3 as the radius of the contact area with either compression plate. is the contact angle of the aggregate with the plate. Shape equilibrium in the compressed aggregate becomes apparent by the stabilization of the recorded force (red line in (A″)) and aggregate shape. The geometric characteristics of the aggregate and the force at equilibrium (F_{eq}) are used to determine the surface tension by applying Laplace's law. (B) Centrifugation/gravitational force. Rounded aggregates (B) are placed in a tube on an agar base and are exposed to a centrifugal force of $100 \times g$ (B′), after which the aggregate is fixed immediately and photographed. Laplacian curves with different surface tensions are fitted to the aggregate outline, using the Asymmetric Drop Shape Analysis for Imperfect Profiles (ADSA-IP) algorithm. The generated profiles are superimposed on the aggregate (dashed lines) (B″), and the best fit is used to determine the aggregate surface tension. (C) Micropipette aspiration: Rounded aggregates with radius R_A are aspirated using a pipette with radius R_P (C). Cells from the aggregate start to flow into the pipette after applying $\Delta P > 0$ (blue arrow in (C′)) and flow back into the aggregate after release of the applied pressure (red arrow in (C″)). The length L_t of the tongue in the pipette is recorded over time, from which the aspiration and retraction curves of the cells in the pipette are obtained. The slopes of these curves are used to determine the critical pressure ΔP_c, which is used to determine the surface tension by applying Laplace's law (see also main text). (For interpretation of the references to color in this figure legend, the reader is referred to the Web version of this chapter.)

that the use of this formula may introduce inaccuracies in the computation of TSTs as in particular R_3 is difficult to measure. Alternative methods were therefore introduced where TST is computed on the basis of R_1 and R_2 only, with $F_{eq} = \pi R_1(R_1/R_2 - 1)\sigma$ (Mgharbel *et al.*, 2009; Narotte *et al.*, 2008).

In addition to the TST, the aggregate compression can also be used to calculate tissue viscosity. In analogy with the behavior of viscous liquids, tissue viscosity can be estimated by analyzing aggregate shape recovery after long-term compressions (Marmottant *et al.*, 2009). Since tissue viscosity is an important factor influencing cell-sorting dynamics, and the dynamics of the viscous–elastic biphasic response in an aggregate compression experiment, determining it will facilitate the interpretation of the data obtained in TST measurements.

3.2. Centrifugation/gravitational force

Initial attempts to measure TST were centered around experiments, which analyzed the resistance of cell aggregates to flattening in response to a known centrifugal force (Phillips and Steinberg, 1969). More recently, similar methods of deforming tissues by a known centrifugal force (g) have been used to determine the TST of *Xenopus* explants (Kalantarian *et al.*, 2009; Fig. 6.1B). Small rounded-up tissue fragments were exposed to a predetermined centrifugal force to deform the aggregate. The resulting deformed shape of the aggregate was used to calculate the TST by fitting a set of Laplacian profiles to its altered shape using an Axisymmetric Liquid–Fluid Interfaces (ALFI) computer algorithm (Ninomiya and Winklbauer, 2008). This method has most recently been optimized to be able to cope with the analysis of tissue explants with irregular surfaces (Kalantarian *et al.*, 2009). The disadvantage of the centrifugation method is that the tissue needs to be fixed after centrifugation, before the shape analysis, which possibly influences the aggregates shape and thereby might modulate the obtained TST values. In addition, this method does not allow for the analysis of the dynamics of tissue compression and relaxation. Tissue fixation can be avoided by analyzing tissue deformation at normal gravitation ($1 \times g$) for tissues, such as *Xenopus* explants, which will deform by their own weight under normal gravitation. However, this approach does not allow for the analysis of the viscoelastic properties of the tissue, as tissue relaxation will not occur.

3.3. Pipette aspiration

Micropipette aspiration represents another more recently developed technique to determine the surface tension (σ), elastic constant (E), and viscosity (η) of cell aggregates (Guevorkian *et al.*, 2010; Fig. 6.1C). Micropipette aspiration was developed to evaluate the viscoelastic properties of single cells

(Evans and Yeung, 1989), the stiffness of tissues upon small deformations (Von Dassow and Davidson, 2009), and the separation force of cell doublets (Chu *et al.*, 2006; Paluch and Heisenberg, 2009; Sung *et al.*, 1986). By scaling the pipette size so that its diameter does not prohibit individual cells from entering the pipette, a recent study has shown that micropipette aspiration can also be used to measure the mechanical properties of multicellular aggregates (Guevorkian *et al.*, 2010). In short, spherical cell aggregates are deformed by aspirating cells into the pipette under a constant known pressure ΔP (constant stress), while recording the deformation of the aggregate (strain). The viscoelastic properties are determined by plotting the flow of cellular material into the pipette during aspiration under constant ΔP. The observed flow behavior is usually biphasic for cell aggregates, with an initial fast viscoelastic deformation, followed by a viscous flow. The cell aggregate is modeled as a viscoelastic Newtonian fluid, with a surface tension σ, and assuming volume conservation, the critical pressure ΔP_c is thus related to the surface tension by the Laplace's law with $\Delta P_c = 2\sigma (1/R_p - 1/R_A)$, where R_p and R_A are the radii of the pipette and the aggregate, respectively. The critical pressure ΔP_c can be deduced from the aspiration and retraction flow rates $\Delta P_c = \Delta P \dot{L}^r_\infty (\dot{L}^r_\infty + \dot{L}^a_\infty)$, where $\dot{L}^a_\infty$ and $\dot{L}^r_\infty$ are the aspiration and retraction flow rates, respectively, which can be determined by plotting the deformation of the aggregate into the pipette over time.

Interestingly, the results of the pipette aspiration experiments suggest that TST may be strain-dependent (Guevorkian *et al.*, 2010), indicative of an active response of the tissue surface to the applied deformation. No such dependence was found in TST measurements using a PPCA. This difference is likely to be due to differences in deforming the cell aggregate by these two methods: while constant stress (force per area) is applied in the micropipette aspiration experiments, constant strain (deformation) is induced in the PPCA experiments. Moreover, while in the PPCA experiments the whole aggregate is probed, only a small area of the aggregate is analyzed by micropipette aspiration. Future experiments directly comparing TST measurements in experiments using PPCA and micropipette aspiration, and comparing constant strain versus constant stress measurements, will be required to determine the influence of stress versus strain and local versus global effects on TST.

4. Cell Sorting in Development

Bona fide cell sorting is a rare event in development as tissues containing random mixtures of different cell types are difficult to find. Instead, tissues usually contain partially segregated populations of cells with different induced fates, which are separated into distinct domains by changing their relative positions to each other during the course of development. These processes contain some aspects of cell sorting, but also involve distinct processes such as tissue border formation, border maintenance, and border maturation. In the following, we will describe developmental processes in which some aspects of cell sorting are implicated.

4.1. Induced cell-sorting processes

Initial evidence for cell sorting in development came from classical experiments on cell sorting *in vitro*, in which whole embryos were dissociated into single cells, and their reaggregating and segregation into distinct tissues were documented (Townes and Holtfreter, 1955). Prominent examples of this include the reaggregation and segregation of germ layer progenitor cells from gastrulating *Rana pipiens*, *Xenopus*, and zebrafish embryos (Davis *et al.*, 1997; Klopper *et al.*, 2010; Ninomiya and Winklbauer, 2008; Townes and Holtfreter, 1955). These experiments revealed important information about the sorting potential of different cell mixtures. However, they do not provide evidence for a critical involvement of cell sorting *in vivo*. One reason for this is that, in reaggregation experiments, cells are brought into contact, which do not necessarily contact each other within the developing embryo (Foty *et al.*, 1996). *In vitro* cell-sorting experiments were also used to determine the specific contribution of various molecular and cellular features, such as adhesion molecule expression and cell motility, to cell sorting. By changing the activity and/or expression of specific proteins through the application of pharmacological drugs or the use of genetically modified cells, the expression and/or activity of cell–cell adhesion molecules and cytoskeletal elements were shown to be capable of influencing the outcome of cell sorting (Duguay *et al.*, 2003; Foty and Steinberg, 2005; Krieg *et al.*, 2008).

The development of various genetic tools also allowed for the analysis of induced cell sorting *in vivo*. To date, this has mainly been achieved by generating genetic mosaic embryos through the transplantation of genetically modified cells or the stochastic/targeted induction of gene expression in a subset of cells within the embryos. These mosaic embryos can then be used to observe alterations in cell behavior/sorting associated with the induced change in the expression of specific genes. In *Xenopus* embryos for example, injection of single blastomeres with mRNA and/or antisense oligonucleotides to induce mosaic gene expression was used to demonstrate that cell–cell and cell–matrix adhesion molecules, such as NCAM (Kintner, 1988), Cadherins (Detrick *et al.*, 1990), Protocadherins (Chen and Gumbiner, 2006a), and Integrins (Marsden and DeSimone, 2003), can induce cell sorting *in vivo*. Likewise in *Drosophila* and mouse, the induction of clonal gene expression in a spatially and temporally controlled manner by transgenesis was employed to investigate the ability of different genes to induce cell sorting within specific embryonic and adult tissues, such as the eye (Hayashi and Carthew, 2004) and imaginal disc in *Drosophila* (Wei *et al.*, 2005) and the blastocyst in mouse (Moore *et al.*, 2009; Plusa *et al.*, 2008; Rossant *et al.*, 2003).

4.2. Endogenous cell sorting and segregation processes

During development, progenitor cells of different tissues and organs, which are initially specified in partially overlapping domains, eventually need to segregate into separate compartments, such as germ layers and organ primordia. Differences in adhesive, mechanical, and motile cell properties are thought to drive the initial "sorting out" of different progenitor cell populations into segregated compartments.

In the following, we will summarize the roles of cell sorting and segregation in three exemplary developmental processes: germ layer formation during zebrafish and *Xenopus* gastrulation, blastocyst development in mouse, and limb bud formation in chicken.

4.2.1. Cell segregation during zebrafish and *Xenopus* gastrulation

Gastrulation is the first major morphogenetic process in the development of most metazoans, leading to the formation of the three germ layers: ectoderm, mesoderm, and endoderm. During gastrulation, specification of the different germ layer progenitor cell types is accompanied by their segregation into distinct cell layers, the prospective ectoderm, mesoderm, and endoderm germ layers. Eventually, this leads to the formation of a multilayered gastrula consisting of endoderm in the interior, enveloped by mesoderm, which is itself surrounded by ectoderm at the exterior of the gastrula. A stereotypic and evolutionarily conserved set of morphogenetic cell and tissue movements governs germ layer formation and organization: *epiboly,* the spreading and thinning of the germ layers; *internalization*, the movement of mesoderm and endoderm progenitors to the inside of the gastrula; and *convergence and extension* movements, the narrowing and extension of the forming body axis (Solnica-Krezel, 2005).

Germ layer progenitor cell segregation is triggered by the internalization of mesoderm and endoderm progenitors. Distinct types of progenitor cell internalization movements were observed in different organisms. While in mouse and chick gastrulation, single mesoderm and endoderm progenitor cells delaminate from the epiblast (single cell ingression; Bachvarova *et al.*, 1998; Vasiev *et al.*, 2010), progenitors internalize in a coordinated manner in zebrafish (coordinated cell ingression; Arboleda-Estudillo *et al.*, 2010; Montero *et al.*, 2005; Warga and Kimmel, 1990) and as a sheet of coherent cells in *Xenopus* (involution; Shook and Keller, 2008; Winklbauer and Kwang, 2009; Fig. 6.2A and B). Although the mode of mesoderm and endoderm cell internalization is different between different organisms, in all

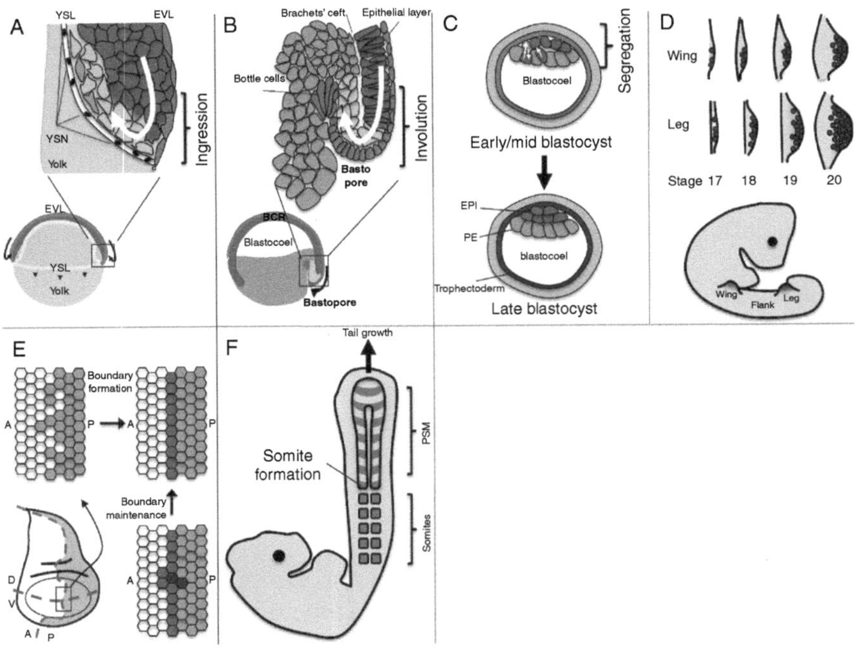

Figure 6.2 Developmental processes involving cell segregation and boundary formation. (A) Mesendoderm progenitor cell internalization during zebrafish gastrulation. Mesendoderm progenitor cells (green) specified at the germ ring margin ingress in a coordinated manner below the epiblast (ectoderm progenitors, red). Upper panel is a magnified representation of the boxed area in lower panel. (B) Mesoderm and endoderm progenitor cell involution during *Xenopus* gastrulation. Mesoderm progenitor cells (green) involute as a coherent cell sheet at the blastopore. Upper panel is a magnified representation of the boxed area in lower panel. (C) Progenitor cell sorting during mouse blastocyst formation. During early blastocyst formation (upper panel), the inner cells can be subdivided in epiblast cells (red) and primitive endoderm cells (blue), which segregate into epiblast cells at the interior surrounded by endoderm cells in the mature blastocyst (lower panel). (D) Limb bud formation in chicken. FGF-induced proliferation of mesoderm cells from the flank region lead to increased cell density in the forming limb bud (purple), which in turn increases TST promoting bud formation. (E) Boundary formation and maintenance in the *Drosophila* wing imaginal disc. Top left panel depicts a magnification of the anterior (A, white) and posterior (P, gray) cells before A/P boundary formation/maturation. Tension at the forming A/P boundary is needed to straighten the boundary and resist challenges of boundary integrity by divisions of boundary cells (boundary cells = red; dividing boundary cells = purple). (F) Somite boundary formation in chicken, mouse, and zebrafish. The presomitic mesoderm extends as the tail grows. The forming somites separate from the presomitic mesoderm by periodically forming boundaries. A, anterior; BCR, blastocoel roof; D, dorsal; EPI, epiblast; PSM, presomitic mesoderm; PE, primitive endoderm; P, posterior; V, ventral; YSL, yolk syncytial layer; YSN, yolk syncytial nuclei. (For interpretation of the references to color in this figure legend, the reader is referred to the Web version of this chapter.)

cases mesoderm and endoderm progenitor cells, once segregated from the epiblast (ectoderm) progenitor cells, acquire a more mesenchymal and migratory character and lose their ability to intermix with the epiblast cells above. This change in the character and affinity of the internalizing cells is thought to be triggered by TGFβ-type Nodal signals, key inducers of mesoderm and endoderm cell fate in vertebrates (Schier, 2003,2009).

To elucidate the effector mechanisms by which Nodals change the character of mesoderm and endoderm progenitors, their segregation behavior from ectoderm progenitors was reconstituted by cell-sorting experiments *in vitro* (Davis *et al.*, 1997; Klopper *et al.*, 2010; Montero *et al.*, 2005; Ninomiya and Winklbauer, 2008; Townes and Holtfreter, 1955). Such sorting experiments are performed by setting up heterotypic cell or tissue cocultures either by homogeneously mixing primary progenitor cells from different (induced) germ layers or placing germ layer tissue fragments in close proximity. In either case, the different progenitor cells/tissues rearrange into a final configuration in which the ectoderm is surrounded by mesoderm or endoderm cells. This stereotypical sorting behavior can be explained by differences in the surface tension between the different germ layer tissues. Measurements of germ layer TST showed that ectoderm tissues, which sort to the middle of heterotypic aggregates, display higher surface tensions than mesoderm and endoderm tissues, consistent with the assumption that tissues with higher surface tensions are surrounded by those with a lower TST (Davis *et al.*, 1997; Schötz *et al.*, 2008).

In addition to ectoderm, mesoderm and endoderm several other tissues are known to display sorting behavior during gastrulation. In *Xenopus*, sorting was reported for cell mixtures from the animal and vegetal poles at mid-blastula stage (Turner *et al.*, 1989), between anterior and posterior chordamesoderm (Ninomiya *et al.*, 2004), and between neural crest cells and epidermal cells (Davis *et al.*, 1997). In most of these cases, TST measurements suggested that differences in TST influence the sorting order of these different heterotypic cell/tissue mixtures, although direct evidence for a critical function of surface tension in these processes is still insufficient.

Although differences in TST were implicated in cell sorting during gastrulation, the molecular and cellular mechanisms determining TST are still debated. Seminal studies by Steinberg and colleagues suggested that differential adhesion determines differential TST. As outlined above (Section 2.2), the DAH proposes that the level of cell adhesion directly scales with the level of surface tension of the tissue formed by these cells and, consequently, that differences in surface tension between different tissues are due to differences in cell adhesion between the respective cell populations (Foty and Steinberg, 2005). Evidence in favor of the DAH is provided by studies in zebrafish, which show that interfering with the endogenous expression level of the cell–cell adhesion molecule E-cadherin induces the engulfment of germ layer tissues expressing higher levels of E-cadherin by those with lower E-cadherin expression (Schötz *et al.*, 2008). While these observations demonstrate that differences in E-cadherin expression between progenitor cells can influence their sorting behavior, they do not prove that differences in the endogenous E-cadherin expression levels are responsible for the sorting order of different cell populations during gastrulation. Moreover, the finding that mesoderm progenitor cells, which in a heterotypic aggregate with ectoderm cells sort to the outside and exhibit lower TST than ectoderm tissues, express higher levels of E-cadherin than mesoderm cells, suggests that TST and the level of E-cadherin expression might not be directly correlated in progenitor cells (Blanco *et al.*, 2007; Montero *et al.*, 2005).

In *Xenopus* gastrulation, C-cadherin is highly expressed in both epiblast and hypoblast cells (Kühl and Wedlich, 1996). Mosaic expression of C-cadherin induces cell sorting both *in vivo* and *in vitro*. Differences in C-cadherin activity between ectoderm and mesoderm progenitors, which are regulated by Paraxial Protocadherin C (PAPC) in a Nodal/TGFb-signaling dependent manner, are required for progenitor cell sorting and germ layer morphogenesis (Chen and Gumbiner, 2006b). Moreover, signaling through the Wnt-receptor Frizzled-7 (Fz7), also controlled by PAPC (Chen and Gumbiner, 2006b), is important for progenitor cell segregation and germ layer separation (Winklbauer *et al.*, 2001). However, the cell-type specific regulation of C-cadherin activity, and the relationship between C-cadherin activity and TST in progenitor cell segregation remains unclear.

It was proposed that cell cortex contraction or tension, in addition to cell adhesion, influences cell sorting (Harris, 1976). By using single cell force spectroscopy (SCFS) to measure cortical tension of individual zebrafish germ layer progenitor cells, it was shown that ectoderm progenitors exhibit higher cortical tension than mesoderm and endoderm progenitors. Furthermore, interfering with cortical tension in germ layer progenitor cells influences their sorting order such that progenitor cells displaying higher cortical tension sort to the inside of heterotypic aggregates (Krieg *et al.*, 2008). In principle, cortical tension of individual cells can influence TST in two ways (see also Section 2.3). At cell–cell interfaces cortical tension will reduce the interfacial cell–cell contact area. Consequently, this leads to reduced cell compaction and reduced TST. By contrast, at cell–medium interfaces, cortical tension will increase TST by building a contractile "scaffold" around the cell aggregate. The observation that differences in cortical tension between the progenitor cell types correlates with high surface tension of germ layer tissues suggests that these differences are particularly pronounced at cell–medium interfaces. Analysis of the subcellular distribution of actin, as readout for cortical tension, in germ layer tissue explants provides experimental evidence that differences in cortical tension between the different germ layer tissue types are predominantly visible at cell–medium and not cell–cell interfaces. This observation has been challenged by a recent study showing that cortical actin in germ layer tissue explants is predominantly localized at cell–cell rather than cell–medium interfaces. However, it remains unclear whether the surface cells analyzed in these explants express the same fate and physical properties than bulk cells (Manning *et al.*, 2010). Interestingly, recent studies in *Drosophila* have shown that high Myosin II coincides with low E-cadherin localization at cell–cell junctions, suggesting a mutual exclusion of E-cadherin and Myosin II at these interfaces (Fernandez-Gonzalez *et al.*, 2009; Rauzi *et al.*, 2010).

4.2.2. Cell segregation during mouse blastocyst formation

Morphogenesis of the early mammalian blastocyst is thought to be triggered by two factors: the unequal distribution of cell fate determinants between the blastoderm cells leading to cell fate specification at different positions within the blastocyst, and the segregation of cells with different fates during blastocyst formation. Cell segregation in the blastocyst is best described between pluripotent epiblast cells and primitive endoderm cells resulting in epiblast cells being surrounded by endoderm cells. Long-term time-lapse analysis combined with cell lineage tracing revealed that epiblast and endoderm cell fates are specified at random positions and subsequently sorted into distinct domains in the early mouse blastocyst (Morris *et al.*, 2010; Plusa *et al.*, 2008). It was proposed that repositioning of cells through sorting feeds back on their cell fate specification, and consequently, that the interplay between cell positioning and specification determines the different cell fates within the blastocyst (Rossant *et al.*, 2003; Fig. 6.2C). The observation that the repositioning of cells during the sorting process influences their cell fate specification has general implications for the analysis of cell sorting *per se*, since most of the present cell-sorting models are based on the assumption that cell fate remains unchanged by sorting (Armstrong, 1989).

Recent studies have shown that differential adhesion is not sufficient to explain segregation of epiblast and endoderm cells during blastocyst formation (Moore *et al.*, 2009). E-cadherin is a prominent cell–cell adhesion molecule expressed during the early stages of mouse development and is equally distributed throughout blastocyst formation. Moreover, in heterotypic aggregates of pluripotent ES cells, resembling epiblast cells and endoderm cells, downregulation of E-cadherin expression in epiblast cells does not affect the sorting of endoderm cells to the outside. This suggests that the sorting order in epiblast–endoderm aggregates is not determined by differential adhesion between these two cell types. Instead, the sorting order appears to be driven by the tendency of surface endoderm cells to polarize along their apical–basolateral axis and assemble into epithelial cell sheets (Moore *et al.*, 2009).

While the surface tension of different developing tissues during blastocyst formation has not yet been measured directly, simulations using assigned surface energies for the different tissues were developed to mimic cell sorting during blastocyst formation. In these simulations, epiblast and endoderm cell sorting *in silico* resembles the actual sorting behavior *in vivo* when the endoderm tissue surface energy is set to be lower than that of the epiblast. In addition, the simulation predicted that outer cells will be larger than inner cells, a feature also experimentally observed using time-lapse recordings of blastocyst development (Honda *et al.*, 2008).

4.2.3. Cell segregation during chick limb bud formation

During chick limb bud development, limb mesoderm cells are thought to segregate from the somatopleural mesoderm from which they arise. It was suggested that limb mesoderm cells become more cohesive than their surrounding somatopleural mesoderm cells and thus form a tissue with higher surface tension. As a result of these disparate surface tensions, limb bud mesoderm cells segregate from the somatopleural mesoderm, leading to the formation of the morphological recognizable limb bud (Heintzelman *et al.*, 1978). This explanation was first experimentally confirmed by showing that limb bud tissues indeed have larger surface tensions than flank mesoderm (Foty *et al.*, 1996). In addition, recent simulations of limb bud development have confirmed that differences in surface tensions between limb bud and somatopleural mesoderm tissues are in principle sufficient to trigger limb bud development (Damon *et al.*, 2008). The increase in surface tension of limb mesoderm tissue was shown to be accompanied by an increase in cell density, suggesting that cell cohesion is involved in this process. FGF signaling, a key pathway implicated in limb bud development, was found to trigger this increase in cell density and TST at least partly by promoting cell proliferation, suggesting that FGF plays a critical role in the segregation of limb bud cells from the somatopleural mesoderm by regulating TST (Fig. 6.2D).

4.3. From cell sorting to boundary formation

Cell sorting and segregation are often accompanied by the establishment of boundary structures which maintain the segregated state. Boundary formation was first described during abdomen (Lawrence *et al.*, 1978) and wing development (Garcia-Bellido *et al.*, 1973) in *Drosophila*, and subsequently analyzed in various other developmental processes in both vertebrates and invertebrates. Most of these processes were already reviewed extensively elsewhere (Cooke and Moens, 2002; McGrew and Pourquiè, 1998; Tepass *et al.*, 2002), and we will therefore only summarize boundary formation briefly, with specific emphasis on the differences in mechanical and adhesive properties of the boundary forming cells, in two exemplary developmental processes: imaginal disc development in *Drosophila* and somitogenesis in vertebrates.

4.3.1. Boundary formation in *Drosophila* wing imaginal disk

Cell segregation and boundary formation are thought to be integral parts of *Drosophila* wing imaginal disc development (Tepass *et al.*, 2002). The imaginal disc comprises a monolayer of columnar epithelial cells. Like in most epithelia, imaginal disc cells are polarized along their apical–basolateral axis and most cell–cell adhesion molecules are localized in juxta-apical junctional complexes, which serve as localized regions of strong adhesiveness to stabilize the epithelial structure (Martin, 2010; Papusheva and Heisenberg, 2010). The processes driving cell segregation and boundary formation within the imaginal disc epithelium are therefore mostly restricted to the plane of the epithelium where the junctional complexes are localized (Fig. 6.2E).

Two major boundaries become apparent during wing imaginal disc development: an anterior–posterior (A/P) boundary, subdividing the disc into anterior and posterior compartment, and a dorsal–ventral (D/V) boundary, oriented perpendicular to the A/P boundary and subdividing the anterior and posterior compartments into dorsal and ventral compartments. Initial studies revealed critical roles for DPP signaling in A/P boundary formation, and Delta–Notch signaling in D/V boundary formation. The mechanism by which these signaling molecules control differences in the physical properties of different cell populations at the boundaries, and thus drive cell segregation and boundary maintenance, is not yet fully understood. E-cadherin, one of the main cell–cell adhesion molecules expressed in wing imaginal disc cells, shows uniform expression throughout the disc and is thus unlikely to drive cell segregation and border formation. However, induction of cell clones over- or underexpressing E-cadherin within the wing imaginal disc can induce ectopic cell segregation and border formation, suggesting that E-cadherin is able to function in these processes (Dahmann and Basler, 2000; Wei *et al.*, 2005). Echinoid (Ed), an immunoglobulin-domain containing transmembrane protein, is another adhesion molecule with a putative function in wing imaginal disc cell segregation (Wei *et al.*, 2005). Clones of imaginal disc cells over- or under-expressing Ed sort from their surrounding cells, suggesting differential Ed expression has the potential to induce cell segregation. It is not clear yet whether Ed functions by regulating differential adhesion (Lecuit, 2005), since Ed mutant cells also display ectopic contraction of their apices, which might contribute to their segregation potential. Importantly, Ed is not differentially expressed along the imaginal disc boundaries, and endogenous Ed is not required for boundary formation, suggesting that Ed is not a critical factor driving wing imaginal disc cell segregation and boundary formation (Major and Irvine, 2005; Rawlins *et al.*, 2003; Wei *et al.*, 2005). Finally, the leucine-rich domain containing adhesion molecules Capricious and Tartan are known to be specifically expressed within the dorsal compartment of the wing imaginal disc. Although their exact molecular function is not yet known, Capricious was found to act as a homotypic adhesion molecule in cell culture (Shinza-Kameda *et al.*, 2006). Expression of either of these proteins in dorsal compartment cells is sufficient to restore D/V boundary formation in mutants for *apterous*, a gene required for dorsal wing imaginal disc cell fate specification and D/V boundary formation (Milan *et al.*, 2001). While this suggests that Capricious and Tartan function downstream of Apterous in D/V boundary formation, their endogenous function does not seem to be required for this process (Milan *et al.*, 2001).

In addition to adhesion, Actin- and Myosin-dependent cell cortex tension is also involved in wing imaginal disc boundary formation (Landsberg *et al.*, 2009; Monier *et al.*, 2010). Maintenance of the A/P boundary in the proliferating fly wing tissue is constantly challenged by cells dividing at the boundary, and thus requires a mechanism to counteract these cellular rearrangements. Recently, Myosin II accumulation at the A/P boundary has been suggested to locally increase boundary cell bond tension, leading to the straightening and stabilization of this boundary. This notion is supported by simulations demonstrating that the increase in boundary cell bond tension is sufficient to counteract division-induced cellular rearrangements at the boundary (Landsberg *et al.*, 2009; Monier *et al.*, 2010).

As for cell-sorting experiments *in vitro*, differential adhesion and/or cell cortex tension are likely to be involved in wing imaginal disc cell segregation and boundary formation. However, it remains unclear whether *bona fide* cell sorting occurs during imaginal disc development. Notably, cultures of mixed anterior and posterior wing imaginal disc cells fail to sort, arguing against cell sorting being the primary mechanism by which cells segregate in the wing imaginal disc (Fausto-Sterling and Hsieh, 1987).

4.3.2. Border formation during vertebrate somitogenesis

During vertebrate development, somites form as paired blocks of mesoderm flanking the notochord and neural tube, giving rise to all skeletal muscles of the body, the axial skeleton, and part of the dermis (McGrew and Pourquiè, 1998). Somites are a common feature of the segmented vertebrate body plan and are formed by segmentation of the presomitic mesoderm (PSM) into somites by a clock and wave front mechanism (Cooke and Zeeman, 1976). Somite segmentation is achieved by the formation of a border separating the newly formed somite from the PSM (Fig. 6.2F). Differences between the TST of the PSM and the forming somite were originally thought to be involved in border formation and somite separation. However, simulation studies showed that given the high bulk viscosities of the somitic tissues, differences in TST measured *in vitro* would not suffice to drive these processes (Damon *et al.*, 2008; Grima and Schnell, 2007).

A key player in somite border formation is Eph–Ephrin signaling. Eph receptors and Ephrin ligands are expressed in opposing cells at the forming border and are thought to trigger border formation by mediating cell repulsion (Durbin *et al.*, 1998; Watanabe *et al.*, 2009). Other genes that are expressed at the forming somite border and required for somite border formation and segmentation in zebrafish and *Xenopus*, are N-cadherin (Warga and Kane, 2007) and PAPC (Kim *et al.*, 1998). Whether any of the genes function in somitogenesis by regulating differential cell adhesion, cortical tension, or other cell properties implicated in cell sorting is not yet fully clear.

5. Outlook

Two major questions remain with respect to cell sorting in development. The first governs the extent to which cell sorting can be observed within a developing organism *in vivo*, and the second, the role that differences in TST play in driving cell sorting *in vivo*. Cell sorting, which starts with a randomly mixed population of cells with different properties and/or fates, is a process which can be initiated and studied *in vitro* by mixing cells in culture and observing their subsequent sorting behavior, but is rarely observed in a developing organism *in vivo*. Usually, cells with similar properties and/or the same fate do not emerge at random positions within a developing tissue, but instead are induced in a certain temporal and spatial pattern within the embryo. Consequently, cells with different fates and/or properties are already in a partially sorted configuration when they are induced within the embryo. This configuration might then be modified or fine-tuned in subsequent stages of development. Therefore, cell sorting *in vivo* is more reminiscent of later stages of cell sorting *in vitro*, during which other processes that are not directly involved in the initial stages of demixing become more important, such as border formation and stabilization. Future studies will have to address whether the key principles of cell sorting identified *in vitro*, such as differences in TST, also drive cell segregation *in vivo*, or whether other processes involved in cell segregation, such as border maturation through extracellular matrix deposition, are of greater importance.

In general, TST has to date been measured for cell aggregates cultured in liquid media. The measured values thus represent a tissue property specific for the interface between cells and a liquid medium. In the developing organism, only the most superficial cells/tissues are in contact with the outside medium, while most other tissues are in contact with each other. To understand how these internal tissues interact *in vivo*, values of TST that are measured once these tissues are taken into culture, only provide limited insights as they do not necessarily correspond to the surface (interfacial) tension of these tissues when surrounded by other cells and/or extracellular matrix. A recent example for this phenomenon is the apparent difference between the sorting behavior of zebrafish germ layer progenitor cells *in vitro* and *in vivo*. A heterotypic aggregate of ectoderm and mesoderm progenitors *in vitro* sorts into a configuration where ectoderm is surrounded by mesoderm, consistent with the observation that ectoderm TST measured in culture is larger than mesoderm TST. However, within the developing embryo, mesoderm progenitors internalize underneath the ectoderm progenitors giving rise to a gastrula in which ectoderm surrounds mesoderm, a configuration opposite to that is predicted by the differences in their TST. The most likely explanation for the different behaviors of germ layer progenitors *in vitro* and *in vivo* is the simple fact that progenitors in the embryos are rarely in direct contact with the outside medium, but rather interact with each other and/or the extracellular matrix. While these observations do not devalue the use of TST in explaining cell sorting and tissue envelopment, they point toward the importance of developing methods for measuring TST *in vivo*. Recent studies in *Drosophila* have used laser systems to cut the border between cells/tissues and measure the resulting relaxation behavior of the border around the lesion (Bertet *et al.*, 2009; Fernandez-Gonzalez *et al.*, 2009; Landsberg *et al.*, 2009). This allows for computation of the interfacial tension at these borders, which serves as a readout for the combined TST at cell/tissue borders. In the future, the advancement of this and other techniques to measure interfacial tension *in vivo* will shed light on the extent to which differences in TST account for cell sorting and segregation in development.

REFERENCES

Adamson, A. C. (1967). Physical Chemistry of Surfaces. John Wiley & Sons, New York.

Arboleda-Estudillo, Y., Krieg, M., Stuhmer, J., Licata, N. A., Muller, D. J., and Heisenberg, C. (2010). Movement directionality in collective migration of germ layer progenitors. *Curr. Biol.* **20,** 161–169.

Armstrong, P. (1989). Cell sorting out: The self-assembly of tissues in vitro. *Crit. Rev. Biochem. Mol. Biol.* **24,** 119–149.

Bachvarova, R. F., Skromne, I., and Stern, C. D. (1998). Induction of primitive streak and Hensen's node by the posterior marginal zone in the early chick embryo. *Development* **125,** 3521–3534.

Bertet, C., Rauzi, M., and Lecuit, T. (2009). Repression of Wasp by JAK/STAT signalling inhibits medial actomyosin network assembly and apical cell constriction in intercalating epithelial cells. *Development* **136,** 4199–4212.

Blanco, M. J., Barrallo-Gimeno, A., Acloque, H., Reyes, A. E., Tada, M., Allende, M. L., Mayor, R., and Nieto, M. A. (2007). Snail1a and Snail1b cooperate in the anterior migration of the axial mesendoderm in the zebrafish embryo. *Development* **134,** 4073–4081.

Borghi, N., and Nelson, W. J. (2009). Intercellular adhesion in morphogenesis: Molecular and biophysical considerations. *Curr. Top. Dev. Biol.* **89,** 1–32.

Brodland, G. W. (2002). The differential interfacial tension hypothesis (DITH): A comprehensive theory for the self-rearrangement of embryonic cells and tissues. *J. Biomech. Eng.* **124,** 188–197.

Brodland, W. G., and Chen, H. H. (2000). The mechanics of cell sorting and envelopment. *J. Biomech.* **33,** 845–851.
Brodland, G. W., Yang, J., and Sweny, J. (2009). Cellular interfacial and surface tensions determined from aggregate compression tests using a finite element model. *HFSP J.* **3,** 273.
Chen, X., and Gumbiner, B. (2006a). Crosstalk between different adhesion molecules. *Curr. Opin. Cell Biol.* **18,** 572–578.
Chen, X., and Gumbiner, B. M. (2006b). Paraxial protocadherin mediates cell sorting and tissue morphogenesis by regulating C-cadherin adhesion activity. *J. Cell Biol.* **174,** 301–313.
Chu, Y., Eder, O., Thomas, W. A., Simcha, I., Pincet, F., Ben-Ze'ev, A., Perez, E., Thiery, J. P., and Dufour, S. (2006). Prototypical type I E-cadherin and type II cadherin-7 mediate very distinct adhesiveness through their extracellular domains. *J. Biol. Chem.* **281,** 2901–2910.
Cooke, J. E., and Moens, C. B. (2002). Boundary formation in the hindbrain: Eph only it were simple. *Trends Neurosci.* **25,** 260–267.
Cooke, J., and Zeeman, E. C. (1976). A clock and wavefront model for control of the number of repeated structures during animal morphogenesis. *J. Theor. Biol.* **58,** 455–476.
Curtis, A. S. G. (1960). Cell contacts: Some physical considerations. *Am. Nat.* **94,** 37–56.
Curtis, A. S. G. (1961). Timing mechanisms in the specific adhesion of cells. *Exp. Cell Res.* **8,** 107–122.
Dahmann, C., and Basler, K. (2000). Opposing transcriptional outputs of hedgehog signaling and engrailed control compartmental cell sorting at the *Drosophila* A/P boundary. *Cell* **100,** 411–422.
Damon, B. J., Mezentseva, N. V., Kumaratilake, J. S., Forgacs, G., and Newman, S. A. (2008). Limb bud and flank mesoderm have distinct "physical phenotypes" that may contribute to limb budding. *Dev. Biol.* **321,** 319–330.
Davis, G. S. (1984). Migration-directing liquid properties of embryonic amphibian tissues. *Am. Zool.* **24,** 649–655.
Davis, G. S., Phillips, H. M., and Steinberg, M. S. (1997). Germ-layer surface tensions and "tissue affinities" in *Rana pipiens* gastrulae: Quantitative measurements. *Dev. Biol.* **192,** 630–644.
Detrick, R. J., Dickey, D., and Kintner, C. R. (1990). The effects of N-cadherin misexpression on morphogenesis in xenopus embryos. *Neuron* **4,** 493–506.
Duguay, D., AFoty, R. A., and Steinberg, M. S. (2003). Cadherin-mediated cell adhesion and tissue segregation: Qualitative and quantitative determinants. *Dev. Biol.* **253,** 309–323.
Durbin, L., Brennan, C., Shiomi, K., Cooke, J., Barrios, A., Shanmugalingam, S., Guthrie, B., Lindberg, R., and Holder, N. (1998). Eph signaling is required for segmentation and differentiation of the somites. *Genes Dev.* **12,** 3096–3109.
Evans, E., and Yeung, A. (1989). Apparent viscosity and cortical tension of blood granulocytes determined by micropipet aspiration. *Biophys. J.* **56,** 151–160.
Fausto-Sterling, A., and Hsieh, L. (1987). In vitro culture of *Drosophila* imaginal disc cells: Aggregation, sorting out, and differentiative abilities. *Dev. Biol.* **120,** 284–293.
Fernandez-Gonzalez, R., Simoes, S. d. M., Röper, J.-C., Eaton, S., and Zallen, J. A. (2009). Myosin II dynamics are regulated by tension in intercalating cells. *Dev. Cell* **17,** 736–743.
Foty, R. A., and Steinberg, M. S. (2005). The differential adhesion hypothesis: A direct evaluation. *Dev. Biol.* **278,** 255–263.
Foty, R. A., Forgacs, G., Pfleger, C. M., and Steinberg, M. S. (1994). Liquid properties of embryonic tissues: Measurement of interfacial tensions. *Phys. Rev. Lett.* **72,** 2298–2301.
Foty, R. A., Pfleger, C. M., Forgacs, G., and Steinberg, M. S. (1996). Surface tensions of embryonic tissues predict their mutual envelopment behavior. *Development* **122,** 1611–1620.

Garcia-Bellido, A., Ripoll, P., and Morata, G. (1973). Developmental compartmentalisation of the wing disk of Drosophila. *Nat. New Biol.* **245,** 251–253.

Glazier, J. A., and Graner, F. (1993). Simulation of the differential adhesion driven rearrangement of biological cells. *Phys. Rev. E Stat. Phys. Plasmas Fluids Relat. Interdiscip. Topics* **47,** 2128–2154.

Grima, R., and Schnell, S. (2007). Can tissue surface tension drive somite formation? *Dev. Biol.* **307,** 248–257.

Guevorkian, K., Colbert, M., Durth, M., Dufour, S., and Brochard-Wyart, F. (2010). Aspiration of biological viscoelastic drops. *Phys. Rev. Lett.* **104,** 1–4.

Harris, A. K. (1976). Is Cell sorting caused by differences in the work of intercellular adhesion? A critique of the Steinberg hypothesis. *J. Theor. Biol.* **61,** 267–285.

Hayashi, T., and Carthew, R. W. (2004). Surface mechanics mediate pattern formation in the developing retina. *Nature* **431,** 647–652.

Heintzelman, K. F., Phillips, H. M., and Davis, G. S. (1978). Liquid-tissue behavior and differential cohesiveness during chick limb budding. *J. Embryol. Exp. Morphol.* **47,** 1–15.

Honda, H., Motosugi, N., Nagai, T., Tanemura, M., and Hiiragi, T. (2008). Computer simulation of emerging asymmetry in the mouse blastocyst. *Development* **135,** 1407–1414.

Jones, B. M., Evans, P. M., and Lee, D. A. (1989). Relation between the rate of cell movement under agarose and the positioning of cells in heterotypic aggregates. *Exp. Cell Res.* **180,** 287–296.

Kalantarian, A., Ninomiya, H., Saad, S. M., David, R., Winklbauer, R., and Neumann, A. W. (2009). Axisymmetric drop shape analysis for estimating the surface tension of cell aggregates by centrifugation. *Biophys. J.* **96,** 1606–1616.

Kim, S. H., Yamamoto, A., Bouwmeester, T., Agius, E., and Robertis, E. M. (1998). The role of paraxial protocadherin in selective adhesion and cell movements of the mesoderm during Xenopus gastrulation. *Development* **125,** 4681–4690.

Kintner, C. (1988). Effects of altered expression of the neural cell adhesion molecule, N-CAM, on early neural development in Xenopus embryos. *Neuron* **1,** 545–555.

Klopper, A. V., Krens, G., Grill, S. W., and Heisenberg, C. P. (2010). Finite-size corrections to scaling behavior in sorted cell aggregates. *Eur. Phys. J. E Soft Matter* **33**(2), 99–103.

Krieg, M., Arboleda-Estudillo, Y., Puech, P. H., Käfer, J., Graner, F., Müller, D. J., and Heisenberg, C. P. (2008). Tensile forces govern germ-layer organization in zebrafish. *Nat. Cell Biol.* **10,** 429–436.

Kühl, M., and Wedlich, D. (1996). Xenopus cadherins: Sorting out types and functions in embryogenesis. *Dev. Dyn.* **207,** 121–134.

Landsberg, K. P., Farhadifar, R., Ranft, J., Umetsu, D., Widmann, T. J., Bittig, T., Said, A., Julicher, F., and Dahmann, C. (2009). Increased cell bond tension governs cell sorting at the drosophila anteroposterior compartment boundary. *Curr. Biol.* **19,** 1950–1955.

Lawrence, P. A., Green, S. M., and Johnston, P. (1978). Compartmentalization and growth of the Drosophila abdomen. *J. Embryol. Exp. Morphol.* **43,** 233–245.

Lecuit, T. (2005). Cell adhesion: Sorting out cell mixing with echinoid? *Curr. Biol.* **15,** R505–R507.

Maeda, T. T., Ajioka, I., and Nakajima, K. (2007). Computational cell model based on autonomous cell movement regulated by cell-cell signalling successfully recapitulates the "inside and outside" pattern of cell sorting. *BMC Syst. Biol.* **1,** 43.

Major, R. J., and Irvine, K. D. (2005). Influence of Notch on dorsoventral compartmentalization and actin organization in the Drosophila wing. *Development* **132,** 3823–3833.

Manning, M. L., Foty, R. A., Steinberg, M. S., and Schötz, E. (2010). Coaction of intercellular adhesion and cortical tension specifies tissue surface tension. *Proc. Natl. Acad. Sci. USA* **107,** 1–6.

Marmottant, P., Mgharbel, A., Käfer, J., Audren, B., Rieu, J., Vial, J. C., van der Sanden, B., Marée, A. F., Graner, F., and Delanoë-Ayari, H. (2009). The role of fluctuations and stress on the effective viscosity of cell aggregates. *Proc. Natl. Acad. Sci. USA* **106,** 17271–17275.

Marsden, M., and DeSimone, D. W. (2003). Integrin-ECM interactions regulate Cadherin-dependent cell adhesion and are required for convergent extension in Xenopus. *Curr. Biol.* **13,** 1182–1191.

Martin, A. C. (2010). Pulsation and stabilization: Contractile forces that underlie morphogenesis. *Dev. Biol.* **341,** 114–125.

McGrew, M. J., and Pourquiè, O. (1998). Somitogenesis: Segmenting a vertebrate. *Curr. Opin. Genet. Dev.* **8,** 487–493.

Mgharbel, A., Delanoë Ayari, H., and Rieu, J. (2009). Measuring accurately liquid and tissue surface tension with a compression plate tensiometer. *HFSP J.* **3,** 213.

Milan, M., Weihe, U., Perez, L., and Cohen, S. M. (2001). The LRR proteins capricious and Tartan mediate cell interactions during DV boundary formation in the Drosophila wing. *Cell* **106,** 785–794.

Mombach, J. C. (1999). Simulation of embryonic cell self-organization: A study of aggregates with different concentrations of cell types. *Phys. Rev. E* **59,** 59–63.

Monier, B., Pélissier-Monier, A., Brand, A. H., and Sanson, B. (2010). An actomyosin-based barrier inhibits cell mixing at compartmental boundaries in Drosophila embryos. *Nat. Cell Biol.* **12,** 60–65.

Montero, J. A., Carvalho, L., Wilsch-Bräuninger, M., Kilian, B., Mustafa, C., and Heisenberg, C. (2005). Shield formation at the onset of zebrafish gastrulation. *Development* **132,** 1187–1198.

Moore, R., Cai, K., Escudero, D., and Xu, X. (2009). Cell adhesive affinity does not dictate primitive endoderm segregation and positioning during murine embryoid body formation. *Genesis* **47,** 579–589.

Morris, S., Teo, R., Li, H., Robson, P., Glover, D., and Zernicka-Goetz, M. (2010). Origin and formation of the first two distinct cell types of the inner cell mass in the mouse embryo. *Proc. Natl. Acad. Sci. USA* **107,** 6364–6369.

Moscona, A. A. (1962). Analysis of cell recombinations in experimental synthesis of tissues in vitro. *J. Cell. Comp. Physiol.* **60,** 65–80.

Narotte, C., Marga, F. S., Neagu, A., Kosztin, I., and Forgacs, G. (2008). Experimental evaluation of apparent tissue surface tension based on the exact solution of the Laplace equation. *EPL* **81,** 46003, pp. 46003-p1–46003-p6.

Ninomiya, H., and Winklbauer, R. (2008). Epithelial coating controls mesenchymal shape change through tissue-positioning effects and reduction of surface-minimizing tension. *Nat. Cell Biol.* **10,** 61–69.

Ninomiya, H., Elinson, R. P., and Winklbauer, R. (2004). Antero-posterior tissue polarity links mesoderm convergent extension to axial patterning. *Nature* **430,** 364–367.

Palsson, E. (2001). A three-dimensional model of cell movement in multicellular systems. *Future Gener Comput. Syst.* **17,** 385–852.

Paluch, E., and Heisenberg, C.-P. (2009). Biology and physics of cell shape changes in development. *Curr. Biol.* **19,** R790–R799.

Papusheva, E., and Heisenberg, C.-P. (2010). Spatial organization of adhesion: Force-dependent regulation and function in tissue morphogenesis. *EMBO J.* **29,** 2753–2768.

Phillips, H. M., and Steinberg, M. S. (1969). Equilibrium measurements of embryonic chick cell adhesiveness. I. Shape equilibrium in centrifugal fields. *Proc. Natl. Acad. Sci. USA* **64,** 121–127.

Plusa, B., Piliszek, A., Frankenberg, S., Artus, J., and Hadjantonakis, A. K. (2008). Distinct sequential cell behaviours direct primitive endoderm formation in the mouse blastocyst. *Development* **135,** 3081–3091.

Rauzi, M., Lenne, P.-F., and Lecuit, T. (2010). Planar polarized actomyosin contractile flows control epithelial junction remodelling. *Nature* **468,** 1110–1114, advance online publication.

Rawlins, E. L., Lovegrove, B., and Jarman, A. P. (2003). Echinoid facilitates Notch pathway signalling during Drosophila neurogenesis through functional interaction with Delta. *Development* **130,** 6475–6484.

Rossant, J., Chazaud, C., and Yamanaka, Y. (2003). Lineage allocation and asymmetries in the early mouse embryo. *Philos. Trans. R. Soc. Lond. B Biol. Sci.* **358,** 1341–1348.

Rowlinson, J. S., and Widom, B. (1989). Molecular Theory of Capillarity. Clarendon, Oxford.

Schier, A. F. (2003). Nodal signaling in vertebrate development. *Annu. Rev. Cell Dev. Biol.* **19,** 589–621.

Schier, A. F. (2009). Nodal morphogens. *Cold Spring Harb. Perspect. Biol.* **1,** a003459.

Schötz, E., Burdine, R. D., Jülicher, F., Steinberg, M. S., Heisenberg, C.-P., and Foty, R. A. (2008). Quantitative differences in tissue surface tension influence zebrafish germ layer positioning. *HFSP J.* **2,** 42–56.

Shinbrot, T., Chun, Y., Caicedo-Carvajal, C., and Foty, R. A. (2009). Cellular morphogenesis in silico. *Biophys. J.* **97,** 958–967.

Shinza-Kameda, M., Takasu, E., Sakurai, K., Hayashi, S., and Nose, A. (2006). Regulation of layer-specific targeting by reciprocal expression of a cell adhesion molecule, capricious. *Neuron* **49,** 205–213.

Shook, D. R., and Keller, R. (2008). Morphogenic machines evolve more rapidly than the signals that pattern them: Lessons from amphibians. *J. Exp. Zool. B Mol. Dev. Evol.* **310B,** 111–135.

Solnica-Krezel, L. (2005). Conserved patterns of cell movements during vertebrate gastrulation. *Curr. Biol.* **15,** R213–R228.

Steinberg, M. S. (1962a). On the mechanism of tissue reconstruction by dissociated cells. I. Population kinetics, differential adhesiveness, and the absence of directed migration. *Proc. Natl. Acad. Sci. USA* **48,** 1577–1582.

Steinberg, M. S. (1962b). On the mechanismof tissue reconstruction by dissociated cells, III. Free energy relationsand the reorganization of fused, heteronomic tissue fragmens. *Proc. Natl. Acad. Sci. USA* **48,** 1769–1776.

Steinberg, M. S. (1996). Adhesion in development: An historical overview. *Dev. Biol.* **180,** 377–388.

Sung, K. L., Sung, L. A., Crimmins, M., Burakoff, S. J., and Chien, S. (1986). Determination of junction avidity of cytolytic T cell and target cell. *Science* **234,** 1405–1408.

Tepass, U., Godt, D., and Winklbauer, R. (2002). Cell sorting in animal development: Signalling and adhesive mechanisms in the formation of tissue boundaries. *Curr. Opin. Genet. Dev.* **12,** 572–582.

Townes, P. L., and Holtfreter, J. (1955). Directed movements and selective adhesion of embryonic amphibian cells. *J. Exp. Zool.* **128,** 128–153.

Turner, A., Snape, A. M., Wylie, C. C., and Heasman, J. (1989). Regional identity is established before gastrulation in the *Xenopus* embryo. *J. Exp. Zool.* **251,** 245–252.

Vasiev, B., Balter, A., Chaplain, M., Glazier, J. A., and Weijer, C. (2010). Modeling gastrulation in the chick embryo: Formation of the primitive streak. *PLoS ONE* **5,** e10571.

Von Dassow, M., and Davidson, L. A. (2009). Natural variation in embryo mechanics: Gastrulation in *Xenopus laevis* is highly robust to variation in tissue stiffness. *Dev. Dyn.* **238,** 2–18.

Voß-Böhme, A., and Deutsch, A. (2009). The cellular basis of cell sorting kinetics. *J. Theor. Biol.* **263,** 419–436.

Warga, R. M., and Kane, D. A. (2007). A role for N-cadherin in mesodermal morphogenesis during gastrulation. *Dev. Biol.* **310,** 211–225.

Warga, R. M., and Kimmel, C. B. (1990). Cell movements during epiboly and gastrulation in zebrafish. *Development* **108,** 569–580.

Watanabe, T., Sato, Y., Saito, D., Tadokoro, R., and Takahashi, T. (2009). EphrinB2 coordinates the formation of a morphological boundary and cell epithelialization during somite segmentation. *Proc. Natl. Acad. Sci. USA* **106,** 7467–7472.

Wei, S.-Y., Escudero, L. M., Yu, F., Chang, L.-H., Chen, L.-Y., Ho, Y.-H., Lin, C.-M., Chou, C.-S., Chia, W., Modolell, J., and Hsu, J.-C. (2005). Echinoid is a component of adherens junctions that cooperates with DE-Cadherin to mediate cell adhesion. *Dev. Cell* **8,** 493–504.

Winklbauer, R., and Kwang, W. J. (2009). Cell adhesion in amphibian gastrulation. *Int. Rev. Cell Mol. Biol.* **278,** 215–275.

Winklbauer, R., Medina, A., Swain, R. K., and Steinbeisser, H. (2001). Frizzled-7 signalling controls tissue separation during Xenopus gastrulation. *Nature* **413,** 856–860.

CHAPTER SEVEN

Embryo Mechanics: Balancing Force Production with Elastic Resistance During Morphogenesis

Lance A. Davidson

Contents

Department of Bioengineering and Developmental Biology, University of Pittsburgh, Pittsburgh, Pennsylvania, USA

Current Topics in Developmental Biology, Volume 95
ISSN 0070-2153, DOI: 10.1016/B978-0-12-385065-2.00007-4

Abstract

Morphogenesis requires the spatial and temporal control of embryo mechanics, including force production and mechanical resistance to those forces, to coordinate tissue deformation and large-scale movements. Thus, biomechanical processes play a key role in directly shaping the embryo. Additional roles for embryo mechanics during development may include the patterning of positional information and to provide feedback to ensure the success of morphogenetic movements in shaping the larval body and organs. To understand the multiple roles of mechanics during development requires familiarity with engineering principles of the mechanics of structures, the viscoelastic properties of biomaterials, and the integration of force and stress within embryonic structures as morphogenesis progresses. In this chapter, we review the basic engineering principles of biomechanics as they relate to morphogenesis, introduce methods for quantifying embryo mechanics and the limitations of these methods, and outline a formalism for investigating the role of embryo mechanics in birth defects. We encourage the nascent field of embryo mechanics to adopt standard engineering terms and test methods so that studies of diverse organisms can be compared and universal biomechanical principles can be revealed.

1. Introduction

If you spend any time watching a construction site or find yourself watching time-lapse sequences of buildings being assembled, you will appreciate the task of constructing an embryo. Watching these events unfold, one often forgets about the long planning sessions off-site before a single item is ordered or workers hired. One often overlooks the events leading to the production of raw materials or the tools that arrive mysteriously at key points in the building of the structure. Similarly, after watching a few cases of building construction, you no longer find it odd that the workers start at the bottom rather than the top or that the building is framed before rooms are finished with electricity and plumbing. Watching embryos develop in time-lapse sequences lead to some of the same oversights and bring many preconceived notions. Assembly of the embryonic axis requires considerable off-site planning as the genome is adjusted to ensure fitness and survival. The work-site is prepared ahead of the arrival of materials in the form of the egg and sperm. Last, material properties of the egg and embryo and the biophysical labor of cells and tissues are coordinated to assemble the structure. Just as there are limits to the size and shape of buildings, we can find similar rules that dictate the limits of embryonic development. If the rules of assembly are followed within certain ranges, the product of development is a healthy larvae, if rules are not followed or environmental conditions are unfavorable, the larva dies. The goal of this chapter is to lay out some of the principles that can guide the mechanical processes that shape embryos, explore how these processes may be coupled to improve the rate of success, and how mechanical processes might break down.

The following sections will review: (1) the mechanical principles of embryonic self-assembly, (2) how these processes may guide development and may fail due to environmental or genetic variation, and (3) how anatomical, cellular, and molecular processes might serve to make embryonic development more robust.

2. Mechanics of Development

Studies on the mechanics of development start with analyses of the movements of cells and tissues with a general goal to understand the tissue-, cell- or molecular-scale origin of those movements. We will first review the general principles of a mechanical analysis of morphogenesis and define important engineering terms that put this process into a common analytical framework that can be applied broadly to many questions into the mechanics of development.

Analysis of tissue movements during early development can take many different forms ranging from traditional embryological mechanics to complex multiscale integrative computer simulations. Traditionally, qualitative mechanical analyses of morphogenesis have been carried out with avian or amphibian embryos. These models have been key to understanding the contribution of a specific tissue or structure to a morphogenetic movement by allowing the researcher to microsurgically ablate or isolate a tissue from its surrounding environment. More recently, tissue contribution has been investigated in models such as mouse or zebrafish where tissues can be ablated genetically.

Microsurgical approaches have been key to understanding the complex interplay of cellular and tissue level processes during amphibian gastrulation. In the aquatic frog *Xenopus laevis*, no less than seven distinct tissue movements contribute to the mechanics of gastrulation including: convergent extension, convergent thickening, tissue separation, radial intercalation, bottle cell formation, mesendoderm migration, and vegetal rotation (Fig. 7.1A). Within the intact embryo, one cannot conclusively identify which of these movements is defective when gastrulation fails; however, each of these movements can be microsurgically isolated and studied in a mechanically less complex environment (Fig. 7.1B). For instance, the movements of convergent extension can be isolated from the embryo during the earliest stages of gastrulation in the sandwich explant (Keller and Danilchik, 1988). Thus, removal of mesoderm in the dorsal marginal zone from the context of other movements has allowed developmental biologists to correlate the function of a wide range of molecular factors with these movements (Keller, 2002; Wallingford *et al.*, 2002). Over several decades, increasingly sophisticated microsurgical manipulations have been developed that allow one to investigate the role of each tissue movement to gastrulation and whether a particular molecular perturbation alters that particular movement. Qualitative mechanical analyses such as these have served to connect many signaling pathways and their downstream effectors to distinct tissue movements during gastrulation. These qualitative studies have been extremely useful in annotating the roles of proteins and genes in many morphogenetic movements across phyla but often do not resolve the relative contribution of each pathway to a particular movement or the relative contribution of each movement to the overall shape changes that drive morphogenesis. More quantitative studies of the mechanics of each movement and the cellular mechanics driving each movement are needed to resolve these roles.

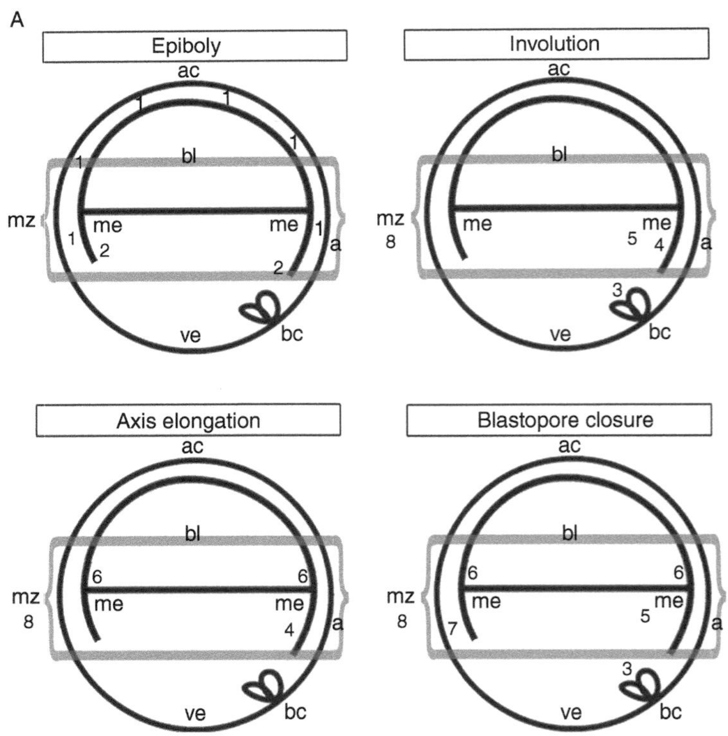

B

Process and Schematic	Where is process active[a]?	Explants[b]	Potential mechanical role
Radial intercalation 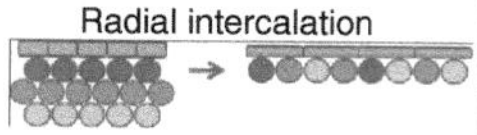	Throughout animal cap ectoderm (1), also immediately post involution (2).	Isolated animal caps cultured over deep cell explants (Marsden and DeSimone, 2001). Deep cell explants from animal cap ectoderm grafted beneath epithelia of whole embryo hosts (Longo *et al.*, 2004).	Release tension?
Bottle cell contraction 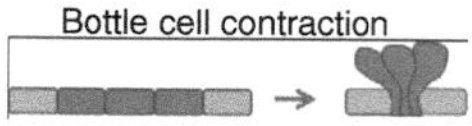	At the boundary (3) between vegetal endoderm and marginal zone.	Extirpated bottle cells (Holtfreter, 1943, 1944).	Increase tension?
Tissue separation 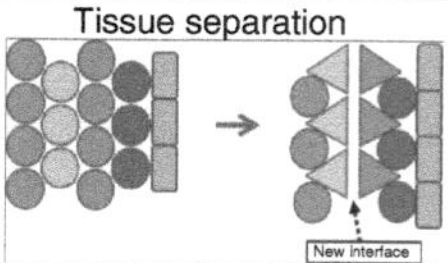	Between vegetal endoderm and overlying mesoderm of marginal zone (4).	Open animal caps combined with "challenge" tissue (Wacker *et al.*, 2000).	New interface may provide a shear surface or allow transmission of shear stress, for example as mesenoderm migrates?
Vegetal rotation 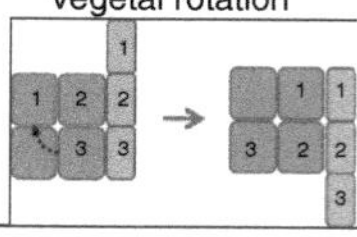	Within the vegetal endoderm beneath the floor of the blastocoel (5).	Whole embryo bisected sagittally (Winklbauer and Schurfeld, 1999).	Shear stress?
Mesendoderm migration 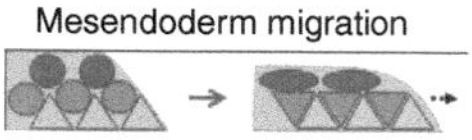	Along the margin of the blastocoel floor. Mesendoderm cells migrate on the blastocoel roof (6).	Fragments of leading edge mesendoderm (Winklbauer and Nagel, 1991). Mesendoderm within larger marginal zone explants (Davidson *et al.*, 2002).	Increase tension within mesendoderm? Shear stress on ectoderm substrate?
Convergent thickening 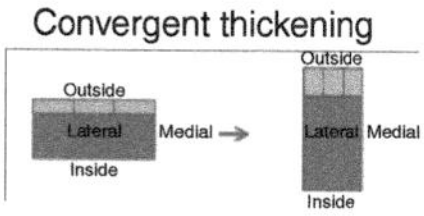	Within the mesoderm of the marginal zone in predominantly lateral and prospective posterior regions (7).	Sandwich explants from the full 360° ofthe marginal zone (Keller and Shook, 2008).	Tension along mediolateral axis or around the margin of the blastopore? Radial compression?
Convergent extension 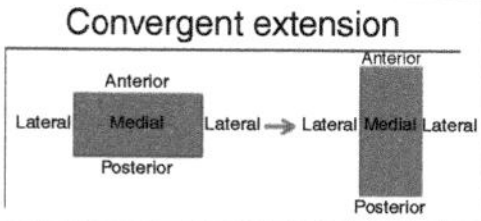	Within the axial and adjacent paraxial mesoderm in the marginal zone (8).	Sandwich explants from anterior-most domain of the marginal zone (Keller and Danilchik, 1988), "open-face" explant (Shih and Keller, 1992) and marginal zone explant cultured on fibronectin (Davidson *et al.*, 2006).	Tension along mediolateral axis and compression along anterior–posterior axis?

[a]Cell- or tissue-scale processes involve localized single cell behaviors or coordinated behaviors of multiple cells within a local cohesive microenvironment.
[b]Each of these processes can be observed and their occurrence assayed using microsurgically prepared tissue explants in organotypic culture.

Figure 7.1 Integrated developmental mechanics of gastrulation. (A) Schematic of an early gastrula stage of *Xenopus laevis* embryo (stage 10) in sagittal section showing morphological landmarks and location of multiple processes that have been proposed to contribute to epiboly, involution, axis elongation, and blastopore closure. These large-scale movements are the product of multiple local cell- and tissue-scale processes that are mechanically integrated. (B) Diagrammatic descriptions of the seven processes that shape the embryo during gastrulation. Processes listed by numbers 1–8 in (B) are active at the approximate locations in the schematic shown in (A). The mechanical role of these local processes, that is, whether a process generates compressive or tensional stress or frees an interface to shear, has not been explicitly tested. bc, bottle cells; ve, vegetal endoderm; ac, animal cap ectoderm; mz, marginal zone; a, anterior most domain of marginal zone; bl, blastocoel; me, mesendoderm, also known as head mesoderm or anterior mesoderm.

3. Morphology and Kinematics: The Foundations of a Biomechanical Analysis

To understand the relative contribution or impact of a particular pathway during a specific morphogenetic movement, developmental biologists have increasingly been turning to more quantitative mechanical analyses of morphogenesis. An experimental mechanical analysis often starts with a detailed anatomical or morphological description of the tissues involved in the movement and then turns to high resolution time-lapse sequences that reveal the kinematics of that movement. Morphological descriptions identify structures or landmarks within the tissue that can be used to create a consensus "staging-table" of key structures and their changes over time in healthy embryos. There can be considerable variation from embryo to embryo (Ewald *et al.*, 2004; von Dassow and Davidson, 2009), but the goal at the earliest stages of analysis is to document a stereotypical description of landmarks and a series of shape changes over a limited time frame. The resulting staging-table describes tissue-scale movements and can be used to generate initial hypotheses regarding the movements of cells within those tissues and how cell–cell boundaries establish tissue compartments. However, more detailed studies of cell movement, cell sorting, and long-range migration can overturn these initial hypotheses.

After a detailed anatomical analysis of tissue movements, the next step in the quantitative mechanical analysis of morphogenesis requires description of the kinematics of cell movements, including information on dynamic events such as the rates of individual cells movements and shape changes within their local microenvironment. The goal is to find and track multiple identifiable landmarks within a tissue over time as the tissue changes shape. It is important that these landmarks and their movements represent physical structures or material points within the tissue. Commonly, these landmarks can be dense regions of pigment or extracellular matrix, cell nuclei, the center of mass of a cell volume or the projected area of a cell, or a larger structure like the blastopore lip, midline of the neural plate, and so on. Fate- or trajectory-maps of cell movements over time provide the raw material for the descriptive mechanics of a morphogenetic movement where engineering properties such as shear, flow, and strain can be quantified. These physical characters of a morphogenetic event are useful for systems-level comparisons, for example, comparing the effects of mutations or other treatments or by contrasting the process between different species or phyla. Properties of flow, strain, and shear (terms defined below) can be reported in units that depend on size so that morphogenetic processes can be compared to *in vitro* studies on single cell mechanics or movements can be reported in units that are independent of size so that mechanical events can be compared between individuals or between organisms that differ in size or morphology. A common language for describing the mechanics of morphogenesis is essential for researchers working on diverse problems so that universal mechanical principles can be extracted from studies of one species or stage and investigated or tested in another.

At this point, we propose the use of a common set of terms derived from physics and mechanical engineering to describe tissue movements during morphogenesis. In the sections below, we briefly outline the limits of such terms, and how biomechanical analysis can be used to understand the molecular and cellular mechanisms responsible for driving morphogenesis.

3.1. Tissue deformation and elementary descriptive kinematics

One of the simplest mechanical statements about morphogenesis equates the direction and magnitude of tissue movement or deformation to the degree and orientation of forces acting on and within the tissue reduced proportionately by the degree and orientation of mechanical resistance of the tissue to those forces. Numerous engineering approaches can be used to quantify the kinematics or deformation that accompanies tissue morphogenesis. Deformation includes any shape change within a tissue or structure and can include translation and rotation (Fig. 7.2A), flow, shear, and strain. Often tissue deformation varies continuously or smoothly from one location to another but may also change suddenly as with the introduction of a crack or cleft or a discontinuous shear zone where cells from one tissue may slide over or move in opposite direction with respect to cells from another tissue. *Flow*, by analogy with the movement of a gas or liquid, describes a bulk movement or translocation of a parcel of material that has no intrinsic coupling between its constituent parts. The kinematics of a flowing material or fluid can be described completely with relatively simple gas laws or the physics of fluid dynamics. *Strain* involves the dilation or shrinkage of a tissue along a particular direction (Fig. 7.2B). *Shear* involves a graded change in tissue movement along parallel lines (Fig. 7.2B). Shear can occur in both solid and fluid materials. The kinematic description of flow, shear, and strain can vary smoothly from point to point but may also change sharply or discontinuously. Often discontinuities revealed in a kinematic analysis of tissue movements not only correlate to already recognized tissue boundaries, such as the notochord-somite boundary or margin of the neural plate but can also reveal previously unrecognized boundaries within tissues. In the study of mechanics, the descriptive terms of flow, shear, and strain allow the complex movements to be broken down into these unitary deformations. However, these terms are useful when discussing small changes in the material or body from an initial shape (e.g., the spatial or Eulerian description of deformation); large changes, however, require more sophisticated

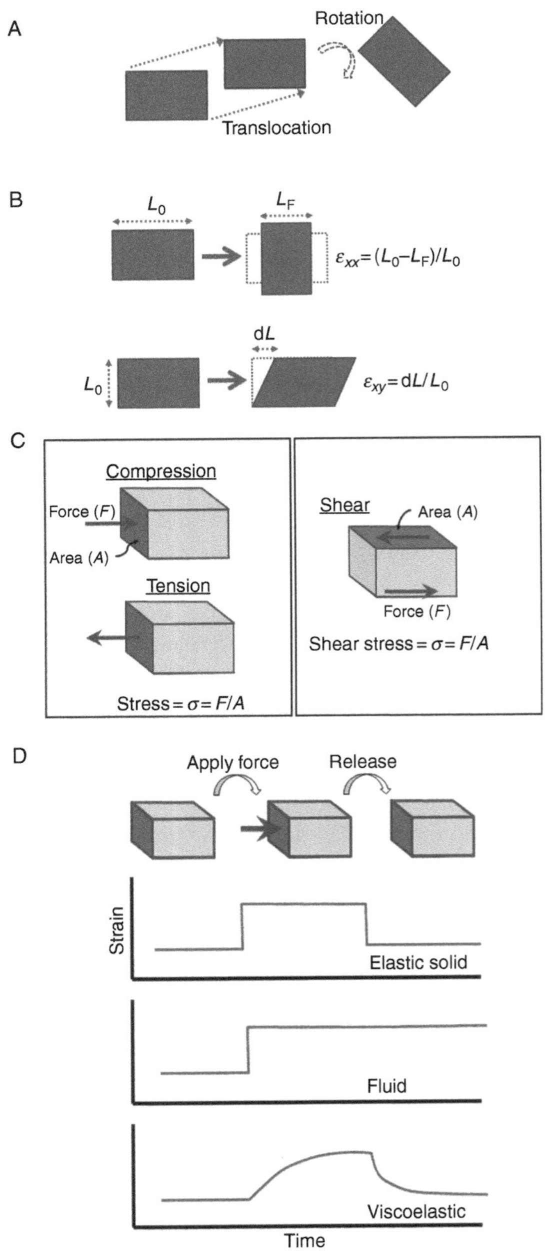

Figure 7.2 Definitions and diagrams for engineering terms describing deformation and stress needed for a biomechanical analysis of morphogenesis. Schematics showing different types of deformation in two dimensions: (A) solid body translation and rotation and (B) strain (ε_{xx}) and shear strain (ε_{xy}). Schematics showing different types of stress that can act within a body: (C) compressive stress, tensile stress, and shear stress. (D) Schematic showing the strain response of a solid, fluid, and viscoelastic material to the application and removal of a force or load.

analyses where deformation is tracked within each parcel of material (e.g., the material or Lagrangian description of deformation). Such distinctions may not be important when looking at small changes but morphogenetic movements can typically involve deformations of 10–20% strain per hour.

A common language to describe the shape changes that occur during morphogenesis is critical to comparative studies of morphogenesis and is increasingly being adopted by developmental biologists seeking to simplify massive data sets collected from full 3D confocal sections collected over the course of morphogenesis (Blanchard *et al.*, 2009; Zamir *et al.*, 2005). Kinematic descriptions and qualitative embryological mechanics are central tools used to investigate how molecular and genetic factors generate the forces and mechanical resistance that direct morphogenetic movements. Kinematic descriptions do not *a priori* require a hypothesis, and it has been suggested that this type of data should be deposited online-like genomic sequences or the results of high-throughput screens or microarray data. The utility of such data will require its presentation in both the raw form of positional data and the processed form of displacements and deformations.

3.2. Morphogenetic kinematics versus morphogenetic dynamics

Once a kinematic analysis or a method for robust kinematic analysis has been devised, a dynamic analysis to investigate the role of force and mechanical resistance can be carried out. Whereas a kinematic analysis can be made with minimally invasive procedures like time-lapse confocal microscopy, measurements of force production and mechanical resistance demand invasive procedures to apply known forces or to place cells and tissues into a context where tissue responses to applied strains and forces can be measured. Methods rooted in the qualitative embryological study of mechanics and biophysical studies of cell mechanics can be adapted to investigate the quantitative generation of force and the response of actively deforming tissues. Before we describe these methods, we will first define some engineering terms relevant to biomechanical dynamics.

In the discussion of kinematics, we defined terms such as flow, shear, and strain; here we will define terms such as stress, tension, and compression and how they are related to the forces generated by cells in the embryo. We typically envision molecular processes such as actomyosin contraction generating discrete sites of force that are *point forces* of tension, transmitted through focal adhesions or sites of cell–cell adhesions (Beningo *et al.*, 2001). Recent *in vitro* work with cultured cells and confluent multicellular sheets has identified forces produced by individual cells through cell adhesions (Trepat *et al.*, 2009). The magnitude of these forces, measured in the range of nanoNewton (nN; a Newton is 1 kg m/s^2 in SI units) to microNewtons (μN), depends strongly on the composition of the adhesion site, cell type, cell shape, and how cells are organized into a sheet. At large distances from these adhesion sites, point forces are distributed or spread over cell surfaces (measured in μm^2). The distribution of a force directed perpendicularly over an area is measured as *stress* (Fig. 7.2C, left panel; force/area; N/m^2 or Pascal or Pa in SI units). The units of stress are identical to the units of pressure. In some instances, due to lack of information about the surface area or due to technical limits, force can be thought to be distributed over a line. In such cases, the distribution of a force over a line is measured as a line stress (force/length; N/m). In this case, the units of line stress are similar to the units of a spring constant but should instead be thought of as (force/area) × (thickness of the area). Forces or stresses exert *tension* when they act in opposite directions that tend to lengthen the structure, whereas forces that are directed toward each other and tend to shorten the structure exert *compression*. Forces do not need to be coaligned with each other and can instead be offset (Fig. 7.2C, right panel). These types of forces induce *shear stress*. Shear stress is also measured as a force distributed over a surface, but instead of being directed perpendicularly with respect to the surface, shear stress is the result of a forces directed within the plane of the surface. The units of shear stress are the same as stress (force/area; Pa) and can be considered tensional or compressive shear depending on whether the forces are directed outward or apposed to one another, respectively.

To understand how forces spread beyond the adhesive contacts between cells and how they can direct shape change, we have to consider the mechanical resistance of tissues to those forces and must define the terms elastic modulus, viscosity, and compressibility. All objects deform in response to applied forces but the degree of that deformation, how much and how fast, depends explicitly on the mechanical properties of the object. The deformation of a multicellular tissue in response to any force or stress depends on the mechanical properties of the tissue. How the tissue responds to applied force determines whether a tissue can be considered an *elastic solid* or a *viscous fluid* or some combination of the two. First, consider the response of an elastic solid; if a tissue is an elastic solid, it will completely recover its original shape after the driving force or stress is removed (Fig. 7.2D). At the other extreme, consider the response of a viscous fluid; if a tissue is a viscous fluid, it will deform under force and simply cease deforming once the force or stress is removed. A viscous fluid will not recover any of its original shape once the driving force or stress is removed (Fig. 7.2D).

It is a useful theoretical exercise to represent embryonic tissues as one of these two extremes. Consider first the example of an embryo composed of a purely elastic solid. Tissue growth and local force production would generate higher and higher stresses in the embryo as development progresses. At any point, microsurgical removal of a piece of tissue would release these stresses and restore the tissue to its original shape before any forces were applied. Similarly, small incisions would result in large wounds "snapping" open wherever tensions were released. Next, consider the example of an embryo composed of a viscous fluid. Tissue growth and local force production would drive movements without any accumulation of stress. Such tissues would be incapable of transmitting force to shape nearby or surrounding tissues. Even the smallest externally applied force would easily reshape the tissue; for instance, small surface tensions could drive all embryos to adopt spherical shapes. Small incisions would never gap or open. In practice, most embryonic tissues should be considered a combination of both elastic solids and viscous fluid, that is, viscoelastic (Fig. 7.2D) and are capable of generating and transmitting force over a large distance, yet unable to accumulate large stresses.

To understand the complex viscoelastic properties of embryonic tissues requires us to define the engineering terms modulus, viscosity, and compressibility that relate the mechanical resistance of embryonic tissues to applied forces and how experimental methods can be used to quantify these properties. One of the most important factors characterizing the response of an embryonic material to an applied force or stress is the tissue's *elastic modulus*, E. The elastic modulus, sometimes referred to as stiffness, Young's modulus, or rigidity, relates the degree of strain or deformation produced in a material by a given stress or force as related by the *constitutive equation*:

$$\varepsilon = \frac{\sigma}{E}, \quad \text{or alternatively,} \quad \sigma = \varepsilon E \tag{1}$$

Since strain has no units the elastic modulus has units of stress (N/m^2, Pascal, Pa). This constitutive equation represents a central physical law of mechanics and in its tensor form relates all forms of strain including shear to all forms of stress by a single array representing the elastic modulus. Other constitutive equations can relate applied forces and resulting movements of a viscous fluid with a characteristic *coefficient of viscosity*, η:

$$\sigma = \eta \frac{\partial \varepsilon}{\partial t} \tag{2}$$

Where the stress is proportional to the time rate of change in the strain. The coefficient of viscosity η has units of stress $\times$ time (Pa $\times$ s, or poise). Additional physical laws such as the influence of gravity or centrifugal forces or an electromagnetic field can be combined with these equations to relate tissue deformation in terms of other sources of applied force. These equations represent the instantaneous responses of elastic solid and viscous fluid tissues to applied forces. A material exhibiting the simplest combination of elastic and viscous responses is called linear viscoelastic material. The response of linear viscoelastic tissues to applied stress requires the use of complex *equations of motion* or of more complex integral forms of the constitutive equation relating time-dependent processes and the "stress- and strain-history" of the tissue. Engineering materials that exhibit time-dependent responses such as these as well as most biological materials are referred to having *nonlinear material properties*.

Along with this more complex relationship between stress and strain, we need to consider other potentially complicating factors such as the compressibility of a tissue and the role of *structure* within embryonic tissues. Tissues may also lose volume as they are subjected to external forces. *Poisson's ratio* (ν; dimensionless units) relates the degree of volume change from one spatial dimension into another. A Poisson's ratio of 0.5 is incompressible and 0.0 is completely compressible. There have been no direct measurements of the Poisson's ratio in embryonic tissues. Most theorists and experimental analyses assume ν ranges from 0.5 to 0.3 but these assumptions often do not introduce more than 10% error into estimated parameters (von Dassow and Davidson, 2009). The complex dependence of the modulus of embryonic tissues on its stress- and strain-history makes direct measurements of compressibility technically challenging.

4. Materials and Structures Responsible for the Mechanical Resistance of Embryonic Tissues

As we consider the kinematic and dynamic mechanics of embryonic development, we need to describe the static structures of the embryo, its component parts, and the key features of its design. It is helpful to consider embryos as being built from materials that are assembled into a structure. A structure is an assemblage of shaped parts, bonded or held together in a specific fashion that are subjected to a defined series of forces or stresses. The term "part" described the unique form of a structural subelement that is formed from a single material. Like a larger structure that is composed of many parts, the deformation of a single part depends critically on the orientation, magnitude, and location of the forces applied to it. Multiple parts can often be combined into a superstructure whose deformation also depends on the orientation, magnitude, and location of an applied force. Structures can change shape in unexpected ways, for instance, by buckling or failing. In this case, a structure may act like a simple elastic structure until applied stresses or loads pass a critical level; once forces exceed that level, the structure may bend, fold, or collapse. Once a structure behaves in this way, it is said the structural response exhibits *nonlinear geometry*. As we use the term structure in the context of an embryo, we may break the embryo into parts like germ layers, cells, and cell–cell walls. In reality, these superstructures are composed of many parts down to the level of protein subdomains (Davidson *et al.*, 2009). However, in practice, we would use the term "material" to describe the composition and behavior of the smallest uniform portion of a larger structure. This is not an atomistic definition, based on the ultimate composition of matter but rather a practical one constrained by experimental methods used to investigate mechanical behaviors of a structure. Tools for biomechanical studies of embryos have very practical limits on the size of samples that can be subjected to experimental tests. For instance, a tissue indenter or microaspirator may sample the mechanical properties of 10–100 cells making this the smallest part of a tissue. We will discuss these tools in the following section, but these limits can guide researchers to conceptually break an embryo into a superstructure assembled from individually shaped pieces that are themselves formed from different materials.

4.1. Experimental approaches to embryo mechanics: Simple geometry and simple analysis

In this section, we will discuss experimental methods for directly measuring the mechanical resistance of embryonic tissues and, to a more limited extent, methods for measuring the forces generated by these tissues. Thus far, we have passed over experimental methods and have focused on the general principles of describing kinematic changes in deforming embryonic tissues (deformation, strain, and flow), the principles of describing the dynamic changes in tissue shape (forces and stress), and outlined general categories of tissue responses to applied stresses (modulus, viscoelasticity). Our coverage of these topics is by no means complete but stops at a convenient point with respect to the limited experimental approaches for directly measuring the quantities required by the theoretical principles we have introduced. The greatest challenge in analyzing the biomechanics of morphogenesis is in quantifying the most essential physical properties of embryonic tissues due to their small size, their ultrasoft mechanical properties, and the difficulties of attaching grips or holders onto sensitive embryonic tissues. In the section below, we will cover three methods in detail and highlight how each technique overcomes these challenges. There have also been many *ad hoc* qualitative and semiquantitative techniques for evaluating the mechanical response of tissues to applied force or strain; however, these approaches rarely provide data that allow direct comparison between different studies. Experimental approaches need to be developed that allow the application of defined forces, stresses, or strains; the measurement of resulting deformations; and the development of interpretive tools to extract quantitative estimates of the mechanical resistance of tissues in common engineering parameters like those provided by the standard linear solid model of a viscoelastic material. In the absence of such standard approaches, it will be difficult to identify universal mechanical principles of morphogenesis and shared mechanisms of mechanotransduction and mechanosensation. In support of this goal, we focus on well-tested methods that have been used to measure the mechanical properties of embryonic tissues during morphogenesis, but at the conclusion of this section, we will discuss briefly several other semiquantitative techniques and what efforts are needed so they can be used to estimate standard engineering parameters.

4.2. Mechanical resistance measured with uniaxial compression or tension

Uniaxial compression or tension offers many advantages for the direct measurement of tissue stiffness. Briefly, a uniform block of embryonic tissue must be microsurgically isolated from the embryo. The block of tissue is either subjected to compression (Davidson and Keller, 2007; Zhou *et al.*, 2009, 2010) or tension (Wiebe and Brodland, 2005) applied by trapping the tissue block between parallel plates, one of which is a force transducer, and moving either of the plates to deform the tissue to a known strain. To estimate standard engineering parameters of mechanical resistance requires time-dependent reports of reactive forces, the degree of axial strain, and the cross-sectional area of the tissue transverse to the direction of applied axial strain (Fig. 7.3A, uniaxial compression). In the case of compression, there is no need to grip the tissue; in contrast, to apply tension, both the force transducer and a fixed-position plate must be attached to the tissue sample (Wiebe and Brodland, 2005). Common biocompatible "glues" include various formulations of cyanoacrylate or "superglue," and fibronectin, a biological adhesive substrate for cells. These glues can be used to affix tissues directly to the parallel plates or force transducer or to fabricated plastic grips. In compression, as long as the tissue does not bend, buckle, or shift, and in tension, as long these attachments hold and the tissue maintains a uniform cross-section anatomy between the parallel plates, mechanical properties such as the time-dependent Young's Modulus can be estimated (Zhou *et al.*, 2009) and compared to tissues and synthetic materials from a variety of sources (Levental *et al.*, 2007).

Challenges to using uniaxial compression or tension range from the destructive nature of microsurgery and the limited size of the tissue sample to the need for uniform anatomy in the cross-sectional area (Zhou *et al.*, 2009). The microsurgical operations needed to isolate tissue fragments are highly invasive and have been restricted to studies of amphibian embryos including *Xenopus* and *Ambystoma* species. Historically, embryos from avians, mouse, teleosts, and nematodes have been the subject of microsurgical methods and should make these systems amenable to mechanical testing by uniaxial compression or tension. Wound healing due to

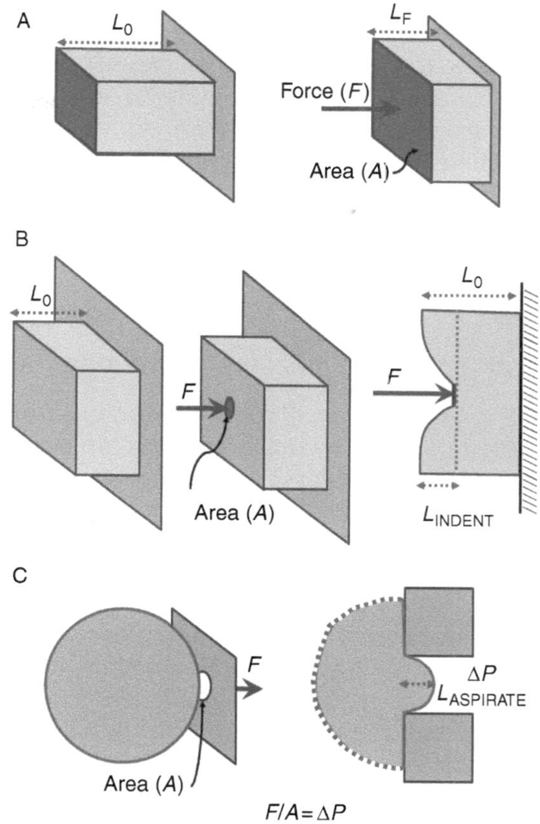

Figure 7.3 Methods for measuring the mechanical resistance of embryonic tissues. (A) Uniaxial compression or stress-relaxation test: force is applied to the shaded surface area causing the sample to undergo strain. Strain is measured by the geometric change in length in the same direction as the force is applied. (B) Indentation test: force is applied through an indenter (rod or shaped silicon cantilever tip) across a small contact area. Deformation is measured by the distance the surface is indented from the original surface. (C) Microaspiration test: force is applied using regulated pressure to a small surface area. Deformation is measured by the distance the sample moves toward the source of low pressure. Each test requires application of a defined force (*F*) over a defined area (*A*). In addition to the force and area, the geometry of the embryonic tissue sample before (L_0) and after deformation of the tissue (L_F, uniaxial compression; L_{INDENT}, indentation; and, $L_{ASPIRATE}$, microaspiration) are needed to calculate the strain and time-dependent Young's Modulus.

microsurgery can complicate interpretations of this test; however, these effects can be addressed using "sham-operated" controls where samples with and without incisional wounds are compared. Simple tension or compression tests of tissue fragments can yield semiquantitative measures of mechanical resistance producing values with apparent units of "modulus" or "stiffness." However, when testing tissues without a uniform cross-sectional area or uniform anatomy, there are no simple methods to estimate standard engineering parameter of modulus and results cannot be compared with other tissues or materials. Such parameters can be estimated, but significant investment is required to develop simulations or theoretical models of the unique geometry and anatomy. Thus, while there are considerable limits to the types of tissues that can be tested with uniaxial compression or tension techniques, the simplicity of their interpretation makes these some of the most robust methods for measuring standard engineering parameters of tissue material properties.

4.3. Mechanical resistance measured with indentation

The two methods of indentation and microaspiration can be used noninvasively to measure standard engineering parameters but have required more investment in the theoretical analysis of the interaction of the indenting probe or microchannel with embryonic tissues. Indentation, where a rod or sphere is pushed into the tissue, has been used extensively with cell and tissue level analyses of mechanical resistance. In brief, the indenting rod or sphere is mounted onto a sensitive force transducer and moved using a micromanipulator. The position of the micromanipulator, the reactive force reported by the transducer, and the amount of deformation of the tissue are measured simultaneously (Fig. 7.3B; Zamir and Taber, 2004a,b). To estimate engineering parameters for the mechanical resistance, the researcher must use a specific theoretical model for the contact mechanics, and the geometry and structure of the indented tissues. These models are typically more complex than those needed for uniaxial testing, but they can be validated using synthetic structures and materials (Zamir and Taber, 2004a).

Challenges to using indentation include limited access to tissues of interests, assumptions made by contact models, and the unintentional wounding of indented tissues. Use of the indenter is limited to tissues that provide clear access, so the indenter can be brought into perpendicular contact with the tissue. Typically, clear access restricts the indenter to measuring epithelial sheets that surround the mesoderm and endoderm. To access deeper tissues or to test mechanical properties of forming organs such as the heart requires microsurgical isolation and raises limitations like those mentioned above for uniaxial testing. Knowledge of the anatomy and theoretical models of tissue response to indentation are critical to using this method quantitatively. As with uniaxial testing methods discussed above and microaspiration discussed below, indentation used without models can provide semiquantitative parameters with apparent units of stiffness or rigidity, but without a model or validation of model assumptions, these parameters cannot be compared across multiple studies or between organisms or organ systems.

4.4. Mechanical resistance measured with microaspiration

Like indentation, microaspiration has a long history of being used in studying cell mechanical properties (e.g., Cole, 1932; Mitchison and Swann, 1954). Microaspiration involves using defined pressure gradients to suck a small patch of embryonic tissue into a narrow channel. Glass capillary tubes or polished glass needles have been used to aspirate small surface patches or whole intact cells. Typically, pressure gradients have been controlled with manometer techniques consisting of large loops of tubing. In contrast, devices that our group has developed for testing the mechanical resistance of multicellular embryonic tissues use very large channels cast into polydimethylsiloxane (PDMS) blocks (von Dassow and Davidson, 2009). Instead of high-pressure gradients produced by manometers needed to aspirate cell surfaces, we use much smaller gradients generated by 20–200 μm differences in aqueous reservoir heights (Fig. 7.3C). In brief, a target patch of tissue is brought into contact with a microchannel opening in the absence of a pressure gradient. A small holding pressure gradient is applied to seal the contact between the tissue and the aspirator microchannel. The pressure gradient is then increased to a fixed level, and deformation of the tissue into the microchannel is observed. Semiquantitative measures of mechanical resistance can be made based on the pressure gradient and movement down the microchannel. Quantitative estimates of engineering parameters characterizing mechanical properties can be obtained using models that incorporate the structural details of the embryonic tissue, contact with the microchannel face, corners, and walls, and the temporal profile of pressure changes. Such models have been previously developed for cell microaspiration and have been extended in recent years by our group to describe the response of more complex multicellular embryonic tissues (von Dassow *et al.*, 2010).

Challenges to using microaspiration parallel those of indentation and include limited access to tissues of interest, assumptions made by aspiration models, and the unintentional wounding of indented tissues. Microaspiration can only be used with tissues that can be brought into clear contact with the microchannel. Detailed microanatomical descriptions of tissue architecture are also key factors when building models and are required before estimations of quantitative engineering parameters can be made. Like indentation, analysis is simplified when tested samples are thick and mechanical resistance due to structural bending can be ignored. Like the two methods listed above, microaspiration can be used semiquantitatively but to compare measurements of mechanical resistance from one organism to the next, or from one organ to another requires the consistent use of validated theoretical models of microaspirator geometry and interactions with microanatomy of the embryonic tissue.

4.5. *Ad hoc* approaches to assessing mechanical resistance

In addition to semiquantitative application of the above methods, a wide variety of *ad hoc* approaches have been developed to estimate mechanical resistance of embryonic tissues to deformation including: centrifugation and shape analysis of sessile drops (Kalantarian *et al.*, 2009), microrheology (Daniels *et al.*, 2006; Panorchan *et al.*, 2007), atomic force microscopy (Krieg *et al.*, 2008; Puech *et al.*, 2005), and tissue wounding (Beloussov *et al.*, 1975; Kiehart *et al.*, 2000). These are extremely useful approaches for applying strain to embryonic tissues or comparing mechanical properties of embryonic tissues under different conditions, generally within the same study, but are of limited use when comparing mechanical properties between organisms or comparing mechanical properties across different length scales. To extract estimates of engineering parameters from these methods, the *ad hoc* experiments need to be simulated in more detail so the contribution of both structural and material mechanics can be understood. Several of these *ad hoc* approaches have been studied in detail; for instance, laser drilling or microdissection has been analyzed theoretically (Hutson *et al.*, 2009; Ma *et al.*, 2009) and has been used to assess the relative contribution of subcellular structures to the recoil-response to laser wounding. More complete theoretical analyses of these *ad hoc* methods are essential to extend findings from mechanics of invertebrate embryos to vertebrate or mammalian embryo mechanics.

5. Approaches for Measuring Stress and Force Production

Tissues movements during morphogenesis are due to transient imbalances in the stresses within the embryo. These stress imbalances may be due to changes in local mechanical resistance or through changes in local force production. Furthermore, these changes may be due to altered mechanics at a number of different length scales, from changes in the activity of motors such as myosin at the protein level, to the frequency and duration of cellular protrusions at the cell scale, to altered anatomy at the tissue level. The major challenge to the experimentalist is to measure these changes at scales that are relevant to the processes driving tissue movement, for instance, measurements at both the protein level and the bulk tissue level. A wide range of quantitative biophysical approaches to estimate stresses at the subcellular scale have been developed for studies of the cytoskeleton, microrheology, and whole-cell mechanics; however, only a few approaches have been developed for quantitatively estimating stress within embryonic tissues.

Laser wounding is used to investigate the balance of stresses within epithelial sheets in the early embryo and later larval stages of the *Drosophila* embryo (Butler *et al.*, 2009; Farhadifar *et al.*, 2007; Fernandez-Gonzalez *et al.*, 2009; Hutson *et al.*, 2003; Rauzi *et al.*, 2008). Laser wounding (also known as laser ablation or laser drilling) is used in much the same semiquantitative fashion as microsurgical incision (described above) to investigate the balance of tensional and compressive stresses within the apical surface of an embryonic epithelium. In brief, a tightly focused laser beam, ideally focused to a diffraction limited spot, is pulsed onto a cell or cell–cell junction. The rapid delivery of heat and light within the epithelium results in rupture of the cell cortex or cell–cell junction. Cell and tissue movements, termed "recoil-response," observed with time-lapse confocal microscopy over the course of the next few seconds can be used as an indication of the relative magnitude and principle directions of stresses that exist within the epithelium. First, introduced as an *ad hoc* tool, laser wounding and theoretical tools describing the recoil-response have been adopted by many research groups to estimate epithelial stress within developing *Drosophila* embryos.

The advantages of laser wounding lie in the optical methods of delivering a rapid incisional wound and in describing the kinematics of the recoil-response; however, there are several limitations to this approach. The first limitation is that the recoil-response may be the product of both actively generated forces and actively generated mechanical resistance. Thus, reduced or increased rates of deformation could be due to changes in either of these mechanical properties of the epithelium. For instance, a low rate of opening of a wound could be due to increased stiffness of the epithelial sheet as well as a decrease in force production. Most computational simulations of *Drosophila* epithelia focus on the changing role of force production and overlook the changing role of mechanical resistance. A notable exception is the recent study which explicitly investigated the changing role of the microarchitecture of epithelial cells in shaping the recoil-response to laser wounding (Hutson *et al.*, 2009). Additionally, the recoil-response after laser wounding may reflect other high-speed cellular responses such as those that have been observed in laser-wounded vertebrate epithelia (Clark *et al.*, 2009; Joshi *et al.*, 2010) or after rapid mechanical stimulation of cultured mammalian cells (Na *et al.*, 2008). Lastly, laser wounding has not been able to provide comparative data on tissue mechanical properties using standard engineering parameters making it difficult to extend the findings in *Drosophila* to studies of vertebrate epithelial morphogenesis. However, even with these limitations, the use of laser wounding has extended quantitative biomechanical analyses of epithelial morphogenesis to one of the premier genetic models of development.

6. Mechanical Hypothesis Building from Histology and Compositional Analysis

There are at least three principle roles for mechanics during morphogenesis: (1) to drive tissue movement, (2) to provide feedback for robust programs of development, and (3) to provide positional cues for cell differentiation. We have been focusing on defining the terms and concepts in biomechanics and methods to measure mechanical resistance and stress within embryos to help guide efforts to test these distinct roles. All mechanical analyses of morphogenesis build on existing descriptive histology and compositional analyses of morphogenesis. Descriptive histology provides information on the size and shape of cells and tissues and the changing localization of intracellular and extracellular proteins; descriptive kinematics provide information on the rates and directions of cell and tissue movements within the embryo. Hypothesis building begins with two possibilities: either movements or shape changes occur autonomously within the cells and tissues, or the movements are driven by forces originating outside the region that is deforming. Most studies focus on autonomous movements, yet it is likely that most of the movements within intact embryos are the product of multiple forces including those acting autonomously, those due to mechanical resistance of surrounding tissues, and forces produced by shape changes in other parts of the embryo (Davidson *et al.*, 2009; Kiehart *et al.*, 2000).

Compositional analysis includes studies of gene expression, protein localization, and protein activity within tissues that are actively changing shape and the temporal changes in these factors before, during, and after movements have occurred. In the past, compositional analysis served to focus research efforts on *morphogenetically active* factors such as proteins involved in the cytoskeleton, extracellular matrix, and signaling factors like the Rho-family of small GTPases. Increasingly, a growing list of *morphogenetically involved* proteins are being included in compositional analysis after being discovered from high-throughput methods based on microarrays, deep sequencing, and proteomics. During the course of histological description and compositional analysis, the researcher commonly constructs one or more "just-so-stories" with plausible, biomechanically phrased hypotheses. Rigorous efforts to disprove the mechanical basis of these hypotheses require application of the tools described in the sections above combined with genetic, molecular, and cell biological tools to perturb or alter the composition or microanatomy of the embryo during morphogenesis (Zhou *et al.*, 2009, 2010).

7. Challenges in Integrative Biomechanics: From Molecular to Whole Organism

Modern cell and molecular biology have succeeded in defining the gene regulatory networks and signaling cascades that pattern and initiate cell behaviors accompanying early development and are increasingly turning toward the question of how these processes are integrated to drive morphogenesis. The goal of modern studies of developmental mechanics is to complement these molecular studies with biomechanical analyses and to identify physical principles and constraints that lie outside the conventional molecular-genetic viewpoint. For instance, biomechanical analysis has identified actomyosin as a key regulator of the mechanical resistance of amphibian embryonic tissues to applied stresses. However, the idea that mechanical processes are *all* under direct molecular control is likely to be misleading. Consider the transmission of stress or force throughout an embryonic tissue. We envision tensional stresses, where cells are actively being pulled apart from one another, will be regulated by molecular processes within cell–cell adhesions involving cadherins and the actin cytoskeleton. However, just because complexes are present does not mean they are actually transmitting force. Once the tissue is under compression force, transmission may not require specific molecular components. What does this mean for morphogenetic movements? Embryonic tissues appear incompressible. An incompressible material means that those tissues do not change in volume upon compression or tension. This simple principle means that compressive forces applied in one direction produce tensional forces in the perpendicular direction. Consider how incompressibility may alter the mechanical response of a tissue subjected to compressive force along its anterior–posterior (AP) axis and as compressive stress is transmitted across the cells' AP faces. Transmission of these compressive stresses need not be transmitted through junctional complexes or for that matter deform the cell cortex along that direction. Since cells are incompressible they must change shape in their mediolateral (ML) or dorsoventral (DV) axes. Thus, molecular regulation of force production and mechanical resistance can integrate with the purely physical properties of multicellular tissues to transmit stress and drive cell and tissue shape change.

One of the greatest challenges to understanding the role of mechanics in development is the careful design of rigorous biophysical and biomechanical experiments. These experiments must aim to directly measure force production at the subcellular, cellular, and tissue scales and to measure the physical material properties of cell- and tissue-scale structures that deform under realistic applied forces. The small size of embryos, the extremely small forces that shape them, and their low stiffness viscoelastic and viscoplastic properties make these measurements challenging. Surprisingly, there are few if any direct measurements of mechanical properties or force production in many of the commonly used developmental model systems.

Work from a small group of labs studying avian and amphibian embryos have begun to make experimental progress needed to understand the mechanics of early morphogenesis. Our group has refined tools to measure the mechanical properties of *X. laevis* embryonic tissues (Davidson and Keller, 2007; von Dassow and Davidson, 2009). We have used these tools to describe variation in the mechanical properties with stage and germ layer (Zhou *et al.*, 2009), to describe variation in mechanical properties between individuals and the correlation of mechanical properties with the success of gastrulation (von Dassow and Davidson, 2009), and to test the role of mechanical feedback during development (von Dassow *et al.*, 2010). Combined with cell and molecular biology techniques, we are using these tools to explore the molecular and genetic control of developmental mechanics, finding that signaling pathways mediated by RhoGTPase can directly regulate the stiffness of gastrulating embryos (Zhou *et al.*, 2010). Avian embryos afford similar access to the mechanical properties of early development (Agero *et al.*, 2010), formation of the heart (Taber *et al.*, 2010; Zamir and Taber, 2004a) and the nervous system (Xu *et al.*, 2009). Tools to measure force production during development are considerably more challenging; however, our group has adapted traction force techniques (Beningo *et al.*, 2002) to observe cell-generated traction during convergence and extension (Zhou *et al.*, 2010) and has begun development of tools to measure bulk force production during axis elongation in *X. laevis* embryos (Zhou and Davidson, unpublished).

More advanced experimental tools will require integrated theoretical models of morphogenesis. A number of theoretical and computational efforts have been started to combine the mechanics of development with signaling factors and complex cell and tissue microanatomy of the early embryo. Most of these efforts are conducted in an *ad hoc* manner to aid the experimentalist in interpreting experiments within a single study. Currently, such efforts are guided by the goal of recapitulating deformations observed during morphogenesis. Commonly, these models present static mechanical structures of an embryo or tissue and test the effectiveness of applied forces or stresses in reproducing observed movements. The process of convergent extension has been extensively modeled using a variety of theoretical approaches including finite element models (Brodland, 2006), elastic vertex network models (Weliky *et al.*, 1991), and Cellular Potts models (Zajac *et al.*, 2000) (for review, see Davidson *et al.*, 2010). As yet, none of these models or other similar models describing morphogenesis consider the dynamically changing mechanical properties of embryonic tissues and lack quantitative estimates of the mechanical properties of embryonic tissues including viscoelastic properties, stress production, and force generation.

Theoretical mechanical representations of embryonic tissues should be viewed as a set of *constitutive equations* and evaluated by standard engineering principles. Future theoretical studies of morphogenesis need to be formulated so their results can be compared from one study to the next and so their results can be extended to understanding morphogenesis in other organisms. The most basic mechanical response of many simulated embryonic tissues is untested. For example, these models should be subjected to biomechanical analyses including quantitative analysis of their responses to applied stress and strain, subject to simulated forces and their elastic or viscoelastic responses reported in standard engineering terms. Establishment of a common biomechanical framework will greatly extend the utility of future modeling efforts.

8. Discussion and Conclusion: Biomechanical Variability, Robustness, and the Origin of Birth Defects

Multicellular organisms live and reproduce within a world where resources may be sparse and environmental conditions are constantly varying. Eggs are formed and embryos develop against a background of these changing conditions making use of inherited genetic programs that have been tuned for success over the course of evolution. However, successful programs may incorporate noise in anticipation of the variable conditions developing embryos encounter, essentially hedging their bets against these random conditions. Such noisy programs may serve two goals: (1) to allow as many embryos as possible to survive under optimal conditions and (2) to ensure some embryos survive even under the most extreme conditions. Our observations of the mechanical properties and forces produced during morphogenesis suggest some of this noise may be in the form of biomechanical variation. We propose an extension of the multifactorial threshold model for the prevalence of birth defects (Boyles *et al.*, 2005; Carter, 1976) to include mechanical variation and biomechanical thresholds for the onset of birth defects.

Many groups have observed that key morphogenetic movements operate as bottlenecks during embryogenesis. Early developmental defects in embryogenesis first appear when embryos arrest or appear blocked at these bottlenecks. Several bottlenecks exist in early development of the frog including: (1) failure to initiate archenteron formation during mid-gastrulation (stages 11–11.5), (2) failure to close the mesendoderm mantle (stages 12–13), (3) failure to close the blastopore (stage 13), and (4) failure to close the neural folds and form the neural tube (stages 19–20). We propose that these bottlenecks occur at biomechanically sensitive phases in development where one tissue movement must be completed successfully before subsequent movements or patterning events can start. Delays or blocks in movements initiate catastrophic cascades when other programs are initiated before the embryo has passed the bottleneck. For instance, if blastopore closure is blocked or delayed too long, the process of convergent extension starts in lateral and posterior paraxial mesoderm. The resulting convergent extension movements of the left and right sides of the presomitic mesoderm are never joined to the midline through the notochord and the resulting embryo elongates forming neural folds that cannot close over the intervening open blastopore.

We propose developmental defects result from specific mechanical failures in morphogenesis at these bottlenecks, either forces are insufficient to drive the needed movement or mechanical resistance is too high for force-generating tissues to overcome. Defects in mechanosensing or mechanotransduction pathways could also short circuit the programs maintaining a balance of force production with mechanical resistance. There are many experimental and theoretical challenges to testing this proposal including the ability to measure and experimentally manipulate both force generation and mechanical resistance. The experimental techniques to measure mechanical resistance and stress described above are often incompatible with the complex microanatomy of embryos at these bottleneck stages, so new experimental tools along with engineering-based theoretical simulations will be needed for their interpretation. Lastly, new multiscale modeling at the level of both cell- and tissue-biomechanics will be needed to investigate the robustness of morphogenetic movements and identify the physical limits of these robust programs.

ACKNOWLEDGMENTS

I would like to thank Michelangelo von Dassow, Sagar Joshi, Hye Young Kim, and Jian Zhou for their helpful discussions. This work would not have been possible without the pioneering embryological studies of Johannes Holtfreter, John Trinkaus, Ray Keller, Doug DeSimone, Rudy Winklbauer, and many others; we greatly admire their determination and are forever indebted for their continuing guidance. Support for this work was provided by grants from the NIH (HD044750 and ES019259) and the NSF (IOS-0845775).

REFERENCES

Agero, U., Glazier, J. A., and Hosek, M. (2010). Bulk elastic properties of chicken embryos during somitogenesis. *Biomed. Eng. Online* **9,** 19.

Beloussov, L. V., Dorfman, J. G., and Cherdantzev, V. G. (1975). Mechanical stresses and morphological patterns in amphibian embryos. *J. Embryol. Exp. Morphol.* **34,** 559–574.

Beningo, K. A., Dembo, M., Kaverina, I., Small, J. V., and Wang, Y. L. (2001). Nascent focal adhesions are responsible for the generation of strong propulsive forces in migrating fibroblasts. *J. Cell Biol.* **153,** 881–888.

Beningo, K. A., Lo, C. M., and Wang, Y. L. (2002). Flexible polyacrylamide substrata for the analysis of mechanical interactions at cell-substratum adhesions. *Methods Cell Biol.* **69,** 325–339.

Blanchard, G. B., Kabla, A. J., Schultz, N. L., Butler, L. C., Sanson, B., Gorfinkiel, N., Mahadevan, L., and Adams, R. J. (2009). Tissue tectonics: Morphogenetic strain rates, cell shape change and intercalation. *Nat. Methods* **6,** 458–464.

Boyles, A. L., Hammock, P., and Speer, M. C. (2005). Candidate gene analysis in human neural tube defects. *Am. J. Med. Genet. C Semin. Med. Genet.* **135C,** 9–23.

Brodland, G. W. (2006). Do lamellipodia have the mechanical capacity to drive convergent extension? *Int. J. Dev. Biol.* **50,** 151–155.

Butler, L. C., Blanchard, G. B., Kabla, A. J., Lawrence, N. J., Welchman, D. P., Mahadevan, L., Adams, R. J., and Sanson, B. (2009). Cell shape changes indicate a role for extrinsic tensile forces in *Drosophila* germ-band extension. *Nat. Cell Biol.* **11,** 859–864.

Carter, C. O. (1976). Genetics of common single malformations. *Br. Med. Bull.* **32,** 21–26.

Clark, A. G., Miller, A. L., Vaughan, E., Yu, H. Y., Penkert, R., and Bement, W. M. (2009). Integration of single and multicellular wound responses. *Curr. Biol.* **19,** 1389–1395.

Cole, K. S. (1932). Surface forces of the *Arbacia* egg. *J. Cell. Comp. Physiol.* **1,** 1–9.

Daniels, B. R., Masi, B. C., and Wirtz, D. (2006). Probing single-cell micromechanics in vivo: The microrheology of *C. elegans* developing embryos. *Biophys. J.* **90,** 4712–4719.

Davidson, L., and Keller, R. (2007). Measuring mechanical properties of embryos and embryonic tissues. *Methods Cell Biol.* **83,** 425–439.

Davidson, L. A., Hoffstrom, B. G., Keller, R., and DeSimone, D. W. (2002). Mesendoderm extension and mantle closure in *Xenopus laevis* gastrulation: Combined roles for integrin alpha5beta1, fibronectin, and tissue geometry. *Dev. Biol.* **242,** 109–129.

Davidson, L. A., Marsden, M., Keller, R., and Desimone, D. W. (2006). Integrin alpha5-beta1 and fibronectin regulate polarized cell protrusions required for *Xenopus* convergence and extension. *Curr. Biol.* **16,** 833–844.

Davidson, L., von Dassow, M., and Zhou, J. (2009). Multi-scale mechanics from molecules to morphogenesis. *Int. J. Biochem. Cell Biol.* **41,** 2147–2162.

Davidson, L. A., Joshi, S. D., Kim, H. Y., von Dassow, M., Zhang, L., and Zhou, J. (2010). Emergent morphogenesis: Elastic mechanics of a self-deforming tissue. *J. Biomech.* **43,** 63–70.

Ewald, A. J., Peyrot, S. M., Tyszka, J. M., Fraser, S. E., and Wallingford, J. B. (2004). Regional requirements for Dishevelled signaling during *Xenopus* gastrulation: Separable effects on blastopore closure, mesendoderm internalization and archenteron formation. *Development* **131,** 6195–6209.

Farhadifar, R., Roper, J. C., Aigouy, B., Eaton, S., and Julicher, F. (2007). The influence of cell mechanics, cell–cell interactions, and proliferation on epithelial packing. *Curr. Biol.* **17,** 2095–2104.

Fernandez-Gonzalez, R., Simoes Sde, M., Roper, J. C., Eaton, S., and Zallen, J. A. (2009). Myosin II dynamics are regulated by tension in intercalating cells. *Dev. Cell* **17,** 736–743.

Holtfreter, J. (1943). A study of the mechanics of gastrulation (part 1). *J. Exp. Zool.* **94,** 261–318.

Holtfreter, J. (1944). A study of the mechanics of gastrulation (part 2). *J. Exp. Zool.* **95,** 171–212.

Hutson, M. S., Tokutake, Y., Chang, M. S., Bloor, J. W., Venakides, S., Kiehart, D. P., and Edwards, G. S. (2003). Forces for morphogenesis investigated with laser microsurgery and quantitative modeling. *Science* **300,** 145–149.

Hutson, M. S., Veldhuis, J., Ma, X., Lynch, H. E., Cranston, P. G., and Brodland, G. W. (2009). Combining laser microsurgery and finite element modeling to assess cell-level epithelial mechanics. *Biophys. J.* **97,** 3075–3085.

Joshi, S. D., von Dassow, M., and Davidson, L. A. (2010). Experimental control of excitable embryonic tissues: Three stimuli induce rapid epithelial contraction. *Exp. Cell Res.* **316,** 103–114.

Kalantarian, A., Ninomiya, H., Saad, S. M., David, R., Winklbauer, R., and Neumann, A. W. (2009). Axisymmetric drop shape analysis for estimating the surface tension of cell aggregates by centrifugation. *Biophys. J.* **96,** 1606–1616.

Keller, R. (2002). Shaping the vertebrate body plan by polarized embryonic cell movements. *Science* **298,** 1950–1954.

Keller, R., and Danilchik, M. (1988). Regional expression, pattern and timing of convergence and extension during gastrulation of *Xenopus laevis*. *Development* **103**(1), 193–209.

Keller, R., and Shook, D. (2008). Dynamic determinations: Patterning the cell behaviours that close the amphibian blastopore. *Philos. Trans. R. Soc. Lond. B Biol. Sci.* **363,** 1317–1332.

Kiehart, D. P., Galbraith, C. G., Edwards, K. A., Rickoll, W. L., and Montague, R. A. (2000). Multiple forces contribute to cell sheet morphogenesis for dorsal closure in *Drosophila*. *J. Cell Biol.* **149,** 471–490.

Krieg, M., Arboleda-Estudillo, Y., Puech, P. H., Kafer, J., Graner, F., Muller, D. J., and Heisenberg, C. P. (2008). Tensile forces govern germ-layer organization in zebrafish. *Nat. Cell Biol.* **10,** 429–436.

Levental, I., Georges, P. C., and Janmey, P. A. (2007). Soft biological materials and their impact on cell function. *Soft Matter.* **3,** 299–306.

Longo, D., Peirce, S. M., Skalak, T. C., Davidson, L., Marsden, M., Dzamba, B., and DeSimone, D. W. (2004). Multicellular computer simulation of morphogenesis: Blastocoel roof thinning and matrix assembly in *Xenopus laevis*. *Dev. Biol.* **271,** 210–222.

Ma, X., Lynch, H. E., Scully, P. C., and Hutson, M. S. (2009). Probing embryonic tissue mechanics with laser hole drilling. *Phys. Biol.* **6,** 036004.

Marsden, M., and DeSimone, D. W. (2001). Regulation of cell polarity, radial intercalation and epiboly in *Xenopus*: Novel roles for integrin and fibronectin. *Development* **128,** 3635–3647.

Mitchison, J. M., and Swann, M. M. (1954). The mechanical properties of the cell surface. I. The cell elastimeter. *J. Exp. Biol.* **31,** 443–461.

Na, S., Collin, O., Chowdhury, F., Tay, B., Ouyang, M., Wang, Y., and Wang, N. (2008). Rapid signal transduction in living cells is a unique feature of mechanotransduction. *Proc. Natl. Acad. Sci. USA* **105,** 6626–6631.

Panorchan, P., Lee, J. S., Daniels, B. R., Kole, T. P., Tseng, Y., and Wirtz, D. (2007). Probing cellular mechanical responses to stimuli using ballistic intracellular nanorheology. *Methods Cell Biol.* **83,** 115–140.

Puech, P. H., Taubenberger, A., Ulrich, F., Krieg, M., Muller, D. J., and Heisenberg, C. P. (2005). Measuring cell adhesion forces of primary gastrulating cells from zebrafish using atomic force microscopy. *J. Cell Sci.* **118,** 4199–4206.

Rauzi, M., Verant, P., Lecuit, T., and Lenne, P. F. (2008). Nature and anisotropy of cortical forces orienting *Drosophila* tissue morphogenesis. *Nat. Cell Biol.* **10,** 1401–1410.

Shih, J., and Keller, R. (1992). Cell motility driving mediolateral intercalation in explants of *Xenopus laevis*. *Development* **116**(4), 901–914.

Taber, L. A., Voronov, D. A., and Ramasubramanian, A. (2010). The role of mechanical forces in the torsional component of cardiac looping. *Ann. NY Acad. Sci.* **1188,** 103–110.

Trepat, X., Wasserman, M. R., Angelini, T. E., Millet, E., Weitz, D. A., Butler, J. P., and Fredberg, J. J. (2009). Physical forces during collective cell migration. *Nat. Phys.* **5,** 426–430.

von Dassow, M., and Davidson, L. A. (2009). Natural variation in embryo mechanics: Gastrulation in *Xenopus laevis* is highly robust to variation in tissue stiffness. *Dev. Dyn.* **238,** 2–18.

von Dassow, M., Strother, J. A., and Davidson, L. A. (2010). Surprisingly simple mechanical behavior from a complex embryonic tissue. *PLoS ONE* **5**(12), e15359.

Wacker, S., Grimm, K., Joos, T., and Winklbauer, R. (2000). Development and control of tissue separation at gastrulation in *Xenopus*. *Dev. Biol.* **224,** 428–439.

Wallingford, J. B., Fraser, S. E., and Harland, R. M. (2002). Convergent extension: The molecular control of polarized cell movement during embryonic development. *Dev. Cell* **2,** 695–706.

Weliky, M., Minsuk, S., Keller, R., and Oster, G. (1991). Notochord morphogenesis in *Xenopus laevis*: Simulation of cell behavior underlying tissue convergence and extension. *Development* **113**(4), 1231–1244.

Wiebe, C., and Brodland, G. W. (2005). Tensile properties of embryonic epithelia measured using a novel instrument. *J. Biomech.* **38,** 2087–2094.

Winklbauer, R., and Nagel, M. (1991). Directional mesoderm cell migration in the *Xenopus* gastrula. *Dev. Biol.* **148,** 573–589.

Winklbauer, R., and Schurfeld, M. (1999). Vegetal rotation, a new gastrulation movement involved in the internalization of the mesoderm and endoderm in *Xenopus. Development* **126,** 3703–3713.

Xu, G., Bayly, P. V., and Taber, L. A. (2009). Residual stress in the adult mouse brain. *Biomech. Model. Mechanobiol.* **8,** 253–262.

Zajac, M., Jones, G. L., and Glazier, J. A. (2000). Model of convergent extension in animal morphogenesis. *Phys. Rev. Lett.* **85,** 2022–2025.

Zamir, E. A., and Taber, L. A. (2004a). Material properties and residual stress in the stage 12 chick heart during cardiac looping. *J. Biomech. Eng.* **126,** 823–830.

Zamir, E. A., and Taber, L. A. (2004b). On the effects of residual stress in microindentation tests of soft tissue structures. *J. Biomech. Eng. Trans. ASME* **126,** 276–283.

Zamir, E. A., Czirok, A., Rongish, B. J., and Little, C. D. (2005). A digital image-based method for computational tissue fate mapping during early avian morphogenesis. *Ann. Biomed. Eng.* **33,** 854–865.

Zhou, J., Kim, H. Y., and Davidson, L. A. (2009). Actomyosin stiffens the vertebrate embryo during critical stages of elongation and neural tube closure. *Development* **136,** 677–688.

Zhou, J., Kim, H. Y., Wang, J. H.-C., and Davidson, L. A. (2010). Macroscopic stiffening of embryonic tissues via microtubules, Rho-GEF, and assembly of contractile bundles of actomyosin. *Development* **137,** 2785–2794.

CHAPTER EIGHT

Mechanotransduction in Development

Emmanuel Farge

Contents

Abstract

Biochemical patterning and morphogenetic movements coordinate the design of embryonic development. The molecular processes that pattern and closely control morphogenetic movements are today becoming well understood. Recent experimental evidence demonstrates that mechanical cues generated by morphogenesis activate mechanotransduction pathways, which in turn regulate cytoskeleton remodeling, cell proliferation, tissue differentiation. From *Drosophila* oocytes and embryos to *Xenopus* and mouse embryos and *Arabidopsis* meristem, here we review the developmental processes known to be activated *in vivo* by the mechanical strains associated to embryonic multicellular tissue morphogenesis. We describe the genetic, mechanical, and magnetic tools that have allowed the testing of mechanical induction in development by a step-by-step uncoupling of genetic inputs from mechanical inputs in

Mechanics and Genetics of Embryonic and Tumoral Development Group, UMR168 CNRS, Institut Curie, Paris, France

Current Topics in Developmental Biology, Volume 95
ISSN 0070-2153, DOI: 10.1016/B978-0-12-385065-2.00008-6

embryogenesis. We discuss the known underlying molecular mechanisms involved in such mechanotransduction processes, including the Armadillo/β-catenin activation of Twist and the Fog-dependent stabilization of Myosin-II. These mechanotransduction processes are associated with a variety of physiological functions, such as mid-gut differentiation, mesoderm invagination and skeletal joint differentiation in embryogenesis, cell migration and internal pressure regulation during oogenesis, and meristem morphogenesis. We describe how the conservation of associated mechanosensitive pathways in embryonic and adult tissues opens new perspectives on mechanical involvement, potentially in evolution, and in cancer progression.

1. Introduction

After the Newtonian revolution of physics in the eighteenth century, embryonic observations were naturally interpreted by some of the scientists following that period as flows governed by the Newtonians laws of hydrodynamics (His, 1875; Leduc, 1912). One of these embryologists, D'Arcy Thompson, extended such vision in proposing the emergence of the variety of living forms as the results of a continuous distortion of common shapes, privileging a view of development and of evolution of shapes apparently driven by modifications of passive-like mechanical properties of tissue deformation (Fox-Keller, 2003; Thomson, 1917). The discovery of the genome and the emergence of molecular biology and the genetics of developmental biology in the middle of the twentieth century revealed the existence of primordial factors of morphogenesis, both genetic and biochemical in nature (Garber *et al.*, 1983; Lewis, 1978; Nusslein-Volhard and Wieschaus, 1980; St. Johnston and Nusslein-Volhard, 1992). In the meanwhile, our understanding of the molecular mechanism linking patterning gene expression to the production of the mechanical forces that shape the embryo increasingly progressed (Martin, 2009; Sweeton *et al.*, 1991). More recently, reverse signals were discovered, showing the mechanical control of the expression of developmental and patterning genes as well as of Myosin-II behavior patterning by the morphogenetic movements, based on biochemical mechanotransduction processes. Here, we review such processes of mechanotransduction involved in development, after a short description of the primary findings having first suggested mechanical factors to play a major role in proliferation, cytoskeleton reaction, or differentiation in cell culture, with the most recent findings in the underlying molecular mechanisms that translate mechanical signals into biochemical cues leading to the activation of transduction pathways.

2. Mechanotransduction in Cultured Cells

2.1. Cell response to external forces

The cell interaction with the substrate or with the extracellular matrix has long been proposed to play a critical role in controlling cell growth. For instance, corneal epithelial cells are highly sensitive to the FGF growth factor, but not to the EGF growth factor, when cultured on plastic substrate, with a reversed behavior when cultured on collagen substrate (Gospodarowicz *et al.*, 1978). The mechanical shape of cells was here proposed to play a direct role in modulating the response of the cells to external mechanical stimuli. Much experimental evidences supported these models, showing that regions of high tensional stress in epithelial monolayers correlate with increased proliferation *in vitro*, and that inhibition of myosin generated tension relaxes these regions of stress and leads to the inhibition of proliferation (Ingber *et al.*, 1981; Nelson *et al.*, 2005). Multicellular events recapitulating morphogenetic processes in cell culture, such as mammary gland morphogenesis, were also found to be dependent on the extracellular nature of the substrate (Ghajar and Bissell, 2008).

In vivo, endothelial cells are constantly submitted to shear stress due to the flow of blood and are thought to react by expressing genes that remodel cell structure. *In vitro*, studies aimed at understanding how these cells respond to such stimuli provided seminal insights in understanding the downstream mechanotranscriptional processes (Brouzes and Farge, 2004; Resnick *et al.*, 2003). Several genes have primarily been shown to be differentially expressed under static and shear stress conditions within endothelial cells, by activating the so-called promoter shear stress responsive elements (SSREs), thought to mediate some shear stress transcriptional responses (Resnick *et al.*, 1993). Different transcription factors involved in proliferation and cytoskeleton elements expression, including NF-kB, Egr-1, Sp1, fos, jun, and SREBP1 (sterol regulatory element-binding protein1), were found to be activated by laminar shear stress and were able to bind to the SSREs of some of these genes. However, the activation cascades of these transcription factors triggered by laminar shear stress remained poorly understood.

In addition to the control of cell proliferation, cell differentiation was found to be modulated by the nature and the rigidity of the extracellular matrix substrate. Specifically, cultured stem cells were found to differentiate as a function of the rigidity of their substrate, from neuronal differentiation on soft substrates to myoblast and osteoblast differentiation on more rigid substrates (Engler *et al.*, 2006). This suggested that stem cell differentiation might be under a combinatorial control of surrounding tissue rigidity with morphogen secretion, *in vivo* (Discher *et al.*, 2009).

Interestingly, the probing of the topology of the substrate by cells has also been documented to modulate their migration ability, thanks to innovative technologies controlling roughness through deformable micro-pillars with variable densities (Ghibaudo *et al.*, 2009). The actomyosin complex in itself has been proposed as a sensor allowing the probing by the cell of the rigidity of the adhesive substrate, possibly regulating cell migration along rigidity gradients (Mitrossilis *et al.*, 2009; Spudich, 2006). During neural cone growth, local actin accumulation strengthens nascent N-cadherin contacts in response to externally applied mechanical strain, supporting a direct transmission of actin-based traction forces to N-cadherin adhesions through catenin partners, driving growth cone advance, and neurite extension (Bard *et al.*, 2008).

2.2. The mechanosensors of the cell

The identification of mechanosensors that are able to transduce mechanical stimuli into a biochemical activity remains a largely opened question of mechanobiology (Brouzes and Farge, 2004). The first models proposed that transduction events take place in the vicinity of the plasma membrane (Ali and Schumacker, 2002). These models suggested that mechanosensitive ion channels, tyrosine kinase receptors, caveolae, and G proteins work as mechanosensors (Ali and Schumacker, 2002; Resnick *et al.*, 2003). Potential shear stress receptors, like integrins αVβ3, FAK (focal adhesion kinase), c-Src, and the VEGFR2 (vascular endothelial growth factor receptor)–VE-cadherin–β-catenin complex, were shown to be involved upstream of transcription factors or transcriptional activation events (Resnick *et al.*, 2003). Interestingly, the mechanical activation of a Src target at membrane surface, p130Cas, was directly observed in FRET. This was achieved by monitoring the fluorescence signal when the p130 peptide is tagged at the C-terminus and N-terminus: in the open and activated phosphorylated state, there is little FRET signal, whereas in the closed and compact conformation characterizing the unphosphorylated inactive state, the FRET signal is strong (Wang *et al.*, 2005). Regarding cytoskeleton rearrangements, mechanical strains locally applied at the plasma membrane induced focal contact growth in a Rho- and Myosin-independent process and suggested the existence of integrin-containing focal adhesion complexes as mechanosensors (Geiger *et al.*, 2009; Riveline *et al.*, 2001). However, in any case, the primary mechanosensor translating mechanical signals into biochemical signal activation remained to be discovered.

The plasma membrane coupled to secreted ligands expression was first experimentally proposed as a primary mechanosensor entity. Effectively, mechanical membrane tension can block the endocytosis and prevent the degradation of the receptor–ligand complex within endosomes (Rauch and Farge, 2000; Raucher and Sheetz, 1999), leading to the enhancement of downstream transcriptional events (Rauch *et al.*, 2002). Indeed, the blockade of BMP2 (bone morphogenetic protein 2) endocytosis induced by plasma membrane mechanical tension dramatically enhanced the BMP2-dependent myoblastic/osteoblastic transdifferentiation of murine cultured cells (Rauch *et al.*, 2002) (Fig. 8.1A). In this case, the membrane itself is the sensor, by regulating very sensitively the efficiency of specific ligands endocytosis in response to mechanical strain, thanks to its remarkable deformability due to its soft matter physics properties. A delocalized model of mechanotransduction also suggested that forces applied at the cell surface are transmitted, through integrins and via the cytoskeleton, to other locations within the cell and potentially to the nucleus, where they could trigger transcriptional events (Ingber, 2003; Mazumder *et al.*, 2010).

More recently, progress in the understanding of how junctional mechanosensors could translate mechanical signals into biochemical signals at the molecular level came from *in vitro* and cell culture experiments having shown

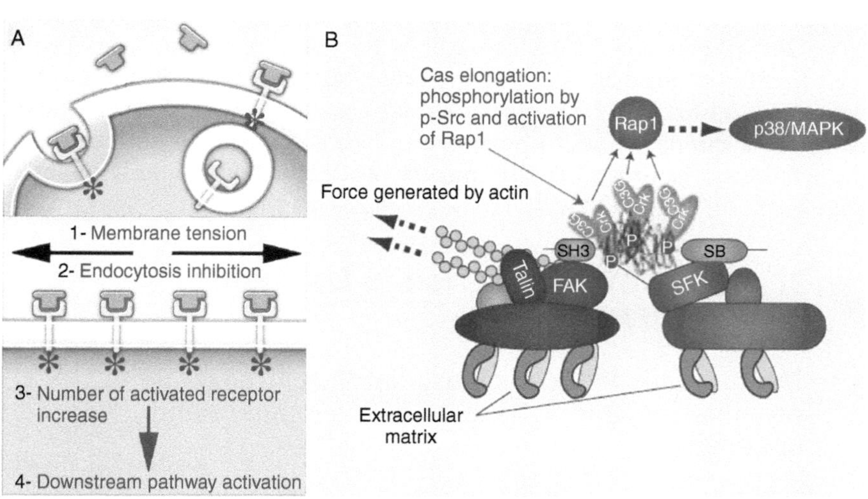

Figure 8.1 Underlying molecular mechanisms of mechanotransduction. (A) Mechanical modulation of signaling protein trafficking. Membrane mechanical tension inhibits endocytosis by membrane flattening. Without tension (top), endocytosis is allowed, with fewer receptors activated at the plasma membrane (red stars) and more degradation of their interactions with the signaling protein (in green) in the cytosolic acidic endosomes, than for the membrane mechanically flattened by tension blocking of endocytosis (bottom). Adapted from Fernandez-Gonzales and Zallen. (B) Mechanical induction of protein conformation changes. p130Cas (the gray-coiled domain in between the SH3 and Src SB interacting domain) involved in the focal adhesion protein complex, changes of conformation in response to the strain applied to cells. This opens the site of its interaction with Src. If Src is activated and present in solution, this site is phosphorylated, which activates the enzymatic activity of p130Cas, leading to the activation of the p38/MAPK pathway. (For interpretation of the references to color in this figure legend, the reader is referred to the Web version of this chapter.)

that the direct mechanically induced change of conformation of p130Cas leads to the opening of its site of phosphorylation by activated Src, thereby activating its enzyme activity (Sawada *et al.*, 2006) (Fig. 8.1B). Mechanical reinforcement of focal adhesion and adherent junctions was also suggested to be controlled by such a mechanotransduction process, involving talin (del Rio *et al.*, 2009), or α-catenin (Yonemura *et al.*, 2010) conformation change, respectively, and vinculin recruitment. Application of physiologically relevant forces caused stretching of single talin rods that exposed cryptic binding sites for vinculin (del Rio *et al.*, 2009; Grashoff *et al.*, 2010). In adherens junctions, molecular mechanotransduction can occur by protein binding after exposure of buried binding sites in the talin–vinculin system (del Rio *et al.*, 2009; Liu *et al.*, 2010). These experiments showed that, like the opening of an ionic pore in response to strain (Ali and Schumacker, 2002; Resnick *et al.*, 2003), any protein associated to any deformable structure, such as growth factors interacting directly with the ECM, cytoskeleton, nuclear membrane, or nuclear chromatin, is a potential mechanosensor through conformation change in response to strain (del Rio *et al.*, 2009; Hynes, 2009; Johnson *et al.*, 2007; Schultz and Wysocki, 2009). This not only opens a potentially high number of mechanosensitive molecules but also highlights the need to definitively distinguish between molecular complexes that are activated by mechanical stimuli without downstream physiological significant effect and the cellular mechanosensor(s) that triggers subsequent physiological responses.

3. Mechanotransduction in Embryonic Development

3.1. Mechanically induced cytoskeleton rearrangements, proliferation, and activation in development and oogenesis

Mechanotransduction in embryogenesis was initially proposed to play a major role in active shape change responses of cells to mechanical stimulation by the morphogenetic movements of the environmental tissue. Based on *ex vivo* artificial deformation experiments on *Xenopus* early embryos explants, cells of the ventral ectoderm were found to generically counter-react to stretching by an active contraction leading to an elongation of the tissue in the opposite direction of the stretching (Beloussov and Luchinskaia, 1995). Such a putative mechanotransduction process could potentially target the cytoskeleton directly. It could also target the extracellular matrix, and thus indirectly the cytoskeleton, as suggested by recent experiments showing the increase of fibronectin assembly into a fibrillar matrix in the blastocell roof in response to artificial external mechanical strains locally developed on the entire embryo (Dzamba *et al.*, 2009).

In parallel to animal embryos, plant meristem cells seem to respond to mechanical strain by an orientation of the microtubules along the lines of the tissue mechanical field strain (Hamant *et al.*, 2008). This orientation is driven by the field of mechanical strains developed by the growth pattern of meristem cells, given the geometrical limit conditions of the tissue. It subsequently leads to anisotropic morphogenetic deformation of the meristem respective to the polarity of the tissue stiffness controlled by the alignment. Such an orientation of the microtubules could belong to the activation of a mechanotransduction pathway similar to yeast single cell behavior (Terenna *et al.*, 2008). Alternatively, it could be due to a passive orientation of the rigid microtubules along the direction of the deformation of the cells, possibly driven by the mixed actin/microtubule cytoskeleton passive elongation of the cells along the field lines.

Recent experiments indicate that localized proliferation could be controlled by mechanical stresses during *Drosophila melanogaster* egg chamber morphogenesis (Wang and Riechmann, 2007). In this case, myosin activity is restricted to the epithelium apical surface of the egg chamber, which is facing the growing germline cyst, maintaining surface epithelial shape by balancing the force coming from volumic cyst growth (Fig. 8.2A(a)). Following germline cyst growth, surface epithelial cells divide and increase the surface area of the epithelium. Epithelial cell division was suggested to be induced by endogenous tensional stresses in response to volumic germline cysts growth pressure. Such proposal was based on the fact that epithelial clonal cells not expressing Myo-II show a defect in epithelia cell division, with no more defect after the blocking of germline cyst growth (Fig. 8.2A(b); Wang and Riechmann, 2007). Localized myosin activity at the apical face of the epithelium was proposed to increased tension in the epithelial cells in response to germline cyst, resulting in localized proliferation regulating epithelial tissue growth during oogenesis. A recent mathematical model of cell proliferation in wing development corroborates that this type of mechanical feedback, which prevents cell division after compression and induces it in response to dilation, could stabilize growth to maintain *D. melanogaster* tissue shape and form (Hufnagel *et al.*, 2007).

Interestingly, the oocyte activation is under the initial control of increasing Ca^{2+} concentration into the cytosol, and Ca^{2+} pores are known to be mechanosensitive: recent experiments involving osmotic pressure shocks and hydrostatic pressure showed a Ca^{2+}-dependent mechanical activation of the *Drosophila* embryo oocyte (Horner and Wolfner, 2008). Even though the mechanical stimuli of these experiments are far from physiologically mimicking the mechanical strains developed by ovulation across the oviduct, these observations suggest that the mechanical strains developed by deformation potentially associated to laying might play a role in the process of activation of the oocyte (Horner and Wolfner, 2008).

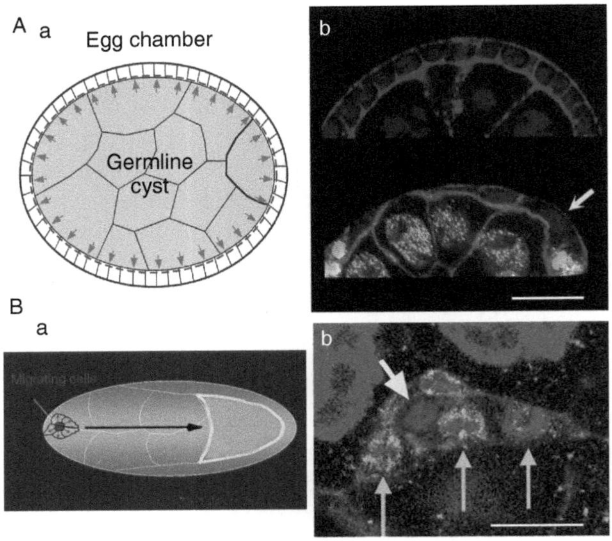

Figure 8.2 Mechanotransduction in oogenesis. (A) Mechanical feedback in oogenesis growth control. (a) The volumic growing germline cyst generates tension of the epithelial follicle cells, which accommodate growth by cell division and accommodate tension by apical actomyosin activities. (b) Blocking Myo-II expression in epithelial clonal populations (bottom, in cells nonmarked in green, yellow arrow) induces defects of cell division, compared to wild-type cells (top). These defects are absent in this clonal population when germ cyst growth is inhibited (not shown). Adapted from Wang and Riechmann (2007). Scale bar is 20 μm. (B(a)) Migrating cells are activated at the anterior pole of the follicular epithelium (in red) and migrate posteriorly following chemoattractants (black arrow). (b) During migration, these cells are deformed by friction forces, which activate some nuclear translocation of MAL-D (in green), necessary for cytoskeleton plasticity accommodating migration. Adapted from Somogyi and Rorth (2004). Scale bar is 10 μm. (For interpretation of the references to color in this figure legend, the reader is referred to the Web version of this chapter.)

3.2. Mechanical activation of transcriptional events involved in cytoskeleton reinforcement

During *Drosophila* oogenesis, a group of eight cells delaminate from the anterior follicular epithelium as a cluster, invade the underlying germline tissue, and migrate posteriorly to the oocyte (Fig. 8.2B(a)). These cells will contribute to the formation of the micropyle of the oocyte, which allows the entry of the spermatozoids for fertilization. The migration ability of these border cells requires the activity of several transcription factors, with guidance by the PDGF/VEGF and EGFR tyrosine kinases (Somogyi and Rorth, 2004). In addition to the transcription factors necessary for migration, Mal-D and SRF cotranscription factors were found to be required for robust F-actin cytoskeleton formation to accommodate migration. Effectively, MAL-D mutants show border cell fragmentation when attempting to migrate. Interestingly, MAL-D nuclear translocation correlates with the amplitude of border cell stretching deformation during migration, suggesting that MAL-D nuclear translocation could be mechanically induced by the tension applied to border cells associated with their movement (Somogyi and Rorth, 2004) (Fig. 8.2B(b)).

The hypothesis of a mechanical induction of MAL-D nuclear accumulation was tested by using the *slbo* mutant genetic tool, which is deficient in cell migration. In this nonmigrative mutant, cells are not stretched and show no nuclear translocation of MAL-D. Interestingly, intermediate mutant situations in which only some of the border cells are *slbo* mutants show that the wild-type border cells, which do migrate, are able to attract *slbo* mutant border cells. The latter are thus stretch-deformed and show a MAL-D nuclear accumulation, suggesting that MAL-D nuclear translocation is not prevented by the *slbo* mutation *per se*, but rather by the lack of cell deformation. Here, mechanical manipulations are indirectly monitored by genetic manipulations but logically predict MAL-D nuclear translocation as mechanically induced by cells stretching (Somogyi and Rorth, 2004). Even though here the underlying molecular mechanotransduction pathway remains to be elucidated, the existence of such a mechanotransduction feedback pathway could be of high importance in other invasive configurations, including cancer invasivity.

3.3. Mechanical activation of developmental patterning gene expression in development

Developmental genes control both the biochemical patterning and the generation of morphogenetic movements that geometrically shape the embryo. Biochemical patterning consists in a cascade of developmental gene expression, which specifies the fate of the cells in the developing embryo according to their position (Gilbert, 1994; Lawrence, 1992; Spemann and Mangold, 1924). In the early *Drosophila* embryo, for instance, anteroposterior patterning begins with the generation of Bicoid and Nanos anteroposterior gradients into the syncytial yolk, in addition with the expression of the terminal Torso-like protein in the two embryonic poles of the embryo. The nuclei of the syncytial embryo are in direct contact with those maternally loaded transcription factor gradients and express different gap genes as a function of the combined concentration of those morphogens. Gradients and gap gene expression subsequently combine to lead to an anteroposterior segmentation of the embryo through the striped expression of genes like of *even-skipped* (Small *et al.*, 1992). Those three patterning systems finally combine, leading to the expression of *Hox* genes that design the fly body plan (Lawrence, 1992). Dorsoventral patterning begins with the maternal activation of the Toll receptor on the ventral side of the embryo, which induces the formation of a ventrodorsal gradient of dorsal protein nuclear localization (Roth *et al.*, 1989). Dorsal acts as a morphogen because high concentrations activate the expression of *twist* and *snail* in ventral nuclei, and represses *dpp* and *zen* to restrict their expression to the dorsal side of the embryo (St. Johnston and Nusslein-Volhard, 1992).

Since 20 years ago, experiments initiated in early *Drosophila* embryos have shown that the morphogenetic movement sequence is tightly controlled by patterning gene expression (Sweeton *et al.*, 1991). For instance, embryonic mesoderm invagination requires the expression of the Fog (expressed under the control of Twist) and Snail zygotic dorsoventral patterning proteins in the mesoderm (Seher *et al.*, 2007), whereas the germ-band extension movement (i.e., the anterior–posterior elongation of the invaginated embryo) requires the expression of the Bicoid, Nanos, and Torso-like maternal proteins, which control the anterior–posterior polarity of the embryo (Irvine and Wieschaus, 1994). Recently, the role of the dorsoventral and anterior–posterior patterning genes was established to induce apico-basal and planar polarities in Myo-II concentration leading to morphogenetic movements (Bertet *et al.*, 2004; Dawes-Hoang *et al.*, 2005).

Is there signaling feedback pathways that modify genome activity depending on progression of morphogenesis? Regarding classical induction, the state of RNA expression of any cell at a given stage of development can be considered as a feedback response of the genome to the pattern of expression of signaling proteins characterizing the previous stage. However, in contrast to RNA or protein patterning, multicellular geometric morphology *per se* has no signaling properties. In that case, the putative existence of such feedback cannot be based on classical signaling protein inductive cues and was proposed to be due to mechanical cues associated to tissue deformation, with morphogenesis-associated mechanical strains as activators of signaling pathways controlling master patterning gene expression (Farge, 2003).

Looking at endogenous morphogenesis in the early *Drosophila* embryo, the expression of the Twist protein is strongly amplified in anterior pole stomodeal cells, after 10–20 min of compression of these cells by the morphogenetic movement of germ-band extension. Twist expression being found to be mechanically induced in early *Drosophila* embryos submitted to global artificial shape changes (Farge, 2003), this suggested that Twist expression might be mechanically induced in the stomodeum at the onset of gastrulation (Fig. 8.3A(a, b); Desprat *et al.*, 2008; Farge, 2003). Demonstrating the existence of mechanical cues leading to Twist mechanical induction in stomodeal cells in response to the endogenous morphogenetic movements of germ-band extension required the elaboration of tools allowing the inhibition and rescue of the Myo-II-dependent germ-band extension morphogenetic movement within the physiological wild-type genetic background, with physiological mechanical strain deformation, *in vivo*. A noninvasive two-photon ablation of the most dorsal part of the

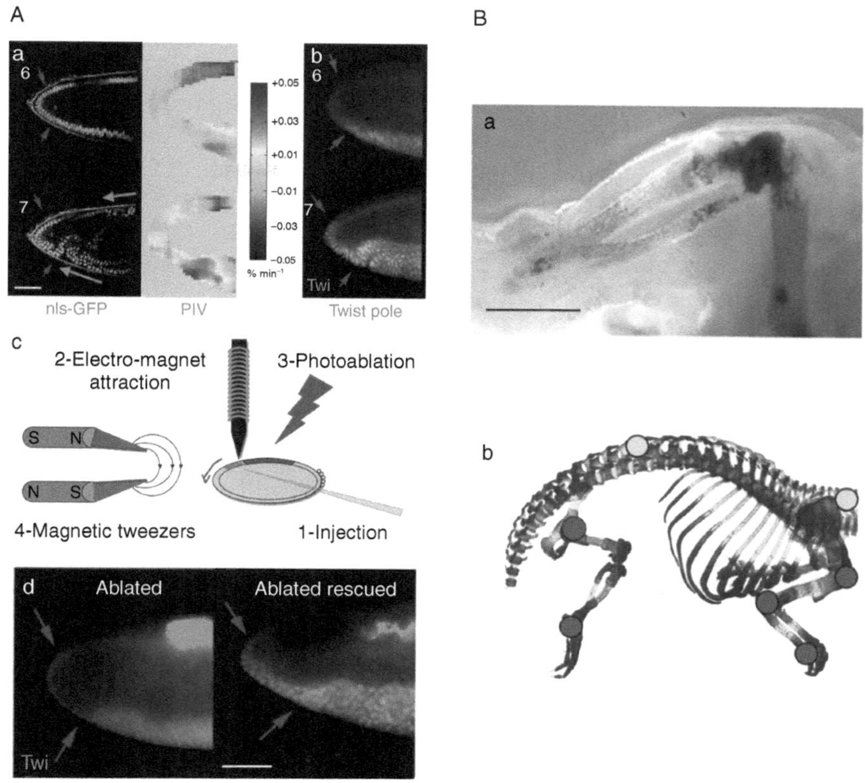

Figure 8.3 Mechanical induction of developmental gene expression in embryogenesis. (A) Twist expression increase is induced in stomodeal cells at the onset of gastrulation by compression by germ-band extension (GBE). (a) Nls-GFP nuclear labeling shows the compression of stomodeal cells (in between red arrows) in between early stage 6 and late stage 7 due to GBE movements (orange arrows). Quantitative PIV analysis shows a 2% min^{-1} dynamics of compression during the first 10 min of GBE, encoded in red. Scale bar is 50 μm. (b) Twist is overexpressed at late stage 7 in compressed stomodeal cells (adapted from Farge, 2003; Desprat *et al.*, 2008 data). (c) The dorsal domain of the embryo is photoablated to block GBE, after injection and concentration of a calibrated ferrofluid of magnetic nanoparticles into anterodorsal cells. Calibrated magnetic tweezers are positioned in order to attract magnetized dorsal cells to compress anterior pole stomodeal cells with a deformation rate that mimics endogenous GBE compression dynamics. (d) Loss of Twist overexpression in noncompressed stomodeal cells in the ablated embryo (in between red arrows), and recovery in the stomodeal cells of the ablated embryos for which the physiological compression is rescued by magnetic manipulation. Adapted from Desprat *et al.* (2008). Scale bar is 50 μm. (B) The β-catenin pathway is mechanically activated during bone morphogenesis in mouse development. (a) The Top-gal promoter, target of the β-catenin, is activated (X-gal blue revelation of the Top-gal expression) into the bone joints submitted to recurrent mechanical shocks due to muscle oscillations, (b) in several bone joint domains of the developing skull. Activation and joint formation were lost in all three muscleless mutants (lacking musculature) and in the *mdg* mutant mice (lacking excitation contraction coupling, leading to paralysis) in red dots, or only in the *mdg* mutant in yellow dots. Adapted from Hens *et al.* (2005) and Kahn *et al.* (2009). Scale bar is 1 mm. (For interpretation of the references to color in this figure legend, the reader is referred to the Web version of this chapter.)

embryo blocks the germ-band extension on the time scale of endogenous stomodeal cell compression. This inhibits both the posterior and anterior extensions that normally compress stomodeal cells (Farge, 2003; Supatto *et al.*, 2005) (Fig. 8.3A(c)). Lacking compression, the stomodeal cells show a lack of Twist expression amplification (Fig. 8.3A(d)). Rescuing the compression of stomodeal cells with physiological dynamics is realized after a magnetic attraction of injected ferrofluid into the anterior dorsal cells neighboring the stomodeal cells in the photoablated embryo (Fig. 8.3A(c)) (Desprat *et al.*, 2008). Once cells are magnetized by ferrofluid insertion, cells are subsequently attracted by a calibrated magnetic tweezers to rescue stomodeal cells compression into the ablated embryo, which leads to the rescue of the strong expression of Twist into stomodeal cells (Fig. 8.3A(d)), demonstrating that Twist overexpression is mechanically induced by stomodeal cell compression due to germ-band extension during endogenous development. Interestingly, the physiological function of Twist mechanical induction in stomodeal cells was suggested to be the Dve-dependent (*defective proventriculus*) and vital functional differentiation of the anterior mid-gut of the embryo (Desprat *et al.*, 2008). The underlying molecular mechanism of Twist mechanical induction belongs to a Src42A-dependent mechanically induced process of release of an Armadillo/β-catenin from the junctions into the nuclei, which subsequently leads to TCF-dependent expression of Twist (Desprat *et al.*, 2008; Farge, 2003).

Interestingly, the mechanical activation of the β-catenin pathway was more recently found to be involved in bone development in mice. During embryogenesis, synovial joints develop from a pool of progenitor cells which differentiate into the cell types that will constitute the mature joint, in parallel to the development of the musculature. The role of musculature formation in joint formation has long been known, with poorly understood underlying molecular mechanisms. However, recent evidence demonstrated the mechanical activation of the β-catenin pathway in the maintenance of progenitor joint cells, in response to the shocks submitted to the joints by adjacent bones, developed by muscle oscillations during development (Fig. 8.3B; Hens *et al.*, 2005; Kahn *et al.*, 2009). The inhibition of muscle contractions in a large series of independent muscleless mutants, including Pax3 and MyoD, leads to the failure of joint formation and to abnormal differentiation of many skull joints, as well as of the failure of β-catenin activation in the related progenitor cells. In this case, mechanical strains associated with the muscle function control the genetically dependent processes of cell differentiation in development.

In addition to embryonic muscle movements, reversing blood flows was also suggested to ensure normal valvulogenesis in the developing heart of zebrafish. In this case, the expression of *klf2A* was found to be mechanically induced by oscillatory shear stress due to heart beating, as well as to be necessary for valve formation (Vermot *et al.*, 2009).

Very recently, *ex vivo* experiments of bending of early *Xenopus* embryonic neuro-mesoderm rudiment explants suggested a role for mechanical strain of the tissue in controlling differentiation states within this organism (Kornikova *et al.*, 2010). In these experiments, the explants were bent at the earliest stages, with a labeling of Sox3 and muscular actin, respectively, monitoring the neural and mesoderm differentiation of the tissues 24 h after perturbation. The compressed concave domain of the bent embryos leads to Sox3 expression and neural differentiation, whereas the dilated convex domain leads to muscular actin expression and mesoderm differentiation. Interestingly, these differentiation events correlate well with the natural morphological bending of the developing embryos, with neural differentiation at the concave and mesoderm differentiation at the convex domains. These observations suggest that mechanical induction in development could be sensitive to the nature of the deformation, with differential responses according to compression versus dilation.

3.4. Mechanotransduction in posttranslational morphogenetic events controlling morphogenetic movements

In addition to controlling the state of expression of the genome, mechanotransduction processes in development can also control posttranslational events directly involved in morphogenesis, involving, for instance, Myosin-II intracellular behaviors at the onset of gastrulation. In *Drosophila* embryos, gastrulation begins by a Snail- and Twist-dependent apical redistribution of Myosin-II that leads to a constriction of apical cell surfaces (Martin *et al.*, 2009). Fog (folded gastrulation) is a secreted signaling molecule that is expressed under the control of Twist in the mesoderm and in the posterior pole, the Fog signaling pathway having been demonstrated to promote the apical accumulation of Myo-II (Dawes-Hoang *et al.*, 2005; Sweeton *et al.*, 1991). In addition to the Fog-dependent activation of the Myo-II apical redistribution process, Snail is also necessary for apical redistribution of Myo-II and mesoderm invagination, through a still unknown molecular mechanism (Martin, 2009; Pouille *et al.*, 2009). Note that *T48*, another gene acting downstream of Twist, has been proposed to cooperate with Fog in the apical attraction of RhoGEF2, a protein also involved in the mesoderm invagination process (Kolsch *et al.*, 2007).

Apical constriction generates a trapezoidal shape change of individual cells, leading to the decrease of the apical surface area of the mesoderm compared to the basal surface area, which induces the inward bending constraints of mesoderm invagination (Sweeton *et al.*, 1991). Different simulations were developed to test whether the apical surface tension increase induced by redistribution of Myo-II would be the only genetically controlled active perturbation necessary for mesoderm invagination, or if the invagination would require additional active movements, such as cell shortening. Whereas simulations describing cells as a continuous viscoelastic medium suggest the necessity of an active shortening of mesodermal cells to accomplish invagination (Conte *et al.*, 2009), hydrodynamical simulations describing the tissue as composed of individual cells with individual plasma membranes characterized by an actomyosin cortical tension and contractile apical rings connected by intercellular junctions suggest that the Myo-II-dependent increase of apical surface tension of mesodermal cells is sufficient to trigger the movements observed during invagination (Fig. 8.4A; Pouille and Farge, 2008). Simulations of mesoderm invagination based on individual apical constriction cell shape changes were also performed in sea urchins embryos (Drasdo and Forgacs, 2000; Odell *et al.*, 1981), which have an extracellular matrix that should be specifically compliant to allow gastrulation (Davidson *et al.*, 1995).

Interestingly, there exists two phases of apical constriction. The first 4-min phase is stochastic, randomly involves the uncorrelated reversible pulses of constriction and relaxation of individual cells, and is unable to trigger mesoderm invagination (Sweeton *et al.*, 1991). These pulses are associated with reversible pulses of apical spots of Myo-II (Martin *et al.*, 2009). The second phase is collective and involves the constriction of all mesodermal cells (Sweeton *et al.*, 1991), through a process of pulsatile constrictions, including a ratchet process progressively stabilizing cell apexes into more and more constricted states. This is associated with the progressive stabilization of the apical spots of Myo-II, leading to apical Myo-II coalescence and redistribution (Martin *et al.*, 2009). Because mutants of *twist* only show the stochastic phase, the collective phase is Twist dependent. In fact, the Fog secreted factor, which is expressed under the control of Twist, is one of the key signaling proteins triggering the collective phase (Costa *et al.*, 1994). However, *snail* mutants are defective in both the stochastic and collective phases, indicating that the stochastic phase is indeed Snail dependent, but also that the two phases interact.

Synthetic analysis of the literature having revealed difficulties in understanding such interaction in a purely genetic or biochemical way, a model of mechanical activation of the collective constriction by the strains developed by the stochastic constriction was tested (Pouille *et al.*, 2009). Rescuing the existence of a mechanical strain in the mesoderm of *snail* mutants by an indent of 5 μm (namely, 30% of the thickness of the ectoderm and 2% of the thickness of the embryo), rescued the apical redistribution of Myo-II and the mesoderm invagination, both of which are missing in these mutants (Fig. 8.4B). Interestingly, the mechanotransduction pathway, in this case, was suggested to be the activation of the Fog signaling pathway, associated with the mechanically induced inhibition of Fog endocytosis by the plasma membrane tension generated by the mechanical strains of the indent or of *snail*-induced stochastic constrictions (Pouille *et al.*, 2009). Interestingly,

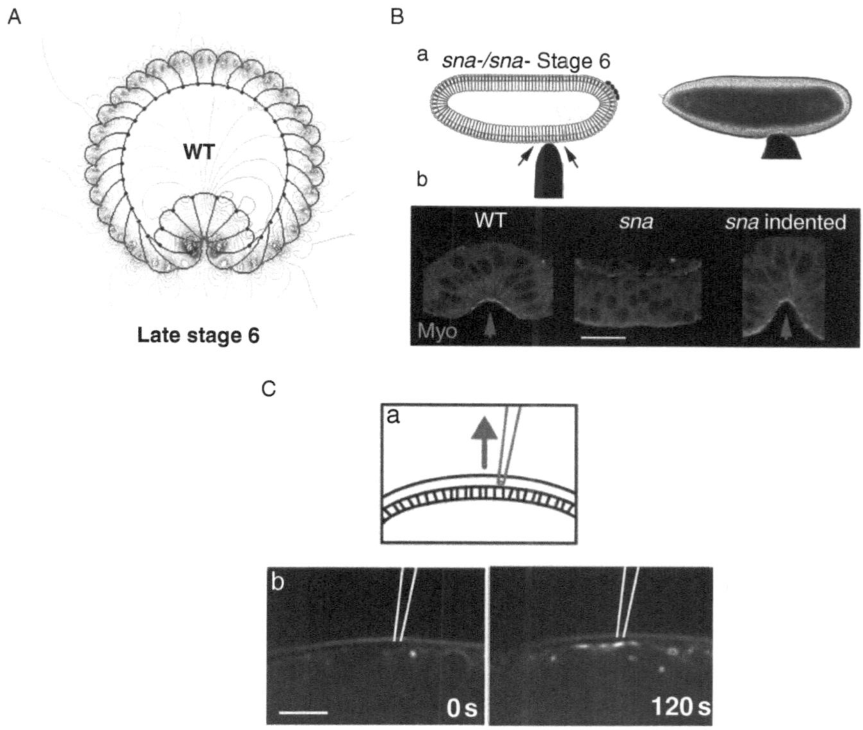

Figure 8.4 Mechanical induction of Myo-II apical redistribution and stabilization at gastrulation. (A) Mesoderm invagination simulation. Simulating an embryo in which the redistribution of Myo-II induces the increase of apexes surface tension into the mesoderm. In response to surface tension increase, the apical surface decreases, leading to the inward curvature of mesoderm invagination. The simulation recapitulates all the cellular and subcellular movements experimentally observed in *Drosophila* embryos (Pouille and Farge, 2008). The *in silico* embryo is an assembly of deformable elastic membranes (in black) connected by junctions, immerged into viscous water, with apical junctions (in red) additionally connected by an elastic actomyosin contractile ring (not drawn). The colored lines are hydrodynamic velocity field lines. Simulations describing cells as a continuous viscoelastic medium require the addition of active cell shortening to get to similar closed invagination (not shown, Conte *et al.*, 2009). Ventrolateral cell intercalation should help in relaxing lateral cells distortions, and trigger lateral forces leading to more profound inward invagination. The movie of the simulation can be seen at: http://u168.inst-curie.lbn.fr/node/104. (B(a)) Indenting *sna* mutants to rescue the lack of mechanical strain of *snail* mutants and to test the mechanical reactivation of the Fog signaling pathway controlling both apical redistribution of Myo-II and mesoderm invagination. (b) Rescuing apical accumulation of Myo-II and mesoderm invagination, lost in the *sna* mutant, after soft indent of the mesoderm of *sna* mutants, in a Fog-dependent process. Adapted from Pouille *et al.* (2009). Scale bar is 20 μm. (C(a)) Micropipette aspiration of gastrulating ectoderm embryos tissues. (b) Apical stabilization of Myo-II in actin cables driving cell intercalation leading to GBE (Bertet *et al.*, 2004) and domain segregation (Monier *et al.*, 2009). Adapted from Fernandez-Gonzales (2009). Scale bar is 10 μm. (For interpretation of the references to color in this figure legend, the reader is referred to the Web version of this chapter.)

such oscillations were also observed during dorsal closure and were in this case suspected to be maintained by mechanical feedback mechanisms (Solon *et al.*, 2009). In mesoderm invagination, mechanotransduction was suggested to lead to a rapid and long distance interaction between all the cells expressing Fog regulating the coordination of apical constriction throughout the complete mesoderm on the typically 300 μm length and 4-min time scales, thereby allowing efficient mesoderm invagination. Mechanical activation of apical redistribution of Myo-II and mesoderm invagination was also suggested to be reminiscent of a putative multicellular primitive motor sensorial feeding response of phagocytosis to touch in ancient organisms, which could have been recapitulated in response to *snail*-dependent mechanical stimulation internally produced during embryogenesis (Pouille *et al.*, 2009).

Interestingly, Twist expression into the mesoderm is known to dramatically decrease at late stage 8, normally at 1 h after mesoderm invagination (Leptin, 1991) in *sna* mutants lacking the morphogenetic movement of mesoderm invagination, suggesting a possible participation of mechanical strains associated to mesoderm invagination at stage 6 in Twist expression at stage 8 (Brouzes *et al.*, 2004). Consistent with this, stage 6-indented *sna* homozygous mutants showed a rescue of Twist expression into the mesoderm at stage 8, both in the pool of embryos having responded by mesoderm invagination rescue (around 70%), and in the pool having not responded that show a typical *snail* mutant noninvagination phenotype at late stage 8 (around 30%) (Fernandez-Sanchez *et al.*, 2010). This suggests the role of Myo-II-dependent tension in mechanical induction of Twist in mesoderm cells in closely coordinating morphogenesis with differentiation of the mesoderm.

Mechanical stabilization of apical Myosin-II in actin cables was also found to be involved in response to the strains developed by the second major morphogenetic movement of gastrulation: germ-band extension. The driving force of germ-band extension is the formation of dorsoventral actomyosin cable structures (Fernandez-Gonzalez *et al.*, 2009), leading to the cell intercalation that generates anteroposterior embryo axis elongation. Photoablation experiments locally deleting the mechanical integrity of the multicellular tissue lead to a rapid release of the Myo-II from the actin cable of neighboring cells, indicating the role of the tissue mechanical tension in the stabilization of Myo-II in these apical actin cables. Consistently, a local aspiration of the epithelial cells leads to a rapid attraction of such apical Myo-II into cables (Fig. 8.4C). In this case, mechanical strains are developed out of the Fog expression domain suggesting an alternative mechanism than the Fog endocytosis blocking by membrane tension as an underlying molecular mechanotransduction process. Interestingly, the Fog-dependent mechanism could stabilize cortical Myo-II under the complete surface of the apical membrane, with a Fog-independent mechanism stabilizing it at cell–cell contacts in cables (Fernandez-Gonzalez and Zallen, 2009) as observed in indented mesoderm cells in the absence of Fog expression (Pouille *et al.*, 2009). Cortical actomyosin cables leading to embryonic elongation might be stabilized in response to their own strains through a positive mechanosensing feedback process (Spudich, 2006).

By integrating the latest findings on the involvement of Myo-II at critical mechanical cross talks between differentiation and morphogenetic patterning, one might speculate that mechanical strains developed by either mesoderm invagination or germ-band extension could stabilize or enhance actomyosin cable tension and therefore participate in anteroposterior differentiated compartmental boundary maintenance having been proposed to be regulated by sorting effects driven by actomyosin cables line tensions (Fernandez-Gonzalez *et al.*, 2009; Landsberg *et al.*, 2009; Monier *et al.*, 2009).

4. Perspectives in Evolution and Cancer

As we saw, the emergence of the Fog/Myo-II mechanosensitive pathway, or a primitive equivalent, might have been at the origin of the generation of a transient primitive gastric organ in response to external stimuli of touch. This reflex response could have been subsequently recapitulated during early embryogenesis in response to *snail*, or primitive equivalent, dependent internal deformations (Fernandez-Sanchez *et al.*, 2010). The emergence of such mechanotransduction pathway has thus been speculated to have been a putative key event leading to the emergence of the first organisms (by definition, a multicellular system with an organ) from the earliest embryos defined as an aggregation of cells without collective functional cell behavior (Fernandez-Sanchez *et al.*, 2010). In addition, the activation of the mechanosensitive β-catenin pathway has been linked to the initial polarity of numerous species (Wikramanayake *et al.*, 2003). It was thus also tempting to speculate that the local response of earliest embryos tissues to mechanical contact with the ground after gravity sedimentation might have participated in the determination of the primary axis formation of earliest embryos through mechanotransduction (Fernandez-Sanchez *et al.*, 2010).

In addition to embryonic development, the activation of the β-catenin pathway is involved in the progression of many distinct tumors, including colon cancer progression (Fodde *et al.*, 1994). Direct uniaxial mechanical deformation of mice colon tissues also showed the β-catenin-dependent activation of the Twist-1 and Myc oncogene expressions, in APC heterozygous mutants but not in wild types (Whitehead *et al.*, 2008). Defects in APC expression are correlated to 70% of the human colon cancer cases having been related to genetic alteration (Fearon and Vogelstein, 1990). APC is known to send cytoplasmic β-catenin to the degradation pathway before it translocates into the nucleus and targets oncogene (Fodde *et al.*, 1994). Consistent with this, tissues defective in one copy of APC appear to not succeed in preventing the nuclear translocation of the β-catenin pool released from the junction after mechanical perturbation. This indicates a potential role of tumor growth pressure, or in the specific case of intestinal transit strains, in tumor gene expression, within the APC-mutated genetic background. Interestingly, the mechanical activation of the β-catenin is today also known to play a critical role in inhibiting adipogenesis and in stimulating bone mass formation in medicine, with mechanical vibrating stimulations beginning to be clinically used to fight against obesity and osteoporosis (Robinson *et al.*, 2006; Rubin *et al.*, 2004; Sen *et al.*, 2008).

In addition, mechanical activation of cytoskeletal elements in response to the variation of the stiffness of cell substrates (related to tumor rigidity) has been systematically studied in breast cancer (Bissell *et al.*, 2005; Ghajar and Bissell, 2008). A correlation between prostate tumorigenicity *in vivo* and disorganized growth depending on vimentin and β1 integrin expression has also been established (Zhang *et al.*, 2009). Further, a novel both chemical and mechanosensitive signaling pathway that controls angiogenesis has been found, whose deregulation contributes to the development of many diseases, including cancer. This pathway involves a direct regulation of p190RhoGAP by growth factors, integrin-dependent ECM binding and mechanical distortion of the cytoskeleton, which in turn controls VEGFR2 (vascular endothelial growth factor receptor 2) expression by modulating the balance between two mutually antagonistic transcription factors, TFII-I and GATA2 (Mammoto *et al.*, 2009).

Of current interest is thus the question of the activation of signaling pathways directly connected to oncogene expression, with future investigations designed to probe the involvement of the mechanical induction process in response to tumor growth pressure during tumor progression (Alexander *et al.*, 2008; Whitehead *et al.*, 2008).

ACKNOWLEDGMENTS

We thank Joanne Whitehead for her reading of the chapter. The author's lab is funded by the ANR (PCV and PiriBio ANR-09-Piri-0013-02), the ARC, Microsoft, NanoldF, the Fondation Pierre Gilles de Gennes, and the HFSP RGP001-14/2006 grants.

REFERENCES

Alexander, N. R., Branch, K. M., Parekh, A., Clark, E. S., Iwueke, I. C., Guelcher, S. A., and Weaver, A. M. (2008). Extracellular matrix rigidity promotes invadopodia activity. *Curr. Biol.* **18,** 1295–1299.

Ali, M. H., and Schumacker, P. T. (2002). Endothelial responses to mechanical stress: Where is the mechanosensor? *Crit. Care Med.* **30,** S198–S206.

Bard, L., Boscher, C., Lambert, M., Mege, R. M., Choquet, D., and Thoumine, O. (2008). A molecular clutch between the actin flow and N-cadherin adhesions drives growth cone migration. *J. Neurosci.* **28,** 5879–5890.

Beloussov, L. V., and Luchinskaia, N. N. (1995). Biomechanical feedback in morphogenesis, as exemplified by stretch responses of amphibian embryonic tissues. *Biochem. Cell Biol.* **73,** 555–563.

Bertet, C., Sulak, L., and Lecuit, T. (2004). Myosin-dependent junction remodelling controls planar cell intercalation and axis elongation. *Nature* **429,** 667–671.

Bissell, M. J., Kenny, P. A., and Radisky, D. C. (2005). Microenvironmental regulators of tissue structure and function also regulate tumor induction and progression: The role of extracellular matrix and its degrading enzymes. *Cold Spring Harb. Symp. Quant. Biol.* **70,** 343–356.

Brouzes, E., and Farge, E. (2004). Interplay of mechanical deformation and patterned gene expression in developing embryos. *Curr. Opin. Genet. Dev.* **14,** 367–374.

Brouzes, E., Supatto, W., and Farge, E. (2004). Is mechano-sensitive expression of twist involved in mesoderm formation? *Biol. Cell* **96,** 471–477.

Conte, V., Munoz, J. J., Baum, B., and Miodownik, M. (2009). Robust mechanisms of ventral furrow invagination require the combination of cellular shape changes. *Phys. Biol.* **6,** 016010.

Costa, M., Wilson, E. T., and Wieschaus, E. (1994). A putative cell signal encoded by the folded gastrulation gene coordinates cell shape changes during Drosophila gastrulation. *Cell* **76,** 1075–1089.

Davidson, L. A., Koehl, M. A., Keller, R., and Oster, G. F. (1995). How do sea urchins invaginate? Using biomechanics to distinguish between mechanisms of primary invagination. *Development* **121,** 2005–2018.

Dawes-Hoang, R. E., Parmar, K. M., Christiansen, A. E., Phelps, C. B., Brand, A. H., and Wieschaus, E. F. (2005). Folded gastrulation, cell shape change and the control of myosin localization. *Development* **132,** 4165–4178.

del Rio, A., Perez-Jimenez, R., Liu, R., Roca-Cusachs, P., Fernandez, J. M., and Sheetz, M. P. (2009). Stretching single talin rod molecules activates vinculin binding. *Science* **323,** 638–641.

Desprat, N., Supatto, W., Pouille, P. A., Beaurepaire, E., and Farge, E. (2008). Tissue deformation modulates twist expression to determine anterior midgut differentiation in Drosophila embryos. *Dev. Cell* **15,** 470–477.

Discher, D. E., Mooney, D. J., and Zandstra, P. W. (2009). Growth factors, matrices, and forces combine and control stem cells. *Science* **324,** 1673–1677.

Drasdo, D., and Forgacs, G. (2000). Modeling the interplay of generic and genetic mechanisms in cleavage, blastulation, and gastrulation. *Dev. Dyn.* **219,** 182–191.

Dzamba, B. J., Jakab, K. R., Marsden, M., Schwartz, M. A., and DeSimone, D. W. (2009). Cadherin adhesion, tissue tension, and noncanonical Wnt signaling regulate fibronectin matrix organization. *Dev. Cell* **16,** 421–432.

Engler, A. J., Sen, S., Sweeney, H. L., and Discher, D. E. (2006). Matrix elasticity directs stem cell lineage specification. *Cell* **126,** 677–689.

Farge, E. (2003). Mechanical induction of twist in the Drosophila foregut/stomodeal primordium. *Curr. Biol.* **13,** 1365–1377.

Fearon, E. R., and Vogelstein, B. (1990). A genetic model for colorectal tumorigenesis. *Cell* **61,** 759–767.

Fernandez-Gonzalez, R., and Zallen, J. A. (2009). Cell mechanics and feedback regulation of actomyosin networks. *Sci. Signal.* **2,** 78.

Fernandez-Gonzalez, R., Simoes Sde, M., Roper, J. C., Eaton, S., and Zallen, J. A. (2009). Myosin II dynamics are regulated by tension in intercalating cells. *Dev. Cell* **17,** 736–743.

Fernandez-Sanchez, M. E., Serman, F., Ahmadi, P., Farge, E., *et al.* (2010). Mechanical induction in embryonic development and tumour growth: Integrative cues through molecular to multi-cellular interplay, and evolutionary perspectives. *In* "Nuclear Mechanics and Genome Regulation," (G. V. Shivashankar, ed.), pp. 295–321.

Fodde, R., Edelmann, W., Yang, K., van Leeuwen, C., Carlson, C., Renault, B., Breukel, C., Alt, E., Lipkin, M., Khan, P. M., *et al.* (1994). A targeted chain-termination mutation in the mouse Apc gene results in multiple intestinal tumors. *Proc. Natl. Acad. Sci. USA* **91,** 8969–8973.

Fox-Keller, E. (2003). *Making Sense of Life.* Harvard University Press, Cambridge, US.

Garber, R. L., Kuroiwa, A., and Gehring, W. J. (1983). Genomic and cDNA clones of the homeotic locus Antennapedia in Drosophila. *EMBO J.* **2,** 2027–2036.

Geiger, B., Spatz, J. P., and Bershadsky, A. D. (2009). Environmental sensing through focal adhesions. *Nat. Rev. Mol. Cell Biol.* **10,** 21–33.

Ghajar, C. M., and Bissell, M. J. (2008). Extracellular matrix control of mammary gland morphogenesis and tumorigenesis: Insights from imaging. *Histochem. Cell Biol.* **130,** 1105–1118.

Ghibaudo, M., Trichet, L., Le Digabel, J., Richert, A., Hersen, P., and Ladoux, B. (2009). Substrate topography induces a crossover from 2D to 3D behavior in fibroblast migration. *Biophys. J.* **97,** 357–368.

Gilbert, S. F. (1994). Specification of cell fate by progressive cell–cell interactions. *In* "Developmental Biology," (S. F. Gilbert, ed.), pp. 575–622. Sinauer Associates Inc., Massachusetts.

Gospodarowicz, D., Greenburg, G., and Birdwell, C. R. (1978). Determination of cellular shape by the extracellular matrix and its correlation with the control of cellular growth. *Cancer Res.* **38,** 4155–4171.

Grashoff, C., Hoffman, B. D., Brenner, M. D., Zhou, R., Parsons, M., Yang, M. T., McLean, M. A., Sligar, S. G., Chen, C. S., Ha, T., *et al.* (2010). Measuring mechanical tension across vinculin reveals regulation of focal adhesion dynamics. *Nature* **466,** 263–266.

Hamant, O., Heisler, M. G., Jonsson, H., Krupinski, P., Uyttewaal, M., Bokov, P., Corson, F., Sahlin, P., Boudaoud, A., Meyerowitz, E. M., *et al.* (2008). Developmental patterning by mechanical signals in *Arabidopsis. Science* **322,** 1650–1655.

Hens, J. R., Wilson, K. M., Dann, P., Chen, X., Horowitz, M. C., and Wysolmerski, J. J. (2005). TOPGAL mice show that the canonical Wnt signaling pathway is active during bone development and growth and is activated by mechanical loading in vitro. *J. Bone Miner. Res.* **20,** 1103–1113.

His, W. (1875). *Unsere Korperform und Das Physiologische problem ibrer entstebung.* F. C. W. Vogel, Leipzig.

Horner, V. L., and Wolfner, M. F. (2008). Mechanical stimulation by osmotic and hydrostatic pressure activates *Drosophila* oocytes in vitro in a calcium-dependent manner. *Dev. Biol.* **316,** 100–109.

Hufnagel, L., Teleman, A. A., Rouault, H., Cohen, S. M., and Shraiman, B. I. (2007). On the mechanism of wing size determination in fly development. *Proc. Natl. Acad. Sci. USA* **104,** 3835–3840.

Hynes, R. O. (2009). The extracellular matrix: Not just pretty fibrils. *Science* **326,** 1216–1219.

Ingber, D. E. (2003). Tensegrity II. How structural networks influence cellular information processing networks. *J. Cell Sci.* **116,** 1397–1408.

Ingber, D. E., Madri, J. A., and Jamieson, J. D. (1981). Role of basal lamina in neoplastic disorganization of tissue architecture. *Proc. Natl. Acad. Sci. USA* **78,** 3901–3905.

Irvine, K. D., and Wieschaus, E. (1994). Cell intercalation during *Drosophila* germband extension and its regulation by pair-rule segmentation genes. *Development* **120,** 827–841.

Johnson, C. P., Tang, H. Y., Carag, C., Speicher, D. W., and Discher, D. E. (2007). Forced unfolding of proteins within cells. *Science* **317,** 663–666.

Kahn, J., Shwartz, Y., Blitz, E., Krief, S., Sharir, A., Breitel, D. A., Rattenbach, R., Relaix, F., Maire, P., Rountree, R. B., *et al.* (2009). Muscle contraction is necessary to maintain joint progenitor cell fate. *Dev. Cell* **16,** 734–743.

Kolsch, V., Seher, T., Fernandez-Ballester, G. J., Serrano, L., and Leptin, M. (2007). Control of *Drosophila* gastrulation by apical localization of adherens junctions and RhoGEF2. *Science* **315,** 384–386.

Kornikova, E. S., Troshina, T. G., Kremnyov, S. V., and Beloussov, L. V. (2010). Neuromesodermal patterns in artificially deformed embryonic explants: A role for mechanogeometry in tissue differentiation. *Dev. Dyn.* **239,** 885–896.

Landsberg, K. P., Farhadifar, R., Ranft, J., Umetsu, D., Widmann, T. J., Bittig, T., Said, A., Julicher, F., and Dahmann, C. (2009). Increased cell bond tension governs cell sorting at the *Drosophila* anteroposterior compartment boundary. *Curr. Biol.* **19,** 1950–1955.

Lawrence, P. A. (1992). *The Making of a Fly. The Genetics of Animal Design.* Blackwell Scientific, Oxford.

Leduc, S. (1912). *La Biologie Synythétique. Etude de Biophysique.* Paris.

Leptin, M. (1991). Twist and snail as positive and negative regulators during *Drosophila* mesoderm development. *Genes Dev.* **5,** 1568–1576.

Lewis, E. B. (1978). A gene complex controlling segmentation in *Drosophila. Nature* **276,** 565–570.

Liu, Z., Tan, J. L., Cohen, D. M., Yang, M. T., Sniadecki, N. J., Ruiz, S. A., Nelson, C. M., and Chen, C. S. (2010). Mechanical tugging force regulates the size of cell–cell junctions. *Proc. Natl. Acad. Sci. USA* **107,** 9944–9949.

Mammoto, A., Connor, K. M., Mammoto, T., Yung, C. W., Huh, D., Aderman, C. M., Mostoslavsky, G., Smith, L. E., and Ingber, D. E. (2009). A mechanosensitive transcriptional mechanism that controls angiogenesis. *Nature* **457,** 1103–1108.

Martin, A. C. (2009). Pulsation and stabilization: Contractile forces that underlie morphogenesis. *Dev Biol.* **341,** 114–25.

Martin, A. C., Kaschube, M., and Wieschaus, E. F. (2009). Pulsed contractions of an actin-myosin network drive apical constriction. *Nature* **457,** 495–499.

Mazumder, A., Roopa, T., Kumar, A., Iyer, K. V., Ramdas, N. M., and Shivashankar, G. V. (2010). Prestressed nuclear organization in living cells. *Methods Cell Biol.* **98,** 221–239.

Mitrossilis, D., Fouchard, J., Guiroy, A., Desprat, N., Rodriguez, N., Fabry, B., and Asnacios, A. (2009). Single-cell response to stiffness exhibits muscle-like behavior. *Proc. Natl. Acad. Sci. USA* **106,** 18243–18248.

Monier, B., Pelissier-Monier, A., Brand, A. H., and Sanson, B. (2009). An actomyosin-based barrier inhibits cell mixing at compartmental boundaries in *Drosophila* embryos. *Nat. Cell Biol.* **12**(60–65), 61–69.

Nelson, C. M., Jean, R. P., Tan, J. L., Liu, W. F., Sniadecki, N. J., Spector, A. A., and Chen, C. S. (2005). Emergent patterns of growth controlled by multicellular form and mechanics. *Proc. Natl. Acad. Sci. USA* **102,** 11594–11599.

Nusslein-Volhard, C., and Wieschaus, E. (1980). Mutations affecting segment number and polarity in *Drosophila. Nature* **287,** 795–801.

Odell, G. M., Oster, G., Alberch, P., and Burnside, B. (1981). The mechanical basis of morphogenesis. I. Epithelial folding and invagination. *Dev. Biol.* **85,** 446–462.

Pouille, P. A., and Farge, E. (2008). Hydrodynamic simulation of multicellular embryo invagination. *Phys. Biol.* **5,** 15005.

Pouille, P. A., Ahmadi, P., Brunet, A. C., and Farge, E. (2009). Mechanical signals trigger myosin ii redistribution and mesoderm invagination in *Drosophila* embryos. *Sci. Signal.* **2,** 16.

Rauch, C., and Farge, E. (2000). Endocytosis switch controlled by transmembrane osmotic pressure and phospholipid number asymmetry. *Biophys. J.* **78,** 3036–3047.

Rauch, C., Brunet, A. C., Deleule, J., and Farge, E. (2002). C2C12 myoblast/osteoblast transdifferentiation steps enhanced by epigenetic inhibition of BMP2 endocytosis. *Am. J. Physiol. Cell Physiol.* **283,** C235–C243.

Raucher, D., and Sheetz, M. P. (1999). Membrane expansion increases endocytosis rate during mitosis. *J. Cell Biol.* **144,** 497–506.

Resnick, N., Collins, T., Atkinson, W., Bonthron, D. T., Dewey, C. F., Jr., and Gimbrone, M. A., Jr. (1993). Platelet-derived growth factor B chain promoter contains a cis-acting fluid shear-stress-responsive element. *Proc. Natl. Acad. Sci. USA* **90,** 4591–4595.

Resnick, N., Yahav, H., Shay-Salit, A., Shushy, M., Schubert, S., Zilberman, L. C., and Wofovitz, E. (2003). Fluid shear stress and the vascular endothelium: For better and for worse. *Prog. Biophys. Mol. Biol.* **81,** 177–199.

Riveline, D., Zamir, E., Balaban, N. Q., Schwarz, U. S., Ishizaki, T., Narumiya, S., Kam, Z., Geiger, B., and Bershadsky, A. D. (2001). Focal contacts as mechanosensors: Externally applied local mechanical force induces growth of focal contacts by an mDia1-dependent and ROCK-independent mechanism. *J. Cell Biol.* **153,** 1175–1186.

Robinson, J. A., Chatterjee-Kishore, M., Yaworsky, P. J., Cullen, D. M., Zhao, W., Li, C., Kharode, Y., Sauter, L., Babij, P., Brown, E. L., *et al.* (2006). Wnt/beta-catenin signaling is a normal physiological response to mechanical loading in bone. *J. Biol. Chem.* **281,** 31720–31728.

Roth, S., Stein, D., and Nusslein-Volhard, C. (1989). A gradient of nuclear localization of the dorsal protein determines dorsoventral pattern in the *Drosophila* embryo. *Cell* **59,** 1189–1202.

Rubin, C., Recker, R., Cullen, D., Ryaby, J., McCabe, J., and McLeod, K. (2004). Prevention of postmenopausal bone loss by a low-magnitude, high-frequency mechanical stimuli: A clinical trial assessing compliance, efficacy, and safety. *J. Bone Miner. Res.* **19,** 343–351.

Sawada, Y., Tamada, M., Dubin-Thaler, B. J., Cherniavskaya, O., Sakai, R., Tanaka, S., and Sheetz, M. P. (2006). Force sensing by mechanical extension of the Src family kinase substrate p130Cas. *Cell* **127,** 1015–1026.

Schultz, G. S., and Wysocki, A. (2009). Interactions between extracellular matrix and growth factors in wound healing. *Wound Repair Regen.* **17,** 153–162.

Seher, T. C., Narasimha, M., Vogelsang, E., and Leptin, M. (2007). Analysis and reconstitution of the genetic cascade controlling early mesoderm morphogenesis in the *Drosophila* embryo. *Mech. Dev.* **124,** 167–179.

Sen, B., Xie, Z., Case, N., Ma, M., Rubin, C., and Rubin, J. (2008). Mechanical strain inhibits adipogenesis in mesenchymal stem cells by stimulating a durable beta-catenin signal. *Endocrinology* **149,** 6065–6075.

Small, S., Blair, A., and Levine, M. (1992). Regulation of even-skipped stripe 2 in the *Drosophila* embryo. *EMBO J.* **11,** 4047–4057.

Solon, J., Kaya-Copur, A., Colombelli, J., and Brunner, D. (2009). Pulsed forces timed by a ratchet-like mechanism drive directed tissue movement during dorsal closure. *Cell* **137,** 1331–1342.

Somogyi, K., and Rorth, P. (2004). Evidence for tension-based regulation of *Drosophila* MAL and SRF during invasive cell migration. *Dev. Cell* **7,** 85–93.

Spemann, H., and Mangold, H. (1924). Induction of embryonic primordia by implantation of organizers from a different species. *In* "Foundations of Experimental Embryology," (B. H. Willier and J. M. Oppenheimer, eds.), pp. 144–184. Hafner, New York, US.

Spudich, J. A. (2006). Molecular motors take tension in stride. *Cell* **126,** 242–244.

St Johnston, D., and Nusslein-Volhard, C. (1992). The origin of pattern and polarity in the *Drosophila* embryo. *Cell* **68,** 201–219.

Supatto, W., Debarre, D., Moulia, B., Brouzes, E., Martin, J. L., Farge, E., and Beaurepaire, E. (2005). In vivo modulation of morphogenetic movements in *Drosophila* embryos with femtosecond laser pulses. *Proc. Natl. Acad. Sci. USA* **102,** 1047–1052.

Sweeton, D., Parks, S., Costa, M., and Wieschaus, E. (1991). Gastrulation in *Drosophila*: The formation of the ventral furrow and posterior midgut invaginations. *Development* **112,** 775–789.

Terenna, C. R., Makushok, T., Velve-Casquillas, G., Baigl, D., Chen, Y., Bornens, M., Paoletti, A., Piel, M., and Tran, P. T. (2008). Physical mechanisms redirecting cell polarity and cell shape in fission yeast. *Curr. Biol.* **18,** 1748–1753.

Thomson, D. A. W. (1917). *On Growth and Form*. Cambridge University Press, Cambridge.

Vermot, J., Forouhar, A. S., Liebling, M., Wu, D., Plummer, D., Gharib, M., and Fraser, S. E. (2009). Reversing blood flows act through klf2a to ensure normal valvulogenesis in the developing heart. *PLoS Biol.* **7,** e1000246.

Wang, Y., and Riechmann, V. (2007). The role of the actomyosin cytoskeleton in coordination of tissue growth during *Drosophila* oogenesis. *Curr. Biol.* **17,** 1349–1355.

Wang, Y., Botvinick, E. L., Zhao, Y., Berns, M. W., Usami, S., Tsien, R. Y., and Chien, S. (2005). Visualizing the mechanical activation of Src. *Nature* **434,** 1040–1045.

Whitehead, J., Vignjevic, D., Futterer, C., Beaurepaire, E., Robine, S., and Farge, E. (2008). Mechanical factors activate beta-catenin-dependent oncogene expression in APC mouse colon. *HFSP J.* **2,** 286–294.

Wikramanayake, A. H., Hong, M., Lee, P. N., Pang, K., Byrum, C. A., Bince, J. M., Xu, R., and Martindale, M. Q. (2003). An ancient role for nuclear beta-catenin in the evolution of axial polarity and germ layer segregation. *Nature* **426,** 446–450.

Yonemura, S., Wada, Y., Watanabe, T., Nagafuchi, A., and Shibata, M. (2010). alpha-Catenin as a tension transducer that induces adherens junction development. *Nat. Cell Biol.* **12,** 533–542.

Zhang, X., Fournier, M. V., Ware, J. L., Bissell, M. J., Yacoub, A., and Zehner, Z. E. (2009). Inhibition of vimentin or beta1 integrin reverts morphology of prostate tumor cells grown in laminin-rich extracellular matrix gels and reduces tumor growth in vivo. *Mol. Cancer Ther.* **8,** 499–508.

Index

N

P

Contents of Previous Volumes

Volume 50

Volume 51

Volume 52

Volume 53

Volume 54

Volume 55

Volume 56

Volume 57

Volume 62

Volume 65

Volume 66

Volume 67

Volume 68

Volume 69

Volume 70

Volume 71

Volume 72

Volume 73

Volume 74

Volume 75

Volume 76

Volume 77

Volume 80

Volume 81

Volume 82

Volume 83

Volume 84

Volume 85

Volume 86

Volume 89

Volume 90

Volume 91

Volume 92

Volume 93

Volume 94

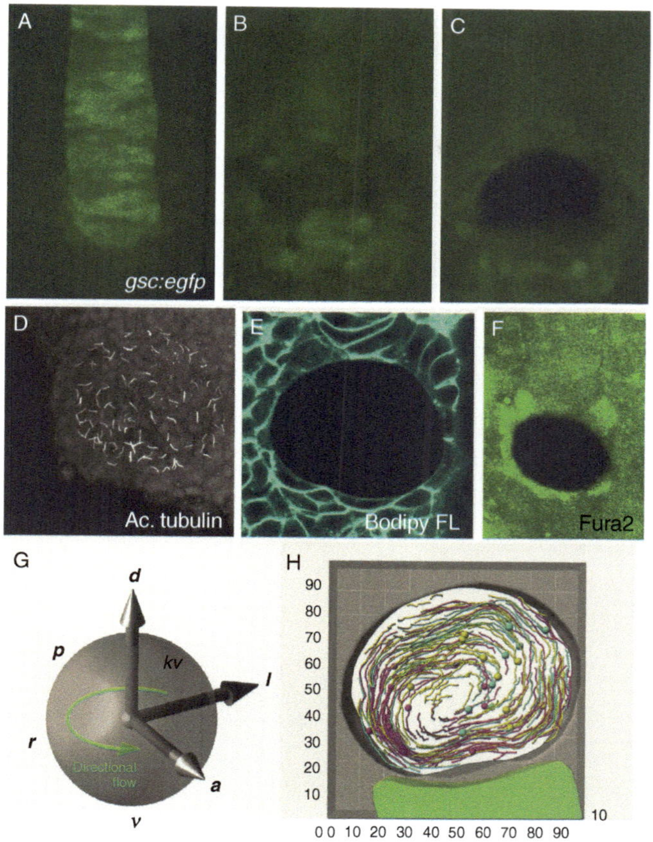

Willy Supatto and Julien Vermot, Figure 2.2 Anatomy of the Kupffer's vesicle. (A and B) Visualization and localization of the Kupffer's vesicle with the gsc:GFP line. Confocal sections at increasing depth showing the localization of the Kupffer's vesicle just below the midline. (D) Labeling using antibodies directed against acetylated tubulin reveals cilia localization in the Kupffer's vesicle. (E) Cell membrane labeling using Bodipy FL staining underlines the spherical shape of the vesicle. (F) Calcium indicator (Fura2) shows asymmetric activation. This asymmetric activation is essential for the left–right embryonic patterning. (G) Scheme representing the Kupffer's vesicle in 3D and the direction of the cilia-driven flow. (H) Particle tracking in the Kupffer's vesicle demonstrates the circularity of the flow *in vivo*. The *Tg(gsc:eGFP)* line is a kind gift from the Houart lab (King's College London, UK). Acetylated tubulin labeling was performed as in Colantonio *et al.* (2009), Bodipy FL labeling as in Hove *et al.* (2003). Panel (H) was obtained as described in Supatto *et al.* (2008).

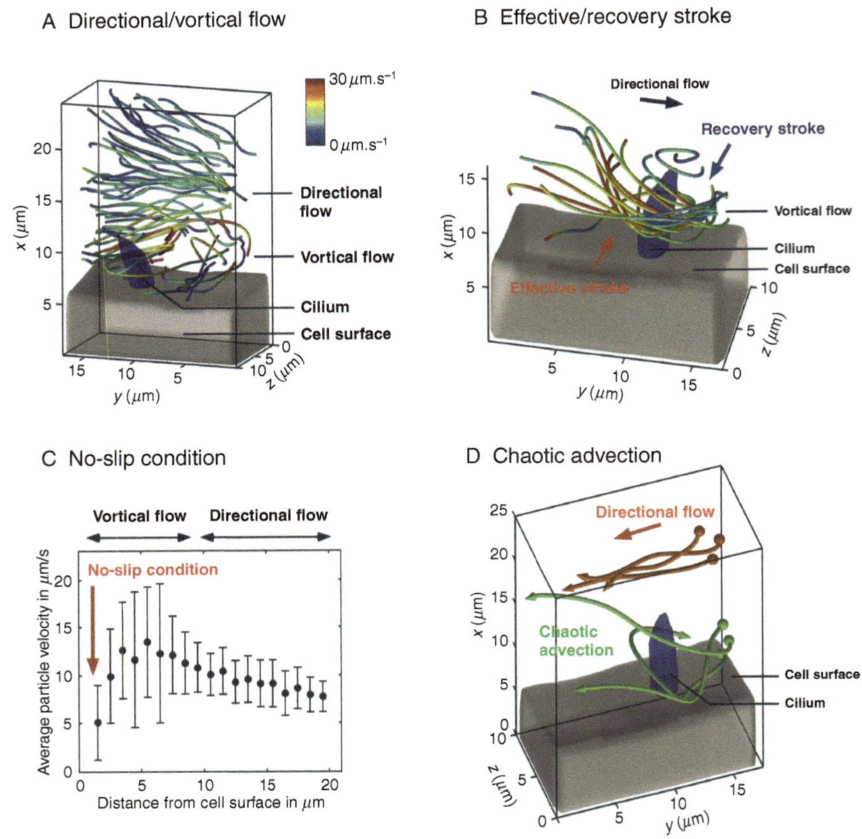

Willy Supatto and Julien Vermot, Figure 2.6 Predicted flow features from simulations are experimentally validated in the zebrafish's left–right organizer. (A) The flow is vortical close to the cilium and the cell membrane and directional far away. (B) Effective and recovery strokes are identified in the flow surrounding a beating cilium with faster velocity on one side and slower velocity on the other. (C) Experimental verification of the no-slip condition: the average flow velocity drop to zero close to the cell membrane. (D) The chaotic nature of the flow surrounding a beating cilium is confirmed experimentally by following particles starting nearby in the vicinity of the cilium and exhibiting drastically different trajectories (green), whereas particle trapped in the directional flow (red) stay close together. Experimental data were obtained using the same procedure as described in Supatto *et al.* (2008). Velocity field display and analysis were done using custom Matlab (The MathWorks) scripts.

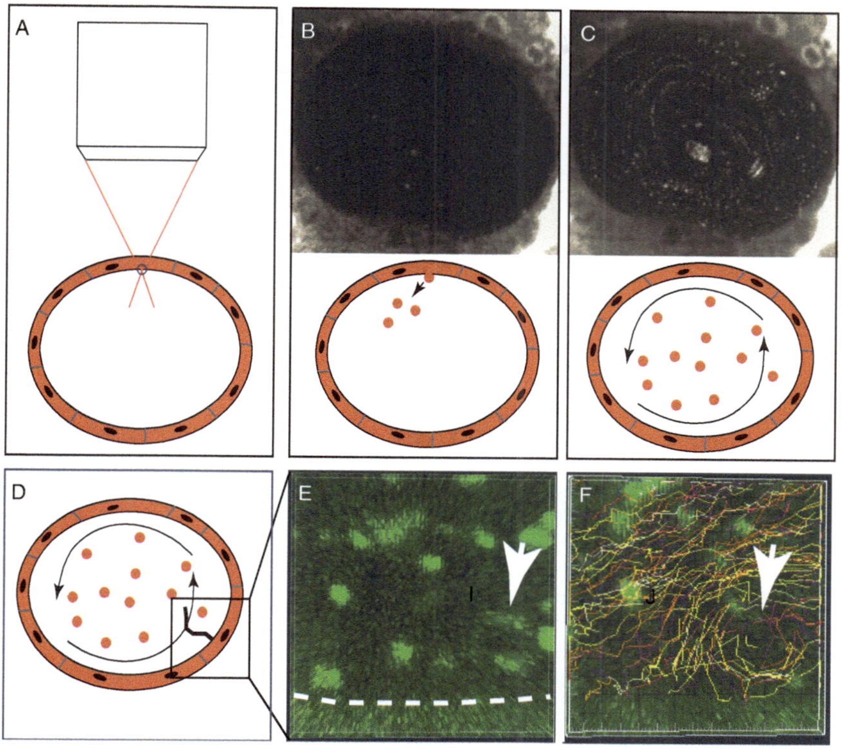

Willy Supatto and Julien Vermot, Figure 2.8 Mapping the cilia-driven flow within the zebrafish Kupffer's vesicle (KV) using femtosecond laser ablation, fast imaging and 3D-particle tracking. The KV cells are labeled with Bodipy TR and a single cell lining the cavity is targeted with tightly focused femtosecond laser pulses at 820 nm wavelength (A). A subcellular femtosecond laser ablation is performed to generate fluorescent microdebris seeding the flow (B). The nonlinear effect used in this technique allows a high spatial confinement and a low invasiveness, thus permitting the targeting of subcellular regions deep inside the embryo. Fast confocal 3D imaging at 4 *z*-stack per second capture the movement of the tracer particles: in (C), time-lapse images are superimposed to show the circular trajectories of the flow around the dorsal–ventral axis of the embryo. Cilia close to the KV surface (D) can be identified by following particles trapped in a vortical flow (arrows in E and F) and mapping the flow using 3D-particle tracking (D). Experimental data were obtained using the same procedure as described in Supatto *et al*. (2008). Image display and analysis were performed using ImageJ (http://rsb.info.nih.gov/ij/) and Imaris (Bitplane).

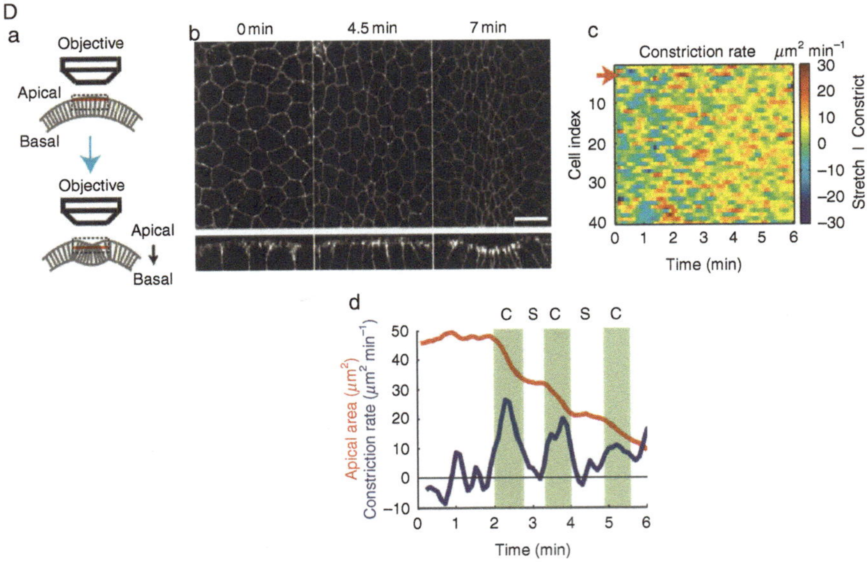

Karsten Kruse and Daniel Riveline, Figure 3.4 Spontaneous oscillations *in vivo*. (D) Apical constriction in developing *Drosophila*: (a) schematic illustration of apical constriction and of the set up, (b) subsequent snapshots of the tissue with membranes stained fluorescently, (c) the apical constriction rate of some cells varies periodically with time, (d) oscillation in the constriction rate associated with a decrease in the apical area (Martin *et al.*, 2009). (E) Dorsal closure in *Drosophila*: (a) Subsequent snapshots of dorsal tissue with membrane stained cells, (b) area of a cell as a function of time shows periodic variations, (c) histogram of recorded pulsation periods (Solon *et al.*, 2009).

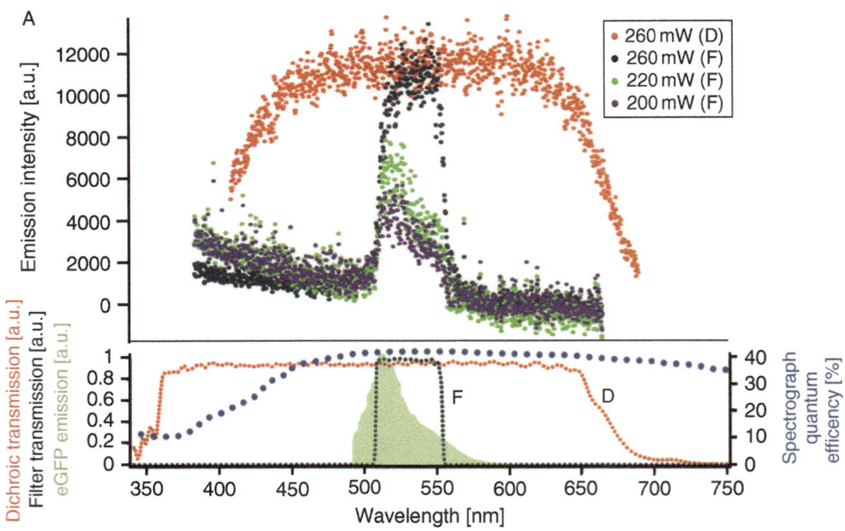

Matteo Rauzi and Pierre-François Lenne, Figure 4.11 (A) Top graph: spectrograph showing light emitted by the ablated cell actomyosin cortex during cell intercalation in the developing *Drosophila* embryo (stage 7). Purple, green, black, and red are scatter plot of the emission curve for exciting power of 200, 220, 260, and 260 mW, respectively. The black and red scatter plots differ in the filters used to acquire photons coming from the sample: the black scatter plot is obtained using a band-pass filter (F) while the red scatter plot is obtained by using a dichroic (D). Bottom graph: spectra of the dichroic and band-pass filter used to detect photon coming from the ablated sample. Also the spectrum of eGFP and the quantum efficiency of the spectrograph are shown.

FSC
www.fsc.org
MIX
Papier | Fördert
gute Waldnutzung
FSC® C083411

Kolibri 360 GmbH
Ferdinand-Jühlke-Straße 7
99095 Erfurt, Deutschland
produktsicherheit@kolibri360.de